ESSAYS *and* REVIEWS
in HISTORY *and*
HISTORY *of* SCIENCE

A la Mémoire de
JEAN DARDE
1920–2006

*Comme un Frère
de l'Autre Côté
de la Mer*

ESSAYS *and* REVIEWS
in HISTORY *and*
HISTORY *of* SCIENCE

—⧈—

CHARLES COULSTON GILLISPIE

American Philosphical Society

Set in Celeste and Aldus by Graphic Composition, Inc., Athens, Georgia.
Text and cover design by Ellen Graben
Printed and bound in the United States of America

Library of Congress Cataloging-in-Publication Data

Gillispie, Charles Coulston.
    Essays and reviews in history and history of science / Charles Coulston Gillispie.
        p.   cm.
    Includes bibliographical references and index.
    ISBN 978-0-87169-965-7 (pbk.)
        1. Science—Historiography. 2. Science—History. I. Title.
Q126.8.G55 2006
509—dc22

                                                              2006052602

Transactions of the American Philosophical Society
Held at Philadelphia
For Promoting Useful Knowledge
Volume 96, Part 5

Table of Contents

—ɯ—

Prefatory Note
and Acknowledgments

—⚶—

Among the various things I have written, the essays and reviews collected in this volume seem to me the most likely to hold a measure of enduring interest. Three referees for the American Philosophical Society generously agreed, for which vote of confidence I am grateful. They further suggested that it would be well to arrange the pieces, not chronologically, but in broad topical categories, and also to provide introductory remarks in order to situate them in the circumstances wherein they were written. I have enjoyed following that advice and hope that having done so does not lend an unduly autobiographical character to the collection.

I am beholden to the American Philosophical Society Held in Philadelphia for Promoting Useful Knowledge for considering that presenting these writings in its *Transactions* conforms to its mission. A special word of thanks is due to the Society's Editor, Mary McDonald, for shepherding them skilfully and tactfully through the press.

Let me further acknowledge the identity of the publishers who hold the copyright to each article and who are cited in the first footnote thereto as follows:

Historically Speaking (The Historical Society), Introduction.

The Journal of Modern History (University of Chicago Press), I, 1; IV, 12.

Archives Internationales d'Histoire des Sciences (Académie Internationale d'Histoire des Sciences), 1, 3.

Isis (History of Science Society), I, 4, 5.

The American Scientist (Sigma Xi), IV, 14, 15; V, 19. *Proceedings of the American Philosophical Society*, II, 7.

Historia Mathematica (Academic Press), IV, 16.

Science (American Association for the Advancement of Science), IV, 17; V, 18.

Cornell University Press, I, 2.

University of Wisconsin Press, II, 6.

Aldus Books, III, 8.

The Doshisha University, III, 9, 10.

American Council of Learned Societies, IV, 13.

Heinemann, V, 20.

In case anyone should wish to locate other of my publications, a complete bibliography appears on my web site, http://www.princeton.edu/~hos/ccgpubs.html, and are also deposited with my papers in the Library of the American Philosophical Society.

A Professional Life
in the History of Science*

—⚀—

It was with some compunction that I acceded to the flattering invitation from Donald Yerxa, editor of *Historically Speaking,* to write of a professional life in the field of my specialty. Reluctance was the greater in that I had already given an account of that career in *Isis* on the occasion of the 75th anniversary of the History of Science Society in 1999.[1] In all probability, however, there is little if any overlap between subscribers to *Isis* and those to *Historically Speaking.* That such should be the case is one of the situations discussed. Anyone who consults the earlier essay will find that it turns on personal and institutional factors. I tried not to repeat myself more than was necessary to make what follows intelligible, and ventured instead to offer some reflections on the context of my work in relation to the development of the historiography of science.

First of all, a word about the subject. The generation to which I have the good fortune to belong is commonly said to have founded the history of science as a professional field of scholarship in the years after World War II. Marshall Clagett, I. Bernard Cohen, Henry Guerlac, Erwin Hiebert, Alistair Crombie, Giorgio di Santillana, Rupert and Marie Hall, Georges Canguilhem, René Taton, Thomas S. Kuhn—those are among the notable names. Having majored in some branch of science as undergraduates or the equivalent, and gone on to graduate school before or just after the war, all of us had somehow developed a strong ancillary taste for history. We came out of service of one sort or another in 1945, dazzled like everyone else by Hiroshima, the Manhattan Project, sonar, radar, penicillin, and so on. Independently of each other, or largely so, we each harbored a sense

* Reprinted from *Historically Speaking: The Bulletin of the Historical Society,* V:3 (January 2004), pp. 2–6.

that science, even like art, literature, or philosophy, must have had a history, the study of which might lead to a better appreciation of its own inwardness as well as its place in the development of civilization.

With a few stellar exceptions, the history of science until that time was the province either of philosophers—Condorcet, Comte, Whewell, Duhem, Mach—each adducing exemplary material in service to their respective epistemologies, or of elderly scientists writing the histories of their science, or sometimes all science, in order to occupy their retirement. Though not written in accordance with historical standards, neither of these bodies of literature is to be ignored. The one is always suggestive and sometimes informative, the other often informative, almost always technically reliable, and rarely of much interpretative significance. Of the two notable scholars who flourished in the 1920s and 1930s, George Sarton was a prophet and scholarly bibliographer rather than a historian, while E. L. Thorndike was a devoted, learned antiquarian riding his hobby horse of magic and experimental science through the library of the Vatican. Though much and rightly respected, neither found a following. Nor did E. J. Dijksterhuis, whose *The Mechanization of the World Picture* (1950) is a classic that will always repay study.

Anticipations of a fully historical history of science appeared in the work of Hélène Metzger on 18th-century chemistry and Anneliese Maier on medieval science. Herbert Butterfield's *The Origins of Modern Science, 1300–1800* (1950) was a godsend both in itself and in that it was one of the few things one could expect undergraduates to read. The same was true of Carl Becker's *Heavenly City of the 18th-Century Philosophers* (1932), a supremely literate essay which (unfortunately in my view) has fallen into disfavor among students of the Enlightenment, and also of Arthur O. Lovejoy's *The Great Chain of Being* (1936), a founding work in the modern historiography of ideas. Two ancillary masterpieces, one from the side of sociology, the other from philosophy, were still more inspirational in exhibiting respectively the social and the intellectual interest that the history of science may hold, namely Robert K. Merton's path breaking *Science, Technology, and Society in Seventeenth-Century England* (1938) and Alexandre Koyré's superb *Études Galiléennes* (1939).

I had read none of these works when, safely out of the army in graduate school at Harvard in 1946–47, I thought to find a thesis subject in what to me was the *terra incognita* of the history of science. My scientific and military backgrounds were respectively in chemistry and a 4.2-inch chemical mortar battalion, but I had taken almost all my electives in history as an undergraduate at Wesleyan, graduating in 1940. The emphasis in the excellent department there was on English history, and my instinct was to look to Britain for a subject, rather than to chemistry. I'm not sure I even knew that there had been a chemical revolution centering on the work of Lavoisier. Darwin was the obvious link between science and intellectual history, but, such was my naiveté, it hardly seemed possible that anything new could be said about the theory of evolution, about science and religion, or about social Darwinism, and I elected to look into the background. That turned out to be in geology, whence my first book, *Genesis and Geology: A Study in the Relations of Scientific Thought, Natural Theology, and Social Opinion in Great Britain, 1790–1850* (1951). It has been in print ever since. Harvard University Press saw fit to put it in a new suit of clothes and reissue it in 1996. A foreword by a scholar of the next generation, Nicolaas Rupke, analyzes the way in which it came to mark a new departure in the historiography of science. He credits me with a novel methodology, first, in consulting, not only the original scientific texts, but the general periodical literature of the time; and second in telling not merely of technical discovery, but of the way in which varying religious views of geologists entered into the formation of their theories, and also the way in which the climate of social opinion entered into the discourse of theology as well as science.

I had no notion of anything of the sort. So far as I was aware, my thesis was a new departure for me, but not for a subject of which I was quite ignorant. Nothing was farther from my thoughts than methodology, something fit for Marxists and sociologists. All that we students of history were taught to do was to go look at the sources, all of them. Perhaps it was lucky that I had never taken a course in geology. Though formally trained in science, I wrote my thesis as someone being trained in history. Had I written it as a scientist, it would have been a chronicle of discovery, a sequence of correct theories displacing incorrect theories, the context being the state of knowledge about the earth in the author's time.

This is not to say that persons trained in a science cannot convert their approach so as to treat its development by historical standards. There are distinguished instances in later years. But I am not among them. Nor is it to deny that it is an advantage, if not quite a necessity, for historians of science to have had scientific training. The reasons are not so much technical as psychological. Except for contemporary or highly mathematical topics, one can always inform oneself about the technicalities, as I was able to do with respect to early 19th-century geology. But it is difficult though not impossible—again there are distinguished instances—to appreciate what it is to know something scientifically without having experienced it.

The department of history at Princeton offered me a job in 1947. Harvard granted me the Ph.D. in 1949, and *Genesis and Geology* appeared to almost inaudible acclaim in 1951. There was no question of my teaching history of science at the outset, and I was quite unprepared to propose any such thing. The curriculum there had the advantage for neophyte faculty that they did not have the labor of preparing courses, and instead led freshman classes and preceptorial discussion groups in the courses taught by senior faculty, whatever the subject. Thus one learned a lot of history while having time to develop one's knowledge and scholarship. When as an assistant professor I had a course of my own, it was modern English history. Only in 1956 did I feel ready to offer history of science. In the interval, I had been able to read all the titles mentioned above and many others. I was informed about courses being offered by Henry Guerlac at Cornell, by Marshall Clagett and Robert Stauffer at Wisconsin, and by Bernard Cohen and others under James B. Conant's leadership in the General Education Program at Harvard. Equally important, and in a personal way more so, I had come to know Alexandre Koyré, who spent half the year annually at the Institute for Advanced Study from 1956 until 1962.

The opportunity to offer an undergraduate course in the history of science opened with the inauguration in the curriculum of an interdisciplinary humanities program. The senior faculty responsible accepted my proposal for a course on the history of scientific ideas from Galileo to Einstein. The notion was to present something that might contribute to the liberal education of students of science and engineering while opening to students in the liberal arts an awareness of the place of science in modern history. Enrollment was nothing of a mass movement, but the undergraduates who did participate in discussion of the material throughout the next three years helped me form a sense of the themes that made for viability. I was thus able to develop the lectures into a book, *The Edge of Objectivity, an Essay in the History of Scientific Ideas* (1960).

The time must have been ripe. That book has been translated into half a dozen languages, beginning with Japanese and ending with Greek. In 1990 Princeton University Press issued a second edition, which is still in print. The preface consists of a review of the thematics of the literature in the intervening thirty years. On its first appearance I had ventured to express the hope that my book might contribute to the development of a professional approach to the history of science.

It would have been more seemly to recognize that *The Edge of Objectivity* was an early instance of such a movement already under way at the hands, largely, of the colleagues mentioned above in the second paragraph. Professional graduate study in history of science was then available only at Wisconsin, Cornell, and Harvard. My book was well enough received that Princeton thereupon agreed to my complementing undergraduate instruction with a graduate program that required additional staff.

In point of content, our attention, like that of colleagues elsewhere, was on the ways in which study of nature reciprocally formed and was formed by the world pictures of classical antiquity, the Middle Ages, the Renaissance, the Enlightenment, and modern times. In point of context, the tendency was to look to philosophy in antiquity, to theology in the Middle Ages, to art and humanism in the Renaissance, to secularism and literature in the Enlightenment, and to industrialization and military technology in modern times. With respect to science itself, the seminal transitions were what attracted scholarship: the Scientific Revolution, mechanization, the Chemical Revolution, the Industrial Revolution, Darwinian evolution. Chronologically, the center of gravity tended to be the 17th century. Other than Darwinism, much else in the 19th century and almost everything in the 20th—relativity, quantum mechanics, and genetics—awaited scrutiny. The narrative line throughout followed the route taken by the creation and transformation of scientific ideas and theories. We wrote, in a word, intellectual history of technicalities with important philosophical overtones. If social, economic, or political awareness crept in, it was around the edges.

The publication of the *Dictionary of Scientific Biography* (1970–1980) affords more objective evidence that a fledgling profession had come into existence by the 1960s, when its preparation began under my direction. The initiative came, not from a historian of science, but from the publisher, Charles Scribner, Jr., who had made a hobby of the history of science since his wartime service in cryptography. Soon after *The Edge of Objectivity* appeared, he asked whether I thought a series of books on the history of science would be viable. I had to say that most of the series known to me started off with one good book by the initiator, and then tailed off into mediocrity since few leading scholars were ever willing to write books on commission. Scribner agreed. His firm was publisher of the *Dictionary of American Biography,* however, and he then had the idea that something of the sort might be feasible in history of science. That, I thought, might work. One could probably persuade first-rate scholars to write, not whole books, but authoritative articles about figures known to them from their own studies.

What had not occurred either to Charles Scribner or myself was that preparation of the *Dictionary of National Biography* and later the *Dictionary of American Biography* had come about at a comparable stage in the formation of a professional discipline of historiography in Britain and the United States respectively. Such, quite serendipitously, proved to be the case with the *Dictionary of Scientific Biography (DSB)*. The quality of the board of editors, of the advisory committee, and of the thousand and more contributors whom it proved possible to enlist from every country with a scientific tradition other than mainland China, then incommunicado, not to mention a large grant from the National Science Foundation and sponsorship by the American Council of Learned Societies—all that succeeded, not only in the main purpose of eliciting over 5,000 articles in sixteen quarto volumes, but also in the unforeseen effect of drawing into a sense of common purpose practitioners dispersed among a miscellany of universities, institutes, national societies, and diverse academies throughout the world.

The *DSB* reflects the time in which it was conceived and composed in another way. The emphasis by design is on the content of the science created—one did not then say constructed—by the men and the few women who are subjects of the articles. The instructions requested authors to keep personal biography and extra-scientific context to the minimum required in order to explicate how the work was possible and wherein it contributed to the development of positive scientific knowledge. It is fair to say that the *DSB* was brought into being by a generation of scholars and scientists who, whatever their other differences, believed in the overall beneficence of science, as by and large did public opinion generally.

The climate of opinion changed amid the seismic shifts in cultural attitudes in the late 1960s and early 1970s. Amid the manifold, largely academic, rebellions of those years, authority became suspect everywhere, including the authority of science. In consequence what had been marginal became central, and social history became the approach of choice in historiography generally, and notably so in history of science. That development bore out a prediction by Robert Merton, to the effect that sociology of science would flourish only if and when the role of science in society should be perceived as problematic.

So it has proved. In consequence, historians of science who came to the forefront in the generation currently in its prime have tended to see sociology, and to a degree anthropology, rather than philosophy as the disciplines with which to link arms. The merit of the approach is not to establish the truism that science is a social and cultural product. No one ever doubted it. But with a few exceptions, the earlier generation never undertook much in the way of analysis of context. We produced little comparable to the fine-grained accounts that distinguish current work by recapturing the actuality of experiment; the life of a laboratory; the labor of field work in natural history and geology: the recalcitrance of instruments; the differences between what scientists say and what they do; the role of research schools; the place of patronage; the occasional cheating; the interplay of professional rivalries, of personal loyalties and hostilities, of institutional standing, of public reputations, of social position, of gender, race, material interest, ambition, shame, guilt, deceit, honor, pride. The practice of scientific research is currently shown to exhibit, in short, the springs of action that make people tick in all walks of life.

All that is to the good. At the same time, the emphasis on the practice, rather than the content, of science may entail certain drawbacks. Current authors often seem to lose interest in science once it is made. Phenomena for which it is difficult to seek any sociological dimension, say the return of Halley's comet, the law of falling bodies, or the fissionability of Uranium 235, are little scrutinized for themselves. What matters is the way they became known. In consequence, or perhaps because of that approach, the fit, if any, with nature is often taken to be ancillary at best, while analysis of the quality of the science under consideration is left aside.

Looking back at my career in the course of writing this essay, I realize that its development might be seen as a set of responses to what was happening in the historiography of science at large. If so, I was a fish in the stream under the impression that the choices were my own. Apart from the *DSB*, an organizational and editorial job, my most considerable effort has been directed toward the material covered in two books, *Science and Polity in France at the End of the Old Regime* (1980) and its sequel, *Science and Polity in France, the Revolutionary and Napoleonic Years* (2004). They are really volumes I and II of a single work. The former is being reissued with the latter, but I did not want to call it Volume I since it could have stood on its own feet if its author had fallen off his in the interval.

That research started, not in response to changing fashion in the historiography of science, but much earlier in consequence of teaching preceptorial discussion groups in Robert Palmer's course on the French Revolution during the academic year of 1951–52. That was the best undergraduate course, including any of my own, in which I have ever participated. *Genesis and Geology* had just appeared. I had begun to feel (no doubt wrongly) that English history, important though it is, held few surprises. It occurred to me that something must have happened to science during the French Revolution, as many things clearly did in this country amid the major events of the last century. The Guggenheim Foundation agreed, and its generosity allowed my wife and me to spend the academic year 1954–55 in Paris, where we have been for part of almost every year until the above work was completed.

That halcyon year was my introduction to archival research. It was clear ahead of time—and this was the attraction of the problem—that the period of French scientific preeminence in the world coincided with that in which political and military events centering in France were a turning point in modern history. The question was: what did these sets of developments have to do with each other? In the process of working that out amid the minutiae of the documents and the magnitude of all that happened in both domains, I came to feel that what I shall call the public history of science may better be elucidated through the medium of events, institutions, and practices than through abstract configurations of ideas and culture. What the relations of science and politics were I shall leave to readers of the books and not attempt to summarize here. Suffice it to say that they turned on the process of modernization in both areas and on the orientation toward the future that is always characteristic of science and was then radically characteristic of politics.

My career, such as it is, has unfolded not in accordance with some agenda, but as a set of responses to a series of lucky accidents—being a historian by nature who happened to study chemistry and mathematics, taking up Charles Scribner's idea for the *DSB*, precepting in Palmer's course on the French Revolution. Personal rather than professional encounters made possible two of the four books that are spin-offs from the research on French science. During our many sojourns in France, my wife and I chanced to meet descendants of two distinguished families, the Carnots and the Montgolfiers. Lazare Carnot has been known to historians only as the "Organizer of Victory" during the revolutionary wars. So he was, but he spent only six years in government during a long life, most of which was occupied with highly original work, not fully appreciated at the time, in mathematics and physics.

Learning of my interest in that aspect of his life, current members of the family arranged for me to spend a summer going through Carnot's papers, which no one had ever seen, in the house in Burgundy where he was born. The result was *Lazare Carnot, Savant* (1971), to which book my esteemed colleague A. P. Youschkevitch of the Soviet Academy of Sciences contributed a chapter. That was another lucky break. He was the only other historian of science who had ever taken an interest in Carnot. In the midst of a discussion about Russian collaboration in the *DSB*, I mentioned a hint in papers I had seen that Carnot had submitted an early draft of his book on the foundations of the calculus to a prize competition set by the Prussian Academy of Sciences. On his way back to Moscow he searched its archives in East Berlin, found it, and contributed a chapter analyzing Carnot's approach.

I knew, of course, that hot-air balloons are called *montgolfières* after the brothers Joseph and Etienne, who invented them in 1783. On meeting Charles de Montgolfier at a wedding reception, I asked whether he was descended from the big balloon. Sure enough, collaterally at least, and since I expressed interest, he invited us to visit in the country house in Annonay, where his ancestors were in the paper business. There he showed me designs, sketches, correspondence, all scattered among drawers and attics in his and his cousins' houses. Thence *The Montgolfier Brothers and the Invention of Aviation, with a Word on the Importance of Ballooning for the Science of Heat and the Art of Building Railroads* (1983). I give the full title (though aeronautics would have been more accurate than aviation) since it suggests, that even like Carnot's work in mechanics, Joseph de Montgolfier's further inventions (which to him were more important than the balloon), along with those of his nephew Marc Seguin, belong to the pre-history of the physics of work and energy.

Two other publications were happenstance in different ways. Firestone Library in Princeton University is fortunate to possess a rare deluxe printing of the *Description de l'Égypte*, this one having been presented by Napoleon to the king of Prussia and bought at auction in 1865 from an impoverished descendant of a Prussian courtier by Ralph Prime of the class of 1843, later one of the founding trustees of the Metropolitan Museum in New York. It had been clear from the outset that a chapter on the scientific component of Bonaparte's Egyptian expedition would be important in my book. While studying the gorgeous plates, I bethought me that a former student who had just started an architectural publishing business might be interested to see them. He turned over a few pages, and said, "Wow, can we do that?" It had never occurred to me to reproduce them, and that was the origin of *Monuments of Egypt, the Napoleonic Edition*, 2 vols. (Princeton Architectural Press, 1987), which I edited in collaboration with Michel Dewachter, an Egyptologist then with the Collège de France.

In like manner, *Pierre-Simon Laplace, a Life in Exact Science* (1997) emerged from an earlier publication, in this case the *DSB*. I had never intended to write a book about Laplace, who lies on the frontier of my ability to follow mathematical reasoning other than qualitatively. Unfortunately, or perhaps fortunately, two colleagues who had successively undertaken to contribute the article on Laplace failed one after the other to keep their commitments. *Faute de mieux* Laplace devolved upon the editor as default author. I worked on him for a year, harder than I have on anything else, and with the collaboration of Robert Fox and Ivor Grattan-

Guinness for particular topics, produced a lengthy article, of which the subsequent book is a revision and enlargement.

Thus, exposure to archives and the close-in research required for these books, as well as editing the articles, many of them very technical, in the *DSB*—these were the experiences that led me to think that limiting one's attention largely to the history of scientific ideas and theories was like following the tips of icebergs, except that the history of science is anything but a frigid subject matter. One might perhaps consider that my individual development exemplifies Auguste Comte's dictum to the effect that, just as every discipline passes through theological and metaphysical stages before becoming positive, so every person is a theologian in infancy, a metaphysician in youth, and a physicist on reaching maturity.

However that may be, the discipline of the history of science has reached maturity. The first meeting of the History of Science Society I attended in 1952 comprised thirty or forty persons, for few of whom was the subject a livelihood. The most recent numbered upwards of 600, the great majority of whom are professional scholars in the discipline. The Society has an endowment and an office with an executive officer. A hundred or more books and collections are reviewed in every issue of the quarterly *Isis*. All that spells success. In only two ways do I feel some slight twinge of regret or disappointment, the first with respect to science and the second with history.

The perception of science as socially problematic in the 1970s and 1980s stemmed in some degree, though by no means entirely, from widespread feelings of anti-scientism in academic and literary circles. In consequence, science studies, whether sociological, political, historical, or a mixture, are often perceived by scientists as hostile enterprises. The most obvious complaint is that critics with no technical qualifications to understand the subjects they discuss are violating the precincts of science. The accusation is nonetheless damaging for being usually, though not always, incorrect or irrelevant or both. The second-order concern among scientists is that the image of science is thus tarnished at a time of weakened political support and stringent restrictions on funding. But the sense of offense goes deeper. While willing to agree that questions of power and advantage are factors both in the macro- and micro-politics of science, scientists resent any implication that their work serves no purpose larger than their own, that they are not in the last analysis investigators of the nature of things, that objectivity is an illusion and rationality a sham. There is the counter-cultural *casus belli* of what journalists have called the science wars.

There was, as well as I can recall, no sense of resentment or hostility to the history of science during the time when our discipline was getting into its stride. On the contrary. We met with every encouragement, institutional and moral, on the part of scientific colleagues. We needed it. I doubt that the discipline could have matured in the face of their enmity and contempt. I do not think that any discipline can flourish in a healthy manner in a mood of hostility to its subject matter. Not that one would argue that prudential reasons should lead historians, or social scientists generally, to refrain from critical and even skeptical scrutiny of the objects of their studies. Still, if we are to recreate the past, the essential matter is to see the subject whole. To set out to see through it is to turn the creatures one studies into specimens. By and large, however, I feel optimistic and think the tide of anti-scientism, if that is what it was, has turned. Much of the work of recent years engages science and scientists on their own terms as well as on the author's.

The slight disappointment has to do with history. It was our hope at the outset, even our expectation, that the historical profession would come to accord the role of science in history a place comparable to that of politics, economics, religion, diplomacy, or warfare. Science after all has been a factor shaping history no less powerfully than have those other sectors. That has not happened. A few departments of history—Princeton's among them—do offer undergraduate and graduate work in the field. But at many, and perhaps most institutions, the subject is taught, if at all, in a separate department or under the aegis of a science and technology studies program. Nor are writings in the history of science as widely read as are those in the conventional fields. The best known, unfortunately in my view, are those written in a more or less iconoclastic vein. Perhaps the barrier is psychological. There may be a fundamental divide between temperaments drawn to history and those drawn to science. At Princeton more of our undergraduate students are majoring in science, engineering, and pre-medical programs than in history or literature. The famous, or infamous, two cultures problem may well be real. Still, we work in hopes that it may be abated.

Note

1. "Apologia pro Vita Sua," *Isis,* 90 Supplement (1999): §84–§94.

PART I

Early Papers

—⁂—

Physick and Philosophy: A Study of the Influence of the College of Physicians of London upon the Foundation of The Royal Society

PERHAPS IT WOULD be well to explain a little more fully than does the foregoing essay how this, my earliest scholarly publication, came to be written and how its author came to be a historian of science at all. For as an undergraduate at Wesleyan University, Class of 1940, I majored in chemistry. From the perspective of Bethlehem, Pennsylvania, in the depths of the depression, companies like Dupont looked to be a better bet than the steel company in which my father was making his career. Chemistry in college was my duty, and I could do it. Most of the courses I took for pleasure, however, were in history. A small, truly liberal arts institution, Wesleyan allowed me to write an honors thesis in history so that I graduated A.B. in chemistry with honors in history.

Graduate study in chemical engineering at MIT in 1940–41 ensued before the Army reached down to me. The military authorities in their wisdom were more impressed with knowledge of chemistry than with a sense of history, however profound, and clapped me into the Chemical Warfare Service. Packed off to Officer Candidate School at Edgewood Arsenal after basic training, I was commissioned and assigned to service in Company A, 94th 4.2″ Chemical Mortar Battalion. The weapons were designed to fire gas shells, but could also be served with high explosives for heavy bombardment and with white phosphorus for incendiary and smoke screening effect. Nothing could have been less pertinent than knowledge of chemistry to my duties, first as platoon leader, finally as company commander in the XV Corps, Third Army, ETO. When the bombing of Hiroshima, as it happened on my 27th birthday, ended all that, to the inexpressible relief of those of us facing the prospect of an invasion, I decided that that was enough duty for one lifetime and that, thanks to the GI Bill, I would do what I liked.

Graduate study in history at Harvard in the spring semester, 1946, was pure joy. The history department at Wesleyan, like many others before the war, had emphasized English history. I gravitated that way at Harvard and enrolled in W. K. Jordan's seminar on the later Stuarts. That was a formative experience. Thinking (as I have explained in the Introduction) somehow to draw on a scientific background in the study of history, but without the slightest idea of how to go about it, I looked for a topic involving science and lighted on the obvious choice of the foundation of the Royal Society. The study that follows was the consequence.

The draft submitted in seminar was grossly overwritten. I had been editor of the college paper at Wesleyan and mistakenly thought I knew how to write in an interesting manner. Jordan's criticism was kindly in manner but relentless about style. "But very badly written" were his words, and I was, therefore, the more astonished when he suggested sending the revision to *The Journal of Modern History.* Imagine my greater astonishment when the editors accepted it! Now that I read it over, sixty years later, I am still embarrassed by the style. As the reader will observe, the paper is full of unnecessary and irritating adjectives and jejune remarks about how things done then would look now.

The substance does hold up, however. The paper has been cited a number of times and does not appear to have been superseded. It brought me to the attention of Dorothy Stimson, then Dean of Goucher College, soon to be President of the History of Science Society, and a great lady whose *Scientists and Amateurs* (1948) is the standard history of the Royal Society. The first of what appear to be positive features of this early article is that I had the wit to perceive that the role of medical people was an interesting problem. Secondly, the amount of counting I did is surprising. If the genre of quantitative history had yet been invented, I was unaware of it. But the treatment does verge on the statistical. What led me to count noses? Was it a scientific background, in which number and quantity do fit naturally? Was it that I had been reading Namier, who did a good deal of it? I simply do not remember. What also strikes me favorably is the choice of quotations, not to mention the stylistic pleasure of rereading 17th-century English. Finally, the conclusion is a more mature piece of writing than are most of the passages throughout.

ONE

Physick and Philosophy:
A Study of the Influence of
the College of Physicians of London
upon the Foundation of
the Royal Society*

—⚊—

This *phisition shall* continually practize together with the naturall philosophor, by the fire and otherwise, to search and try owt the secreates of nature, as many waies as they possiblie may. And shalbe sworne once eu*ery* yeare to deliu*er* into the Treasorer his office, faire and plaine written in Parchment, without *Equiuocacions or Enigmaticall phrases*, under their handes, all those their proofes and trialles made within the forepassed yeare . . . .To thend that their *Successors* may knowe both the way of their working, and the event thereof, the better to follow the good and avoyd the evill, which in time must of force bring great thing*es* to light.
— SIR HUMPHREY GILBERT, *Queene Elizabethes Achademie.*[1]

Queen Elizabeth may have missed a golden opportunity to win distinction as a patron of science, for she never adopted Sir Humphrey's suggestion that an "achademie" of learning be established in her court. No doubt, however, the idea was a bit premature. The founding of a royal society had to wait almost a hundred years until the expansion of experimental knowledge, conventionally described as revolutionary, made the success of such an institution imperative for further progress.[2] In the extensive literature devoted to an event so happily productive of a rapid growth in scientific understanding there has been overly much neglect of the profession to which Sir Humphrey would have looked for the "phisition" on his faculty.[3] Perhaps Sir Humphrey Gilbert

* Reprinted from *The Journal of Modern History* 19, No. 3, September 1947, pp. 210–225.

had never been ill, for he placed no mean valuation upon the intellectual prowess he expected in an Elizabethan doctor. The academy's master in physic would have enjoyed the top stipend of one hundred pounds, no less than that of the instructor in military arts while their colleague, the natural philosopher, would have had to get on with forty.[4] The prospect of so princely a remuneration suggests, perhaps, a contemporary appreciation of what may at first surprise the modern student who develops a nodding acquaintance with the scientific circles of early Stuart England, for, moving in those circles, he will repeatedly encounter a lively set of doctors in physic, enthusiastically engaged in educational and "philosophic" hobbies remarkably remote from medicine itself.

A 20th-century American, eager to improve himself by journeying onward and upward with the arts in his community night school, would, no doubt, think it odd to find himself getting up his astronomy, geometry, music, or rhetoric under the tutelage of the local general practitioner. So his ancestor in Restoration London would have been doing, however, had he chanced, in 1664, to be attending Gresham College, the earliest such institution of adult education; for there, in that particular year, five of the seven teaching posts were held by practicing physicians[5]—the four mentioned and, of course, physic itself. Any student who introduces himself to the profession by way of the misleadingly dreary biographical sketches in *The Roll of the Royal College of Physicians* will continually be coming upon similar instances of extra-medical pursuits.[6] It seems, therefore, worth while to inquire whether the profession as a whole did not develop avocational interest patterns which rendered it a notably dynamic element in the intellectual development of Stuart London. But rather than multiply a wealth of somewhat scattered examples, it may be more conclusive to investigate the influence apparently exerted by the College of Physicians upon the foundation of the Royal Society, for that event marks most conveniently the absorption of the "new philosophy" into London's scientific climate of opinion, and in bringing it about the physician and the natural philosopher did certainly practice together to search and try out secrets of nature in ways more various than Sir Humphrey could have dreamed of.

Ordinarily, historians regard the Royal Society as the earliest scientific body in England, and from the point of view of rigorous definition, so, of course, it was. Almost a century and a half earlier, however, Henry VIII had been prevailed upon to issue letters patent establishing the College of Physicians.[7] The layman might reasonably hope today that a professional organization of physicians would conduct its affairs in a manner at least partially scientific though it may be necessary to demonstrate that such was the case in the time, say, of James I. Unfortunately, we know very little directly of any scientific ferment which may have occurred within the walls of the Royal College under the first Stuarts. There is, however, plenty of evidence of the fascination which the new pursuits of natural philosophy exerted over its individual members.[8] If the college's institutional life was a vigorous one and by no means a merely formalistic affair, it may well be supposed that an opportunity for mutual association in a professional corporation materially advanced the successful prosecution of the fellows' scientific researches. And sufficiently revealing reflections of the college's real vitality may easily be discerned through records of the rather curious quasi-governmental supervision which it exercised over both the medical and the allied professions of pharmacy and surgery.[9]

"By the sentence of the College," flatly declared the president and censors to his lordship of Durham, "Mr. *Lambe* stands convict and guilty of all manner of insufficiency and ignorance in this faculty."[10]Buckingham's really infamous favorite, however, was only the most notorious of the many charlatans whose powerful friends at court were refused the favor of a medical license by which their protégés might profit, and this in an age of fairly venal judicial administration. No person could legally practice medicine in London without the license of the Royal College, and such spectacular cases as John Lambe's were no more than high lights in the college's day-to-day effort to enforce maintenance of proper medical standards. The censors themselves initiated many prosecutions, a time-consuming chore for busy professional men. Offenses were also referred to their jurisdiction by aggrieved, and presumably uncured, victims of quackery or by justices of the peace, in accordance with standing instructions from the privy council. Of course, the college did not succeed in rooting out the quacks who battened on the populace. No doubt, the job would have taxed the resourcefulness of a Gestapo, and London had no police force at all. But successive censors struggled valiantly with the problem under the first two Stuarts. The number of convictions handed down against malpracticing doctors, surgeons, apothecaries, and "empiricks" can at least bear witness to the vigor of an institution which could command of its members so wearisome and unprofitable a service.[11] One runs across nothing of the sort, for example, in *The annals of the Barber-Surgeons,* which preserves the records of a typical craft guild, at this period still more medieval than professional in atmosphere.[12]

Since every doctor legally entitled to nail up his brass plate in London was a licentiate of the Royal College, an inclusive study of the respectable part of the profession is feasible from its roll, lovingly assembled by the somewhat heavy Victorian hand of Dr. Munk.[13] During the reign of James

I a total of eighty-two physicians[14] held the license which entitled them to persuade long-suffering patients that just one final phlebotomy would be judiciously calculated to right the balance of an overcholeric humor. About half the number were practicing at any one time.[15] From this group, one may construct a table reflecting, without undue distortion, the vocational and avocational interests of the profession as a whole. This summary, however, may not be entirely accurate, for some of the source material is little more than an occasional reference to an unpublished manuscript or to a book or pamphlet not available in this country and the title of which may well have been misleading. Seventeenth-century authors ran to full, and even fulsome, titles, often more intriguing than descriptive. Still, the classification seems well worth attempting; and in doing so, a "field of activity" has been ascribed to an individual doctor on the basis of printed information about his manuscript remains, his own publications, and remarks about him in contemporary literature (Table 1).

These neat categories to some extent, of course, read into the past distinctions which it did not make and reflect some overlapping in the extraprofessional exploits of so energetic a group. Nonetheless, taking only what seems to have been the chief outside interest of each, twenty-six of the eighty-two physicians have left distinct traces of their devotion, in their leisure hours, to purely scientific investigations. They may not have added very much to the sum of human knowledge, but then neither did most of the scientists. And, no doubt, later giants like Boyle and Newton would scarcely have risen to such heights except from a social milieu intellectually prepared to smile upon philosophic endeavor.

A number of the doctors whose extramedical pursuits appear in Table I also produced material in their own fields, of course, although in this period laymen published a larger number of books dealing with the plague and with health in general than licensed practitioners did.[16] Only the idea of patents is new in patent medicine. Effusive "empiricks," enthusiastically and profitably peddling *aurum potabile* in various nostrums more romantic and appealing than little liver pills, were responsible for the bulk of the medical pamphlets contributed to the public prints by laymen. Discreetly concerned to publicize their products rather than themselves, most of these authors chose a cautious anonymity as the better part of literary valor.

These medical tracts leave one filled with admiration for the stamina of patients who survived the treatments they describe. Unfortunately, however, no discussion of the literature or more than a hasty mention of the scientific importance of even the more outstanding doctors in the early century is possible here. By and large, respectable how-to-keep-

Table 1
Membership (82) in the College of Physicians, 1603–25

Field of Activity	Number of Participants
I. Extra-medical pursuits:	
A. Natural sciences:	
Astronomy, navigational theory, mathematics	7
Chemistry	7
Physics	5
Mineralogy and meteorology	2
Botany and entomology	5
Psychology	1
	27
B. Nonscientific interests:	
Formal logic—epistemology	1
Music and belles-lettres	5
Ancient and modern languages	3
Military, explorative, or diplomatic expeditions	12
Theology and education	11
	32
II. Medical publications only	9
III. No record of extra-professional activity	18

Note: The following sources have been used: John Aubrey, *Brief lives,* ed. Andrew Clark (Oxford, 1898); C. H. and Thompson Cooper, *Athenae Cantabrigiensis* (Cambridge, 1858–61); *The dictionary of national biography;* R. T. Gunther, *Early science in Cambridge* (Oxford, 1937), and *Early science in Oxford* (Oxford, 1922–34); J. O. Halliwell-Phillipps (ed.), *A collection of letters illustrative of the progress of science in England* (London, 1841); Sir Theodore de Mayerne, "Extrait des œuvres" (title of an unpublished manuscript in Houghton Library at Harvard University); Sir Norman Moore, *The history of St. Bartholomew's Hospital* (London, 1918); Munk; A. W. Pollard, G. R. Redgrave, and Others, *Short-title catalogue of books printed in England, Scotland, and Ireland, 1475–1640* (London, 1926); S. J. Rigaud (ed.), *Correspondence of scientific men of the 17th century* (Oxford, 1841); J. and J. A. Venn, *Alumni Cantabrigiensis* (Part I; Cambridge, 1922–27); Ward; Anthony à Wood, *Athenae Oxoniensis* (3d ed.; London, 1815–20); and Young.

healthy books—that is to say, those written by physicians—reflected a growing reliance upon clinical experience, ill-digested and little understood certainly, but with overtones of a Baconian revolt against Galenic authority. It was only in the literature inspired by the plague that the Royal College's contributions were practically indistinguishable in tone from the most superstitious and obscurantist writings of any "quacksalver" or Puritan divine.

It will be charitable lightly to pass over the profession's record in treating and prescribing for the plague and to suggest that many physicians seem to have thought of themselves less as doctors than as practicing natural philosophers whose field simply happened to be medicine. Harvey's great book, for example, conveys this sense: "For true philosophers, who are only eager for truth and knowledge, never regard themselves as already so thoroughly informed, but that they welcome further information from whomsoever, and from whencesoever it may come."[17] The text of *De motu cordis* ... is impressive as a classic triumph of modern scientific description rather than as a chapter in the art of healing. Though he published nothing until 1628, Harvey had then "for nine years and more" been explaining his researches, formulating his inductions, and testing and demonstrating his hypotheses before the Royal College, where he undertook the Lumleian lectures in 1616.[18] We know little enough about this series, except that Harvey delivered it for forty years. Nor do we have much information about the Goulstonian course, endowed in 1632; but the existence of such lectures indicates that legal prosecutions were by no means the college's sole concern.[19]

In the generation preceding Harvey's, William Gilbert had completed the great researches on magnetism which were for him only a side line to his busy medical practice. He served successively as censor, treasurer, and president of the College of Physicians. He must have found its fellowship rewarding, for he willed his library, globes, instruments, and cabinet of minerals to the college.[20] If his colleagues had not been interested in that sort of thing, Dr. Gilbert would, no doubt, have provided for a more appropriate disposition of objects then so rare and precious. Very likely they were useful to Harvey's contemporaries, in the diverse mathematical, astronomical, physical, and botanical investigations associated with the names of such randomly selected individuals as Dr. Thomas Hood, Dr. John Bainbridge, Dr. John Farmery, Dr. Thomas Fludd, and Dr. Peter Turner. But even without discussing these gentlemen or the chemical researches of Sir Theodore de Mayerne, an unusually able society doctor, it seems abundantly clear that the medical profession as a social group was attracted to the study of extra-medical sciences in early Stuart London.

If this was indeed the case, one would expect to find doctors of physic in sympathetic attendance on the birth pangs of the Royal Society. The story of its foundation is a more than twice-told tale and for the events themselves needs no elaboration here.[21] The Rev. John Wallis has left the most complete accounts of informal gatherings, "about the year 1645," among a coterie of scientific amateurs,[22] whose eager discussions of the new experimental approach to knowledge became the roots of the Royal Society.[23] Meeting weekly, there were, besides Wallis himself, "Dr. [Rev.] John Wilkins, Dr. Jonathan Goddard, Dr. George Ent, Dr. Glisson, Dr. Merret, Mr. Samuel Foster, then Professor of Astronomy at Gresham College, . . . and many others."[24]

In 1648 or 1649 some of the leaders moved to Oxford, and for a time the brilliant scientific lights of that center appear to have thrown what was left of the London circle into shadow. Those in the capital continued their custom of forgathering after Christopher Wren's weekly lecture at Gresham College, until in 1658 their meeting place was taken as a billet for soldiers—a source of "miserable distractions [which] . . . might have made them run the Hazard of the fate of Archimedes."[25] According to Sprat, the London nucleus centered around Lord Brouncker, Sir Paul Neile, John Evelyn, Abraham Hill, Henry Slingsby, Thomas Henshaw, Dr. Timothy Clarke, Dr. George Ent (the most ardent of Harvey's youthful admirers), and Dr. William Croune.[26]

The more active Oxford group assembled habitually in Dr. William Petty's lodgings until their host's departure for Ireland in 1651. In that year under the leadership of John Wilkins, master of Wadham College, the Oxford Philosophical Society organized itself more formally, though it bore only a geographical relationship to the university. Sometime members between 1648 and 1660 were Christopher Wren, Laurence Rooke, Robert Boyle, Robert Hooke, John Wilkins, John Wallis, Seth Ward, Dr. William Petty, Dr. Ralph Bathurst, Dr. Thomas Willis, Dr. Jonathan Goddard, and Dr. Richard Lower.[27] On the evidence of their common enthusiasm for inductive feasts of reason, the individuals mentioned in the last three paragraphs may be regarded as among England's scientific leaders of opinion during the civil wars and interregnum. Ten of the twenty-five either were or soon became fellows of the College of Physicians.

By 1659 the prospect of sharing the fate of Archimedes seemed improbable in London, and most of the more able associates of the Oxford Society were again in town. On November 28, 1660, after about a year of meeting together informally, the group resolved to constitute itself a college "for promoting of physicomathematicall learning."[28] Forty initial members subscribed to the project, and only three dropped out before the new society's incorporation in 1662. Fourteen doctors of physic were among the thirty-seven who were serious enough about their philosophical endeavors to continue as fellows in the Royal Society.[29]

Interest grew very rapidly in the next few years. The "new philosophy" had suddenly become the fashion. King Charles encouraged the budding organization very graciously indeed and equally graciously regretted his inability to afford any marks of his esteem more tangible than a handsome silver mace and permission to use the adjective "Royal." For various obscure legal reasons, the one hundred and fifteen names listed in the Royal Society's second charter, granted on May 20, 1663, constitute the officially recognized original fellows.[30] According to Sprat, writing in 1667, "the farr greater Number are gentlemen free and unconfin'd," vaguely interested in natural philosophy, but not productive. Professional men were the real "Masters and Scholars, some imposing and all the others submitting; and not as equal observers without dependence."[31] The names of twenty-four physicians appear among the professional men. Four of them, Petty, Clarke, Ent, and Goddard, were members of the governing council. To these should be added Dr. Alexander Fraiser, Dr. Thomas Willis, and Dr. Ralph Bathurst, who were out of town at the time of subscription but, having been among the earlier leaders, joined later in the year.[32]

Assured that the king approved its designs, the new society devoted its meeting of December 12, 1660 to framing requirements for membership. No one was to be admitted without scrutiny, "except such as were of, or above, the degree of barons." A further resolution ran: "Any of the Fellows of the said College [of Physicians], if they should desire it, be admitted as supernumeraries upon condition of submitting to the laws of the society, both as to the payments . . . and the particular works, or talks that should be allotted to them." Evidently it was expected that medical practitioners would exhibit a professional aptitude for scientific researchers. An analogous provision recognized the ex officio qualifications of the professors of mathematics, physic, and natural philosophy in Oxford and Cambridge. Only those who merited election as barons or better, however, were free from an obligation to undertake whatever investigations might be assigned to them. The same day someone proposed that the College of Physicians would be a "convenient accommodation for the assemblies of the society."[33] Actually nothing came of the idea, and meetings continued at Gresham College.

The College of Physicians responded cordially to the blanket invitation extended its fellows by the new organization. Its membership numbered sixty-two licentiates at the Restoration, and nearly two-thirds of these joined the Royal Society, twenty-seven as original fellows and twelve more in the ensuing few years.[34] Nor, in the event, did the doctors turn out to be scientific dilettantes. During the year following the meeting of December 12 mentioned above, the Royal Society devoted its attention to one hundred and seventy-one separate projects, some astonishingly trivial, others very profound. Physicians worked on ninety-two of these, either individually or, more often, on committees appointed *ad hoc* to conduct specific researches.[35] Some conclusions about the interests of a social group may certainly be based upon the number of people in it who participate constructively in a dynamic intellectual movement.

And there seems to be tolerably complete statistical evidence of the London medical profession's vital contribution to the foundation of the Royal Society and to the deepening current of scientific curiosity which produced it. But perhaps a brief discussion of a few of the individuals concerned may better illustrate the sources of the intellectual leadership exerted by the College of Physicians.

Sir Thomas Browne, for one, cannot be dismissed simply as a rather charming curio in the history of science. No doubt his mind was credulous, his prose lyrical, and his thought romantic.[36] Probably he never penned a single word in the state of calm, intellectual detachment which would have befitted his idealized empiricism. However engaging it may be, a generously uncritical enthusiasm for scientific caution is likely to defeat its object. But, after Bacon, no one raised so eloquent a tumult in praise of the exciting new experimental approach to nature and to knowledge. Of course, the *Pseudodoxia epidemica* singled out for destruction no more than an illogical fraction of the great body of received myth. Often the book exploded a hoary old legend only to replace it with a wildly original fancy drawn from its author's own rich store of superstition. Nonetheless, in the self-imposed tilt he ran upon philosophic authority, that "mortallest enemy unto Knowledge,"[37] Sir Thomas laid about him shrewdly and exposed a vast collection of "Vulgar Errours" to the salutary ridicule of common experience. And who could be better armed to do so than a doctor? "The course of our Profession . . . leadeth us into many truths which pass undiscovered by others"—though, too, the fact that one's patients were forever falling ill at the most inconvenient moments occasioned troublesome interruptions in the smooth stream of philosophic composition.[38]

Fluent propagandists swell an intellectual current, but expansion of scientific knowledge depends more, no doubt, on the dry and often dusty laboratory technician and the unromantic apparatus he devises. In this direction, the leadership in the nascent days of the Royal Society was assumed by "an admirable chymist," Dr. Jonathan Goddard.[39] Goddard's home in London served the society for a laboratory, and he played host to many meetings from 1645 to 1651 and again from 1659 to 1663, "on occasion of his keeping an Operator in his house for grinding Glasses for Telescopes and Miscroscopes."[40] The doctor took an active part in interregnum politics, rather successfully on the whole.[41] Yet he always found time to tinker with the philosophic devices which he built in his leisure hours. "When any curious experiment was to be done, they [the Royal Society circle] made him their drudge till they could obtain to the bottom of it."[42] To Dr. Goddard, as to Sir Thomas, it seemed that the intellectual development of a respectable physician inherently involved an interest in science. For, though literary style seems, unfortunately, not to have been one of them, a proper professional equipment required many talents of a doctor:[43]

> ... learning of languages; ... then in Arts, some whereof minister advantages to the understanding of the nature and causes of things; all do improve the Mind and Understanding, by exercise at least, ... then to apply his study to Natural Philosophy, such as is more real and solid in this Age, by many happy Experimental discoveries in Nature: and lastly to the Art of Physick, and the knowledge of the Body of Man, by Anatomical Administrations, Experiments, and Observations.
>
> [And it will be necessary for] a Physician to make it his continual Work, to improve in the knowledge of all these (which his interest must incline him to do).

After the Restoration, Goddard and his chief lieutenant, Christopher Merret, triumphantly led the literary forces of the College of Physicians through a virulent battle of the books in which the Apothecaries Company incautiously engaged them. Dr. Merret, like Goddard a gentleman of all-embracing curiosity, had been a member of the Royal Society circle since its earliest days. He had become very depressed indeed about the unreliability of apothecaries, so much so that he resorted to compounding his own prescriptions, which he dispensed gratis, "for saving my Patients' Lives and Purses." And he urged his colleagues likewise to apply the enlightened standards of their age, and to "avoid all pompous, useless, Medicines of the shops, ... the precious Stones, Sapphyres, Emeralds, etc., the high priced Magistrals of Coral and Pearl, as also Unicorns' Horn and Bezoar; ... to lay aside those unintelligible and unreasonable Compositions of Mithridate, Trecle, and the So much Magnified Treacle-Water."[44]

Merret's interest in pharmaceutics carried him into zoology and botany, on which, at the request of the Royal Society, he published his researches.[45] Though scarcely remarkable for original observation, his *Pinax*
employed an unusually efficient scheme of classification. More importantly, perhaps, it was quite free of the teleological speculations characteristic of contemporary excursions into natural history.

In later years he turned his attention to improving the technology
of glass manufacture. His originally annotated edition of the works of
Antonio Neri, the Florentine glassmaker, is available in this country in a
Parisian compilation of 1752. According to his French editor, Merret's
"recherches fastidieuses sur l'origine, les propriétés, et l'excellence du
verre" had won him a "réputation . . . déjà si bien établie parmi les Savans,
qu'il seroit superflu d'insister sur cette matière."[46] More significantly from
the point of view of method, however, Merret's researches into the nature
of cold were conducted simultaneously with those of Robert Boyle. Boyle
has described how, in 1662, "At the same time that the *Royal Society* required of me an Account of what I had observed, or tried, concerning cold,
they recommended the making of trials about that subject to the Learned
Dr. C. Merret." Dr. Merret, "having dispatched what he intended," the experiments which each performed were laid out separately and presented
in a joint publication.[47]

By specifically setting their members just such problems, learned societies acted as essential catalysts in the scientific revolution of which they
were both part and product. In them were pooled the fundamental little
scraps of knowledge, hitherto scattered unproductively among persons
who had had no way of learning that what one sought, another knew. Further than this, these associations facilitated an efficient division of the
scholarly labors devoted to experimental investigations. Perhaps the most
vital stimulus which fellows of the College of Physicians imparted to
founders of the Royal Society lay in setting them striking examples of the
concrete results made possible by such joint efforts. For even before its incorporation, Dr. Francis Glisson, Dr. Thomas Willis, and Dr. William Petty
had headed co-operative research projects of major practical importance.
Their systematized attacks had yielded returns clearly beyond the capacity of any individual to achieve.

Francis Glisson, regius professor of physic in Cambridge from 1636 to
1677, was the chief author of the first complete English account, both
anatomical and clinical, devoted to any single disease. The famous treatise
on rickets still remains a masterpiece of descriptive research.[48] Glisson
had been an enthusiastic member of the "invisible college" ever since its
earliest days in 1645. Like many of his colleagues, he never permitted his

professorial duties unduly to restrict his freedom. Indeed, for many years he lived less in Cambridge than in London.[49] Perhaps there still was such a thing as academic leisure in the seventeenth century, but very likely the modern professor would burn with a less ardent pedagogical zeal if his university casually overlooked the matter of salary payments for five successive years. The College of Physicians engaged the time Glisson had left over from his teaching and his work for the Royal Society. He served as its anatomy reader in 1639, Goulstonian lecturer in 1640, president from 1667 to 1670, and intermittently as elect, censor, and councilor.[50]

The first distinct identification of rickets occurs in the Bills of Mortality for 1634.[51] Glisson had, by then, become interested in the twisted bones, enlarged joints, and stumbling gait of so many unfortunate children in his native Dorset. By 1644 he had amassed a quantity of notes upon his observations there, and he submitted this material to his colleagues in London. A committee of eight doctors was deputized to investigate the disease further under Glisson's direction. In this lies the chief significance of the venture, because the effort to conceive it as a co-operative research project foreshadowed the distinctively modern method of attacking particular scientific problems.

After Glisson himself, George Bate and Ahasuerus Regemorter seem to have been the most energetic of the group. Each doctor was to collect data in a prearranged subdivision of the field and to report the results of his labors. After five years of observational study, they published the treatise under the sponsorship of the College of Physicians. Dr. Glisson, properly enough, received major credit for authorship, since he had undertaken more than the lion's share of the work. But the preface contains acknowledgments of the division of labor among his associates and of their collective contributions.[52] Nor was this the only occasion on which the Royal College played the role of research entrepreneur. Glisson later attributed his discovery of that part of the human organism which medical students still call Glisson's capsule "to the mandate of the College of Physicians . . . [to] remove and study the parenchyma from the liver of a large number of animals," in preparation for a course of public lectures.[53]

Meanwhile in Oxford, Dr. Thomas Willis, Sedleian professor of natural philosophy, struggled with the task of passing through the press his investigations into the anatomy of the brain and nervous system.[54] It may have been this which caused him to miss the opening sessions of the newly incorporated Royal Society. Dr. Willis' earlier intellectual development might be thought almost too pat a case study in the assimilation of modern scientific method were it not so characteristic of his generation. In his younger days he had been guilty of two very silly treatises, on fevers and on fermentations,[55] and in his *Anatomy of the Brain* he laments his youthful philosophic indiscretions,[56] whence he had originally framed

> some not unlikely Hypotheses, which (as uses to be in these kind of businesses) at length accrued into a certain System of Art and frame of Doctrine. But when at last the force of Invention being spent, I had handled each again, and brought them to a severer test, I seemed to myself like a Painter, that had delineated the Head of a Man, not after the form a Master, but at the will of a bold Fancy and Pencil, and had followed not that which was most true, but what was most convenient, and what was rather desired than what was known. Thinking on these things seriously with myself, I awaked at length sad, as one out of a pleasant dream; to wit I was ashamed that I had been so easie hitherto, and that I had drawn out for myself and Auditors a certain Poetical Philosophy and Physick, neatly wrought with Novity and Conjectures and had made a *Fucus*, as it were, with deceits and incantations for either of us.

Wherefore, he resolved to make amends in his next work:

> I determined with myself seriously to enter presently upon a new course, and to rely on this one thing, not to pin my faith on the received Opinions of others, nor on the suspicions and guesses of my own Mind, but for the future to believe Nature and ocular demonstrations: Therefore thenceforward I betook myself wholly to the study of anatomy: and as I did chiefly inquire into the offices and uses of the Brain and its nervous Appendix, I addicted myself to the opening of Heads especially and of every Kind. . . . [By exhaustive study of comparative anatomy,] a firm and stable Basis might be laid, on which . . . a more certain Physiologie . . . and Pathologie of the Brain and nervous stock than I had gained in the Schools might be built.

One man's wit and capacity could scarcely suffice for such a task, however. In blocking out the research, therefore, Willis sought assistance from his friends and scientific associates. Actual dissection was delegated chiefly to Dr. Richard Lower, because of his superior dexterity with the knife. Dr. Thomas Millington contributed his wider anatomical experience, in consultation with Dr. Edmund King and a certain Dr. Masters. And Christopher Wren "was pleased out of his singular humanity, wherewith he abounds, to delineate with his most skillful hands the Figures of the Brain and Skull, whereby the work might be more exact."[57]

By perceiving the latent possibilities of an approach which co-ordinated the observations of specialists in a descriptive synthesis. Dr. Willis was not, of course, being particularly original, although—as so many critics of Bacon have somewhat condescendingly pointed out—perceiving the method itself and implementing it are two rather different things. The ubiquitous Dr. William Petty, that man of an almost frightening variety of talents, had employed a similar technique much more extensively while directing the survey of forfeited Irish estates, a project greatly publicized at the time and one close to the pocketbooks of many.

Possibly Petty drew the importance which his later economic writings attached to subdivision of labor from his own experience in running this "Down Survey." In any case, he procured a leave of absence from the regius chair of physic at Oxford in 1651 to accept appointment as chief surgeon to the army in Ireland. Medical matters occupied the first few years there, but in 1654 Petty happened to see the results of a survey just completed by Benjamin Worsley, upon which were to be based the forfeitures of land designed to finance arrears in soldiers' wages. So impenetrable did he find the obscurity of this document that Petty took the initiative in criticizing it to the supervisory committee appointed by the commissioners for Ireland.[58] Very possibly the state of his own finances stimulated his sense of scientific outrage, but the motives involved in the complicated political negotiations which resulted in his being commissioned to conduct a proper survey need not detain us. Actually, as it turned out, Petty made a rather good thing of it.[59]

The method he employed is the interesting matter. Worsley had never thought to make a map; his "survey" had been merely a conventional specification of tenures and holdings, with brief descriptive notes. Petty, on the other hand, proposed to represent the lands graphically and to scale. To get the job done on time required the most minute organization:[60]

Petty, consideringe the vastnesse of the worke, thought of dividinge both the art of makeing instruments, as alsoe that of usinge them in to many partes, vizt, one man made onely measuringe chaines, vizt, a wire maker; another magneticall needles, with theire pins, vizt, a watchmaker; another turned the boxes out of woode, and the heads of the stand on which the instrument playes, vizt, a turnor; another the stands or leggs, a pipe maker; another all the brasse-worke, vizt, a founder; another workman, of a more versatile head and hand, touches the needles, adjusts the sights and cards, and adaptates every peece to each other.

Each surveying party received instructions to organize itself into measurers, protractors, and checkers, who were to take frequent intersections. All plotters were to plot to the same scale. And, for obvious reasons, Worsley's system of payment by linear distance covered was abandoned. In order to insure accuracy, the checkers had orders to close angles and triangles unknown to the protractors. Even the central office workers performed each a single function. One section valued the land, another determined what districts were most profitable, a third performed the necessary artistic functions, and a fourth fitted together adjacent areas, with a permissible margin of error very strictly defined.[61]

The judgment which a warmly sympathetic student of Irish nationalism might render upon the Down Survey would probably be rather different from that of the historian of science. From the unfeeling point of view of the latter, the project represents an early instance of careful organization of different skills directed toward the solution of a really large problem in applied scientific technique.[62] The whole job was completed with astonishing speed—just over a year—and by 1659 Petty was back in London, renewing his associations with members of the Oxford Philosophical Society, whose host he had been until his departure for Ireland.[63]

Dr. Petty was undoubtedly the most variously acquainted physician among the original fellows of the Royal Society. Neither his purely scientific researches, however, nor his ventures into the field of political economy sapped his interest in his own profession. His writings contain any number of references to the manner in which it had shaped his thought. One may be quoted:[64]

> *Sir Francis Bacon,* in his *Advancement of Learning,* hath made a judicious *Parallel* in many particulars, between the *Body Natural* and *Body Politick,* and between the arts of preserving both in Health and Strength; and it is as reasonable, that as Anatomy is the best foundation of one, so also of the other; and that to practice upon the Politick, without knowing the *Symmetry, Fabrick,* and *Proportion* of it, is as *casual* as the practice of Old-women and Empyricks.
>
> Now, because *Anatomy* is not only necessary in Physicians, but laudable in every Philosophical person whatsoever; I therefore, who profess no Politicks, have, for my curiosity, at large attempted *the first Essay of Political Anatomy.*

Since preparation for and practice of their profession induced among physicians modes of thought which led their minds into wider ranges of natural philosophy, a reverse correlation might be expected to have attracted other scientists to the study of medical problems. Such was, indeed, often the case. Christopher Wren, Robert Hooke, Robert Boyle, and John Wallis are a few obvious examples of laymen who undertook various investigations in anatomy and physiology,[65] but to develop the point would require another article. Science, of course, medical or otherwise, had not yet become departmentalized in the seventeenth century. And natural philosophers were all members of a single fraternity of knowledge, good friends as well as fellow-seekers. Were it feasible here to trace the social contacts of the prominent doctors between 1640 and 1665, their leadership in this circle would be even more apparent; for it was often through the personal and professional associations of physicians—particularly Glisson, Willis, Goddard, Scarborough, and Petty—that younger men were introduced to the fraternity and that different elements within it were kept in touch with one another; those in London, for instance, with those in Oxford.

Both quantitatively and qualitatively, then, the energetic gentlemen who practiced the honorable profession of physic in Stuart London appear to have been instrumental in implementing the "new philosophy." By a mere counting of noses in the sources their quantitative importance is easy to assess. There can be no doubt of their relative productivity as authors in the early part of the century or of their numerical prominence in the coterie which became the Royal Society. Likewise, the percentage of fellows of the College of Physicians who concerned themselves with purely scientific speculation and experiment can be accurately determined. All this, however, does not suggest why it should have happened so or offer a value judgment upon the doctors' collective contribution.

The profession of medicine, like that of law, has achieved an enduring popular reputation for intellectual conservatism. In 17th-century England both lawyers and doctors would seem to have behaved in a rather uncharacteristic fashion. The motives underlying the lawyers' leadership in a political revolution are clear enough, but what were the sources of the physicians' creative interest in the radical innovations of experimental philosophy? Of course, the very fact that a man had successfully taken an M.D. degree indicates that he possessed mental qualities of a fairly distinguished order. He could, if he chose, count himself among the intelligentsia and move in those groups which in any period are the first to feel the impact of a new stream of ideas. Requirements of a medical education—though still notably scholastic in England—presented a hurdle high enough so that only the intellectually agile would be likely to take it in their stride. And, having done so, the better doctors felt profoundly dissatisfied with what they had learned. With Willis, they sought a "more certain Physiologie and Pathologie" than they had gained in the schools. Critical revolts against Galenic authority paralleled in every respect the philosophic protests against Aristotelian physics.

Medicine, however, if it is a science, is a science of life and has necessarily been required to wait upon the development of the basic physical sciences before it could achieve much absolute progress in mastering its materials. Nonetheless, in their professional life, doctors dealt, after all, with physical facts. Physicians did not permit themselves to be discouraged by the insuperable difficulties which barred them from comprehending their own subject. Comte had not yet written, and they were not aware of the logical necessity for their ignorance. Instead, the Baconian optimism of the period bred in them a sense of eager scientific curiosity, and for its immediate satisfaction they seem to have turned to extraneous fields in which the contemporary state of knowledge could promise some results. It is not surprising, therefore, that, long before the age of specialization, physicians attacked scientific investigations with more competence, intensity, and practical aggressiveness than the average, vaguely amateurish, natural philosopher—that gentleman "free and unconfin'd."

Doctors were, moreover, knit together by a professional bond, socially recognized and centuries old. Specifically, the College of Physicians afforded them an institutional vehicle for the corporate development of their avocational interests. By 1660 projects organized by its fellows already offered concrete examples of results made possible only by co-operative investigation. Advantages implicit in the existence of the College of Physicians became the explicit objectives of the Royal Society.

Notes and References

1. Published by Early English Text Society, "Extra series," No. 8 (London, 1869), p. 6.

2. Martha Ornstein, *The role of the scientific societies in the seventeenth century* (New York, 1913). Miss Ornstein's justly famous monograph opened up a field in which later scholarly investigators have toiled most successfully. Dorothy Stimson, in particular, has dealt with various phases of the subject in a series of articles, in which the major ones are: "Dr. Wilkins and the Royal Society," *Journal of modern history*, III (1931), 539–63; "Comenius and the invisible college," *Isis*, XXIII (1935), 373–88; and "Amateurs of science in 17th-century England," *ibid.*, XXXI (1939), 32–47. See also R. P. Stearns, "The scientific spirit in England," *Isis*, XXXIV (1934–35), 293–300; and Fran-

cis R. Johnson, "Gresham College: precursor of the Royal Society," *Journal of the history of ideas*, I (1940), 413–38.

3. Robert K. Merton has suggested that the requirements of medicine were a vital factor in guiding research in the biological sciences and has pointed out the somewhat avocational scientific interests of a number of individual physicians, but the scope of his excellent monograph did not permit him to develop the matter (*Science, technology, and society in seventeenth century England* [Bruges, 1938], pp. 383 and 566).

4. Gilbert, pp. 4–5.

5. These subjects were then taught by Walter Pope, Arthur Dacres, Thomas Baines, William Croune, and Jonathan Goddard, respectively, all practicing M.D.'s (John Ward, *The lives of the professors of Gresham College* [London, 1740], pp. 111–16, 168–69, 227–30, 270–71, and 320–27). Francis R. Johnson (*loc. cit.,* pp. 413–38) has suggested a likelihood that Gresham College constituted an informal focus for interchange of scientific thought in the years prior to the Royal Society. Between its foundation in 1596 and 1664, all the chairs except divinity were held at one time or another by physicians and often by several in succession (Ward, *passim*).

6. W. R. Munk (2d ed.; London, 1878).

7. Text of the letters patent printed in Munk, I, 2–6. The date was Sept. 23, 1518.

8. See below, p. 213.

9. Since its foundation, the college had been entrusted with such functions by statute and by charter (14 & 15 H. 8, c. 5 & 6; 32 H. 8, c. 40; see also, "Confirmation of the Charter to the Royal College," 8 Oct., 15 Jac. I, printed in Charles Goodall [ed.], *The Royal College of Physicians of London founded and established by letters patent, acts of parliament, adjudged cases, etc.* [London, 1684], pp. 37–61; and the charter of James I to the Worshipfull Society of Apothecaries of London, incorporated separately from the Grocers in 1617, C. R. B. Barrett, *The history of the Society of Apothecaries of London* [London, 1905], pp. xix–xxxix).

10. Dec. 19, 1627, Goodall, p. 399. The bishop of Durham had intervened, probably at Buckingham's behest, to secure a license for John Lambe to practice physic.

11. For reports of the cases, see Goodall, pp. 357–472.

12. Sidney Young, *Annals of the Barber-Surgeons of London* (London, 1890), pp. 193–220. Extracts from the minutes of the Court of Assistants from 1600 to 1651—when the minutes were lost—should be compared to the cases tried by the College of Physicians. The apothecaries and the surgeons both shared the outlook, duties, and responsibility to the City of the other companies, but the Royal College is never, so far as I know, listed with them in tax records, accounts of civic processions and pageants, etc.

13. Munk (*passim*) lists every licentiate of the college from its foundation, with such biographical information as was available to the author in the manuscripts preserved in the college's archives. The individual notices are pieced out from the usual biographical sources.

14. *Ibid.,* Vol. I, *passim.* This figure may be slightly in error as it does not include doctors the date of whose death is not known but who had probably died before 1603. Also omitted are some who seem never to have practiced in London.

15. In 1614, e.g., the college was composed of forty-one licentiates, of whom all but a very few were fellows (*ibid.,* p. 101).

16. Pollard and Redgrave. I have used the *Short-title catalogue* for the reign of James I in the chronologically arranged card-index form in which it is available in Houghton Library at Harvard University. Again the occasionally misleading nature of 17th-century titles may have sent me astray in some instances; but as nearly as I can determine, physicians were responsible for only six of the thirty-four volumes listed on the plague and for nineteen of the thirty-nine volumes discussing topics of general health. I have been able to examine about 20 per cent of this material, which should be a representative sampling, but unfortunately very few of the purely scientific books and pamphlets by doctors are available in this country.

17. William Harvey, *Works of William Harvey,* trans. Robert Willis (London, 1847), p. 6. The *De motu cordis* . . . occupies the first half of this edition.

18. *Ibid.,* Dedication to the Royal College, p. 5. Charles Singer (*The discovery of the circulation of the blood* [London, 1922], pp. 50–68) gives an excellent account of Harvey's methodology, though I feel he does an injustice to the impressive clarity of Harvey's own presentation, since he regards the *De motu cordis* . . . as a difficult and occasionally obscure book (pp. 48–49).

19. Munk, I, 127 and 136. The annual Lumleian lectures were established in 1582 by Lord Lumley and Dr. Richard Caldwell (see John Stowe, *A survey of London* [reprinted from the text of 1603; Oxford, 1908], I, 75). The Goulstonian series, endowed by the will of Dr. Theodore Goulston, was devoted to anatomy. It was the founder's wish "to purchase a rent-charge for the maintenance of an annual lecture [course] . . . by one of the four youngest doctors of the College. A dead body, was if possible, to be procured" (quoted in Munk, I, 157).

20. Munk, I, 78–79.

21. The society's contemporary historian does not seem to have known of the earliest meetings (Thomas Sprat, *The history of the Royal Society of London* [London, 1667]). The most complete secondary accounts, both extensively documented, are Thomas Birch, *History of the Royal Society of London* (London, 1756–57); and C. R. Weld, *A history of the Royal Society* (London, 1848). Modern analyses and interpretations rely largely on the original source material available to these authors.

22. For this phrase I am indebted to Dorothy Stimson ("Amateurs of science," *loc. cit.,* p. 32).

23. John Wallis, "Dr. Wallis's account of some passages of his own life" (1696), printed in the preface to Thomas Hearne's edition of *Peter Langtoft's chronicle* (1st ed.; London, 1725), published as Vols. III and IV of *The works of Thomas Hearne* (London, 1810). Citations in this article refer to Vol. III of this edition.

24. Wallis, *loc. cit.,* p. clxii. As a young man, Robert Boyle must also have been a member of the group. See references in his correspondence to the "invisible college" (Thomas Birch, *The life of the Right Honourable Robert Boyle* [London, 1744]), Boyle to Macomber, Oct. 22, 1646, p. 66; to Francis Tallents, Feb. 20, 1646/47, pp. 67–68; and to Samuel Hartlib, May 8, 1647, pp. 78–79.

25. Sprat, pp. 57–58.

26. *Ibid.,* p. 57.

27. Birch, *History of the Royal Society,* I, 2–3; Sprat, p. 55; Wallis, *loc. cit.,* p. clxiv; Andrew Clark, *The life and times of Anthony Wood* . . . *described by himself* (Oxford, 1891–1900), I, 290; and Stimson, "Dr. Wilkins," *loc. cit.,* pp. 539–63.

28. Birch, *History of the Royal Society*, I, 3. The minutes of this meeting, printed in Birch, are the earliest to have survived.

29. The doctors: Francis Glisson, George Bate, George Ent, Charles Scarborough, Alexander Fraiser, Thomas Coxe, Christopher Merret, Daniel Whistler, Timothy Clarke, Thomas Willis, Nathaniel Henshaw, John Finch, Thomas Baynes, and William Croune. Dr. Abraham Cowley (the poet) is also listed. Cowley took an M.D. at Oxford in 1657, though I believe he never practiced, and in any case he appears in none of the later lists (*ibid.,* p. 4).

30. For the charters, negotiations concerning them, and a complete list of fellows from 1663, see *Record of the Royal Society* (4th ed.; London, 1940), pp. 86–88, 215–84, and 375–516.

31. Sprat, p. 67.

32. In addition to Fraiser, Willis, and the twelve others listed in n. 29, the following doctors were original fellows: William Petty, Jonathan Goddard, David Bruce, Walter Charlton, Lord Dorchester, William Hoare, Jasper Needham, Walter Pope, William Quatremaine, George Smyth, Christopher Terne, and Thomas Wren (*Record of the Royal Society,* pp. 375–78).

33. All statements and quotations in this paragraph are from the minutes in Birch, *History of the Royal Society*, I, 5–6.

34. Compiled from Munk, *passim;* and *Record of the Royal Society,* pp. 377–80.

35. Publication of the *Philosophical transactions* did not begin until March 1665, but the four volumes of Birch's *History* print the minutes of the meetings from 1660 until then. These figures are compiled from a survey of the papers, reports of individual researches, and brief notices of experiments assigned to various fellows and committees (Birch, *History of the Royal Society,* I, 5–68). Miss Ornstein regarded these years as the Royal Society's formative period, when the lines it was long to follow were sketched out (p. 123). For an evaluation of the researches see Ornstein, pp. 130–48.

36. For an excellent evaluation see William P. Dunn, *Sir Thomas Browne* (Menasha, Wis., 1926), particularly pp. 1–13 and 74–87.

37. Sir Thomas Browne, *Works*, ed. Charles Sayle (London, 1904–7). The first authorized edition of *Religio medici* appeared in 1643 and the *Pseudodoxia epidemica* in 1646, with the subtitle, *Enquiries into very many received tenets and commonly-presumed truths, which examined prove but vulgar and common errours* (*ibid.,* I, 152).

38. Browne, *Pseudodoxia epidemica, ibid.,* p. 116.

39. Aubrey, I, 268. For examples of the range of his experiments and the apparatus he devised for investigations into problems of chemistry, botany, optics, aerodynamics, and other branches of physics, see Birch, *History of the Royal Society,* I, 248, 255, 270–72, 296–97, 305, 311, 314, 338, and 349; Sprat, pp. 193 and 230; and *Philosophical transactions,* XII, 930 and 953.

40. Wallis, *loc. cit.,* p. clxiii.

41. Goddard served as physician to the parliamentary army and accompanied Cromwell to Ireland as his "great confident." In 1653 he sat in parliament and on the council of state. He was also a member of the parliamentary visitation of Oxford and warden of Merton from 1651 to 1660 (Wood, III, 1029).

42. *Ibid.*

43. Jonathan Goddard, *A discourse setting forth the unhappy condition of the practise of physick* . . . (London, 1670), pp. 12–13.

44. Christopher Merret, *A short view of the frauds and abuses committed by apothecaries* . . . (London, 1669), pp. 5–6 and 33.

45. Christopher Merret, *Pinax, rerum naturalium Brittanicarum, continens vegetabilia, animalia, et fossilia* (London, 1667).

46. *Art de La Verrerie de Neri, Merret, et Kunckel* . . . , trans. from a German edition by "M.D." (Paris, 1752), pp. i and iv. Merret's notes, running throughout the text, seem to me admirable descriptive technology. His preface to the original edition will be found on pp. xix–xlv.

47. Robert Boyle, "Advertisement to the reader," preface to Merret's "An account of freezing made in December and January (*sic*) 1662," which was published as an annex to Boyle, *New experiments touching cold* (London, 1665).

48. Francis Glisson, George Bate, and Ahasuerus Regemorter, *De rachitide, sive morbo puerili, qui vulgo the rickets dicitur, tractatus* (London, 1650). There is an apprecative evaluation in Sir Norman Moore, *The history of the study of medicine in the British Isles* ([Oxford, 1908], pp. 111–13); but in order to grasp the breadth of observation and investigation, the student should turn to the original.

49. Ordinarily, however, he was on hand for medical acts and opponencies and for his scheduled lectures (H. D. Rolleston, *The Cambridge Medical School* [Cambridge, 1932], p. 152; and Gunther, *Early science in Cambridge*, p. 20).

50. Munk, I, 219.

51. John Graunt, *Natural and political observations made upon the Bills of Mortality,* ed. W. F. Willcox (Baltimore, 1939), opp. p. 80.

52. Glisson, Bate, and Regemorter, pp. A3–A7. The other five who contributed were Thomas Sheafe, Robert Wright, Nathan Paget, Jonathan Goddard, and Edmund Trench (*ibid.,* p. A8).

53. Quoted in Gunther, *Early science in Cambridge*, p. 302; see also Francis Glisson, *Anatomia hepatis* (Amsterdam, 1665), pp. *3–*6.

54. Willis' works have been translated and collected in a single ponderous volume, *Dr. Willis's practice of physick,* trans. S. Pordage (London, 1684).

55. *A medico-philosophical discourse of fermentations* (1646) and *A treatise of feavers* (1648), in *ibid.*

56. *The anatomy of the brain and nervous system described* (1664), in *ibid.,* p. 43.

57. *Ibid.,* pp. 43–44; see also Gunther, *Early science in Oxford,* III, 99.

58. Printed manuscripts, including Petty's own account of these negotiations, will be found in Thomas A. Larcom (ed.), *History of the Cromwellian survey of Ireland* . . . *from MSS in the Library of Trinity College, Dublin, the King's Inns, Dublin, and the Marquess of Lansdowne* (Dublin, 1851), particularly pp. 1–3, 13–15, and 18–29.

59. Aubrey, II, 142.

60. Contemporary account from a manuscript in the Record Branch of the Office of the Paymaster of Civil Services in Ireland (Larcom, p. xiv).

61. *Ibid.,* pp. xiv–xvii, 17–18, 46–53, and 110–11.

62. Petty's survey formed the basis for publication, in 1673, of the County and Barony Map of Ireland, which John Evelyn declared the most exact map ever to have appeared (*Diary,* ed. William Bray [London, 1879], II, 307).

63. See above, p. 216.

64. Sir William Petty, "The political anatomy of Ireland" (1672), *Economic writings of Sir William Petty,* ed. C. H. Hull (Cambridge, 1899), I, 129.

65. Gunther, *Early science in Oxford,* III, 96–99 and 135; Wood, IV, 247–49 and 628–29; Birch, *Life of Boyle,* pp. 108–9 and 112, and *History of the Royal Society,* I, 180, II, 187, and III, 77; Ornstein, pp. 140–41; J. F. Fulton, *A bibliography of . . . Robert Boyle* (Oxford, 1932), pp. 94–96, 103–5, and 110–15; Wallis, *loc. cit.,* pp. clxix–cl; Aubrey, II, 112, 114, and 283–85; and Joseph Foster (ed.), *Alumni Oxoniensis* (London, 1891–92), I, 163, and II, 740.

2

English Ideas of the University
in the 19th Century

THE ESSAY THAT FOLLOWS was one of three delivered orally in a session chaired by Margaret Clapp at the annual meeting of the American Historical Association held in Boston in 1949. Paul Farmer of the University of Wisconsin and George Pierson of Yale, gave the other papers, the former on European and the latter on American developments in the same period. George Pierson, one of whose major works is the history of Yale, was the senior member of the panel by far.

Although it was W. K. Jordan's seminar that opened my eyes to the thrill of original research, 19th-century England attracted me more strongly. That had been the field of my mentor at Wesleyan, H. C. F. Bell. At Harvard David Owen was the specialist on the Victorian period. Owen was a fine lecturer and by far the most approachable member of a department that sometimes seemed to beginning graduate students a touch cold, distant, and collectively self-satisfied. Though in no way involved in history of science, he encouraged my idea of looking into science and religion before Darwin and read the successive chapters of *Genesis and Geology* closely, offering not so much guidance as encouraging criticism.

David Owen recommended me to the History Department in Princeton in 1947, and we kept in touch for some years thereafter. A Yale Ph.D., he was probably the one who proposed the unknown young historian I then was to whoever initiated the session on universities, presumably Pierson. If this paper is written in a reasonably straightforward manner, and reading it over, I think it is, I have Owen largely to thank. After reading the draft of the first chapter of *Genesis and Geology*, he gently said, "too many syllables." A fellow instructor at Princeton, the late Walter Woodfill, furthered the cure, observing of the draft of this paper that there were irritating sentences purporting to be jocular. When I finished reading it at the Boston session, Pierson on his way to the podium kindly remarked quietly to me that it was a sprightly paper.

The third of the trio of early papers on English history, the appreciation of the works of Élie Halévy, appears later in this volume in the section on historians. The existence of that paper, too, I owe to a suggestion by

David Owen. When I told him how valuable in preparing for the general examination I had found *England in 1815* and *The Growth of Philosophical Radicalism,* he was interested and said, "Why don't you write that up?" Accordingly I read the rest of Halévy's writings and, as will appear below, I did. Evidently Halévy's sobriety had ended any tendency on my part to ornate writing.

TWO

English Ideas of the University
in the 19th Century*

—⚅—

Oxford and Cambridge over the centuries have traveled as far from their medieval origins as have the European universities. Since the Reformation, however, they have done so more gradually, retaining wherever possible the picturesque externals and ceremonial formalities with which they flavor their traditions.

In England as on the continent the intellectual vitality of the universities reached its lowest ebb during the Age of the Enlightenment, when the world of science, literature, and the arts lay almost entirely outside their walls. It is possible that the familiar picture of academic torpor in 18th-century England, confirmed though it is by the accounts of people as dissimilar as Edward Gibbon, Adam Smith, Lord Chesterfield, and Lord Eldon, has been drawn a little too dark and that it should be touched up here and there with occasional gleams of light and a few real flashes of mental energy. Nevertheless, it would be difficult to show that Oxford and Cambridge at the opening of the nineteenth century gave rise to any real ideas about a university.

The only important thing then coming out of the English universities was the English governing class, and it emerged less the product of education than of a sort of molding process in which the universities finished what the public schools had begun. The single profession for which Oxford and Cambridge pretended to prepare students was the church, but even in theology the formal training was of the sketchiest kind. Law and medicine were in even worse case, and indeed the three higher faculties had become shadows and their degrees formalities. As Sir William Hamilton wrote, without much exaggeration:

* Reprinted from *The Modern University*, ed. Margaret Clapp (Ithaca, N.Y., Cornell University Press, 1950), pp. 27–55.

England is the only Christian country, where the Parson, if he reach
the University at all, receives the same minimum of Theological tuition as the Squire;—the only civilized country, where the degree,
which confers on the jurist a strict monopoly of practise, is conferred without either instruction or examination;—the only country in the world, where the Physician is turned loose upon society,
with extraordinary and odious privileges, but without professional
education, or even the slightest guarantee for his skill.[1]

Regular tuition was offered only in the lowest faculty, that of arts, and
in Oxford this meant chiefly classics, in Cambridge mathematics. Instruction was generally of a perfunctory sort, given not by the professors, who
were university officers and at whose infrequent lectures the attendance
was almost always small and often nonexistent, but by tutors who conducted recitations in the colleges. The colleges, originally no more than
residential foundations, had survived in England, as they had not on the
continent; and, far more richly endowed than the parent corporations,
they had gradually and largely by chance arrogated to themselves almost
all the functions of the university except the granting of degrees. "Alma
Mater," commented the Royal Commission of 1922, "had been devoured
by her own children."[2]

Nor, in the absence of professional training, did the universities or the
colleges encourage devotion to scholarship or the advancement of learning—there was no notion that these pursuits were the business of the universities. If a research scholar was occasionally to be found at either one
of them, it was a happy accident rather than the result of any policy. So far
as Oxford and Cambridge were dedicated to anything, it was to the per-
petuation of themselves and of the type of graduate formed by their peculiar social environment—though even this was simply what they in
fact did rather than a consciously formulated aim.

By the end of the nineteenth century matters were on a very different
footing. In place of the 741 Oxford undergraduates of 1800, there were
over 2,500 in 1900, and at Cambridge the enrollment had grown from 387
to almost 2,800. No longer were degrees available only to members of the
Church of England. Nor were they simply formalities to be approached
through classics or mathematics. The old courses retained great prestige
in 1900, but there were now eight honor schools at Oxford and twelve tri-
poses at Cambridge embracing the natural sciences, the social sciences,
and modern literatures and languages. Seventy-five per cent of the graduates took honors.

The professorial body had grown in proportion to the expansion of studies and the increased number of students. Its members were no longer holders of ill-paid sinecures. They were now overworked—though still ill-paid—and their lectures had a real place in the education of undergraduates. College fellowships had not increased in number. But in 1800 they had been comfortable livings for celibate clergymen, a few of whom were tutors, but most of whom did nothing. In 1900 the majority of resident fellows gave individual tutorial instruction, and most of the remainder, who were now permitted to marry, were on limited tenures engaged in scholarship or preparing for a professional career. Both fellowships and scholarships were open to merit instead of being hedged about by a tangle of local preferences and religious restrictions.

The commanding importance of the colleges was still the major institutional factor which made the system and atmosphere of Oxford and Cambridge unique among European universities, but the colleges had been subordinated in a federal manner to the authority and functions of the universities as a whole and had even been forced to contribute funds to their support.

And, finally, Oxford and Cambridge no longer enjoyed a monopoly. Eight faculties and twenty-four member colleges and institutes of the University of London, linked together and incorporated as a teaching body in 1900, were soon to surpass them as centers of research and instruments of large-scale education. Besides London, there was Durham, founded in 1833, a church university with which were associated a medical school and a school of science in Newcastle. Of the more recently established university colleges, Mason College in Birmingham was raised to the status of a university in 1900. Similar foundations had been established in Manchester, Liverpool, Leeds, Sheffield, Reading, Nottingham, Southampton, and Bristol.

They were not as yet empowered to award degrees; those were conferred by the University of London primarily, or by Victoria University, a sort of academic holding company which had been incorporated in 1880 as the degree-granting authority first for Owens College in Manchester and, a few years later, for the Liverpool University College and the Yorkshire College in Leeds. At the turn of the century these three colleges, following the example of Birmingham, were on the verge of setting up as full-fledged independent universities. In Manchester alone there were about a thousand students at Owens College in 1900; and, if these civic institutions had at first to concentrate chiefly on technical and professional training, their development, like that of the University of London, was certain to enlarge the facilities for research and scholarship in England and to increase the opportunities for higher education.

These 19th-century changes were enormous, and, though some of them were initiated by the older universities themselves, they were effected chiefly as a result of public demand brought to bear in the press, in the political arena, and in Parliament. In the eighteenth century what now seem the glaring scandals tolerated at Oxford and Cambridge were simply examples of the venal condition of most established institutions in that period. "The unreformed House of Commons," wrote Mark Pattison, rector of Lincoln College in Oxford, was "an assembly of gentlemen whose sympathies were always with us."[3] In the nineteenth century, following the reform of Parliament itself, it was inevitable that the universities should not have been spared by the social developments, the political forces, the economic interests, the shifting class relations, and the intellectual and religious influences which modernized the whole fabric of the nation.

Life in Oxford and Cambridge was one of the experiences which helped form the attitudes of the upper classes. Nevertheless, the universities reflected rather than led the major movements of opinion in Victorian England. Until fairly late in the century, the important figures in English intellectual history were for the most part outside the academic fold (with the exception of the leaders of the Oxford Movement), and, although the universities were intimately tied up with the life of the society of which they formed a part, the connection was less simple than in countries where they were government controlled.

In their relationship to the state, the English universities stood midway between continental universities of the nineteenth century and private universities in the United States. Oxford and Cambridge and all of the colleges within them were corporate bodies with the power to regulate their own affairs and manage their own property within the limits imposed by statute, by their charters, and by various provisos in the wills of benefactors who had directed how particular endowments were to be employed. The newer universities were in a similar position. They owed their origin to private philanthropy combined with civic enterprise, and not to the state.

But neither were English universities altogether independent of the sovereign political authority. Oxford and Cambridge were themselves represented in Parliament. Appointment to a number of chairs and college headships was vested in the Crown, and the salaries of a few of the professorships were a charge upon the Treasury. And beginning in 1889, the Committee of Council on Education was authorized to divide an annual subsidy among the newly established university colleges, a sum which, though still only £25,000 in 1900, was growing steadily larger.

After the First World War even Oxford and Cambridge swallowed their pride in order to participate in the government grants. And entirely apart from financial questions, the universities, like all English institutions, were subject to the ultimate authority of Parliament, which could always exercise supreme jurisdiction by means of Royal Commissions and legislative enactments. In practice, however, the universities were left free to administer their own concerns except on the few—though decisive—occasions when the government, usually in response to public pressure, intervened to reform their constitution and define their obligations.

—◊—

By the time of the Parliamentary Reform Bill of 1832, three major currents of opinion had converged in criticism of Oxford and Cambridge—utilitarianism, liberalism, and enthusiasm for natural science. Because the universities were a kind of *ancien régime* in academic microcosm, abounding with anachronisms and charm, Benthamite radicals were bound to attack them. The utilitarian ideal of a university took shape in the plans for the London University, renamed University College when the first attempt to secure a charter carrying authority to award degrees was obstructed by Tory and clerical influence.

This institution, an entirely secular establishment, opened its doors in 1828, and together with King's College, a competing venture founded a few years later under Church of England auspices, formed the nucleus around which the University of London eventually developed. At University College no time was to be wasted on dead languages or collegiate atmosphere. Instead, large classes were to be offered efficient lectures on good solid subjects related to the real world of commercial and professional life. As it turned out, the majority of students elected courses leading to medicine and law.

Unlike utilitarians, who had no use at all for Oxford and Cambridge, Liberals undertook to reform rather than to supersede the existing universities, and their emphasis was political rather than intellectual. They were less critical of Oxford and Cambridge as places of education than as citadels of privilege. Liberal ideas about the content of higher education were never very clear. But the advantages of it, whatever they might be, should go to those who could win them in "unfettered and open competition"—this, according to the Cambridge commissioners of 1850, was "the one good rule."[4]

The intolerable aspect to Liberals was not so much the study of useless classics, as the tissue of restrictions surrounding fellowships and scholarships, the clerical oligarchy which governed the universities in the interest of the colleges, and the whole apparatus of religious tests. The complete secularism of University College in London was the pattern followed later in the century in the foundation of the provincial universities, but the destruction in 1871 of the Anglican monopoly of the resources of Oxford and Cambridge was primarily the consequence of Liberal political principles and Liberal political pressure. In this, of course, Liberals were joined by the dissenting interest, but Nonconformist ideas of education were always complicated by their hostility both to the Church of England and to secularist institutions of learning.

The third main line of criticism arose because the traditional curriculum had not changed despite the growth and increasing social importance of science and technology, the social sciences (particularly political economy), and the medical and legal professions. In view of the achievements and prestige of physical science, the almost total neglect of all its branches in the universities seemed their most glaring omission, and scientists tended to take the lead in demanding that Oxford and Cambridge should offer instruction in science and in all the modern intellectual and professional disciplines. In 1850 parliamentary Liberals secured the appointment of the first Royal Commissions on the universities. By then the campaign of utilitarians, Liberals, dissenters, scientists, and many professional men, manufacturers, and businessmen of the Manchester School variety had created a fairly definite radical ideal of higher education, which, though not fully developed by any one of these groups, may be abstracted from the discussion and summarized in a few sentences.

These critics of Oxford and Cambridge held that the universities belonged to the nation—a point which, denied by their defenders before 1832, was generally admitted by 1850—and that they must not, therefore, be closed to all but members of the established church and the aristocratic class. They must perform a service to the country commensurate with their huge endowments. The business of a university was to supply instruction, to require hard work, and to grant degrees which would be trustworthy evidence of the holder's achievements and ability. The curriculum should meet practical needs, and the graduate should be prepared to advance commerce, manufacturing, or agriculture, to enter public service, or to be a physician, lawyer, engineer, teacher, or scientist. Narrow specialization was not desired, but general knowledge should be taught through subjects relevant to the means by which the country was governed and the enterprises by which it lived: science, with emphasis on applied rather than pure science; political economy; modern law; geography and modern history; modern languages; theology; and moral and political philosophy.

In 19th-century England both the radical and the conservative ideas of a university emerged in the course of disputes about whether training an aristocracy in classics and elementary mathematics served any good purpose in an industrial and commercial society. With a few exceptions, the strength of educational radicalism lay outside Oxford and Cambridge. Necessarily, therefore, the radical idea first appeared as an ideal to be realized, and it came closer to being realized in the later provincial universities than it ever did in Oxford and Cambridge.

The strongholds of educational conservatism, on the other hand, were the older universities themselves, supported by the Church and by Tory interests generally, and in the case of conservatism theory came after practice. Indeed, the theory of a classical education owes its initial elaboration in modern times to the necessity of defending the existing system of Oxford and Cambridge against radical critics. It is true that the rationale was developed while the universities were revitalizing the system on their own account. It is also true that it had many points in common with the Renaissance ideal of humanist education. It would, however, be difficult to trace the survival of any philosophy of education through 18th-century Oxford and Cambridge. Though the system itself had originated earlier, this was the period when it had become fixed (not to say ossified), and this came about chiefly through the accidents of institutional history rather than in consequence of any theory.

One could, indeed, argue that much of the prestige—or at any rate the snob value—of classical studies as a general instrument of education in the modern English-speaking world derives from the fact that in the early nineteenth century Oxford tutors, who were determined to maintain their monopoly of university teaching, did not know anything else to teach. They believed in their wares and valued their positions, and they were forced in self-defense to elaborate a persuasive justification of the manner in which the English upper class was being educated. Publications of Cambridge tutors claimed for the study of mathematics the same pedagogical benefits which in Oxford were attributed to classics—a circumstance which supports the impression that the manner rather than the matter of the education was what ultimately gave rise to the theory developed to defend it.

To the radical contention that the excellence of an education is measured by its utility, conservatives tended to reply that, on the contrary, the value of an education is proportional to its practical uselessness. In Newman's view, a university should dispense liberal knowledge, and he distinguished between liberal and useful accomplishments by quoting Aristotle: "Those rather are useful, which bear *fruit;* those *liberal, which tend to enjoyment.* By fruitful, I mean, which yield revenue; by enjoyable, where *nothing accrues of consequence beyond the using.*"[5] Actually, however, neither the radical nor the conservative argument ever really met the other because the two sides never reached agreement on the purpose of higher education.

For the radical the essence of education was applicability; cultivation of the mind was a by-product. In conservative theory the essence of education was discipline of the personality both in moral and mental qualities, which were thought to be closely associated. The subjects of the curriculum should be laid out, not according to their importance in the humdrum business of getting a living, but according to their value as intellectual gymnastics. Once the mind was trained in processes of logical thought and its powers of reason were properly exercised, once the tastes and values were sufficiently elevated through familiarity with the highest achievements of the human mind and spirit—and experience had proved that the languages, the literature, the mathematics of classical antiquity (plus a little Newton at Cambridge) were the best instruments for these ends—then the student, even if he knew no useful facts, had nevertheless been trained to think, and to think like a gentleman instead of like a manufacturer or mechanic. He would, therefore, be supremely qualified to handle any problems life could set him, and to handle them with grace, poise, balance, and a cultivated judgment. There was no objection to educating only a social elite. Only an elite could (and many people felt should) be educated.

The religious basis of higher education was gone by 1900, as the more subtle class basis was not, but neither had been simply or wholly obstructive. Evangelicalism shared responsibility with the legacy of Newton for the fact that Cambridge stood less in need of reform than Oxford at the opening of the nineteenth century. And in the 1830's Anglican tractarianism shook Oxford out of its aristocratic lethargy and made it the home of vigorous and searching thought. Having done so, the movement performed the university a further, if somewhat negative, service by collapsing amid a revulsion of feeling which had the effect of freeing the majority of Oxford men from their obsession with theological hairsplitting and turning their newly awakened energies outward toward literature, philosophy, and secular learning.

The universities, moreover, were essential in fixing a pattern of service and behavior on the Victorian aristocracy. An influence like Jowett's over the later life of his students could scarcely have come out of any other educational environment than that of an Oxford or Cambridge college. The success of the English governing class may have to be explained on grounds of character rather than intellect, but it would not have distressed the university authorities to have their graduates evaluated in this fashion. In Pusey's view:

> The object of Universities is, with and through the discipline of the intellect, as far as may be, to discipline and train the whole moral and intelligent being. The problem and special work of an University is, not how to advance science, not how to make discoveries, not to form new schools of mental philosophy, nor to invent new modes of analysis; not to produce works in Medicine, Jurisprudence or even Theology; but to form minds religiously, morally, intellectually, which shall discharge aright whatever duties God, in His Providence, shall appoint to them. Acute and subtle intellects, even though well-disciplined, are not needed for most offices in the body politic. Acute and subtle intellects, if undisciplined, are destructive both to themselves and to it, in proportion to their very powers. The type of the English intellectual character is sound, solid, steady, thoughtful, well-disciplined judgment. It would be a perversion of our institutions to turn the University into a forcing-house for intellect.[6]

Opinion on university issues was full of complexities. Before 1830, for example, the device of competitive examinations was developed first in Cambridge and then in Oxford as their major contribution to their own reform. It was the one feature of the universities which was warmly approved by utilitarians. So successful were formal examinations in stimulating hard work and so mechanically satisfactory did they seem that dons fell in love with the system and pointed to it as proof that higher education *was* a serious affair. Eventually the country at large became convinced of the virtue of competitive tests, not only as measures of achievement but even as instruments of schooling. When the state gradually took on responsibility for elementary, secondary, and technical education, it adopted the examination system as the basis of administering its grants, and—a more striking illustration—the methods by which Oxford and Cambridge separated the sheep from the goats were employed by the government in recruiting the civil service both at home and in India.

But while Benthamites and their intellectual successors were enthusiastic about the apparatus, if not the content, of examinations in Oxford and Cambridge, other critics of the universities were hostile to the whole system. Scholars, scientists, and educational liberals like Matthew Arnold and Mark Pattison generally deplored what Americans would call the overemphasis on grades and felt that examinations encouraged the regrettable and philistine English tendency to value the outward badge of learning rather than learning itself, to cram rather than to contemplate, to study only what would pay on examinations and hence lead to prizes and fellowships rather than what was intrinsically true or useful. The professoriate usually agreed with liberals here. (Unlike the material dispensed by college tutors and by the many other fellows who added tidily to their incomes by private coaching, the subjects of professorial lectures were never covered on examinations.)

Although the examination system had been introduced into England by the authorities of Oxford and Cambridge, conservative educational influence was also responsible for much of the dissatisfaction with the University of London so long as that body remained simply an examining mill. Chartered in 1836 to grant degrees to qualified students from University and King's Colleges—and after 1850 from other accredited institutions—the University of London itself had no faculty and no teaching function. In 1858, its facilities were thrown open to all candidates, who, if they passed the questions set them, were awarded degrees regardless of where or how they had studied.

This arrangement was in part an attempt to satisfy the increasing demand for higher education without encroaching on the peculiar and jealously guarded privilege of Oxford and Cambridge. At best, however, the University of London was a workable stopgap. It was never regarded as an altogether happy solution to the problem because, although the restrictiveness of Oxford and Cambridge had been one of the elements which originally forced the adoption of a compromise, their example was also responsible for the fixed English belief that a university should both instruct and examine resident students. The mere possession of information should not in itself carry title to a degree unless accompanied by some assurance that the recipient had acquired with his knowledge the intangible but real advantages given only by life in an academic community.

Even the basic debate as to whether a university should foster professional skills or whether it should forge character on the anvil of the classics was never as clear cut as it seemed. Dons like Pusey and Freeman defended the traditional system on the ground that it trained a class of enlightened and aristocratic public servants, and this is surely one form of professional education. A little later, on the other hand, Huxley urged the study of science because it disciplines the intellect. By the latter part of the century the issue had pretty well subsided, though it was never exactly settled. No one ever replied to Mill's remark that in the ideal university there must surely be room for both literature and science, for both mental training and modern learning. In practice the older universities continued to emphasize the traditional subjects and by their side made room for the sciences and for other modern studies, while the new foundations which eventually developed into the provincial universities offered chiefly practical subjects at first and made subsidiary provision for arts.

Despite many concessions to the spirit and necessities of the times, however, the traditionalists cannot be said to have lost their battle. John Stuart Mill had no reason to feel kindly toward Oxford, but even Mill was strongly if unconsciously influenced by the Oxonian conception of the function of a university. He told the University of St. Andrews:

> Men are men before they are lawyers, or physicians, or merchants, or manufacturers; and, if you make them capable and sensible men, they will make themselves capable and sensible lawyers or physicians. What professional men should carry away with them from an University, is not professional knowledge, but that which should direct the use of their professional knowledge, and bring the light of general culture to illuminate the technicalities of a special pursuit. Men may be competent lawyers without general education, but it depends on general education to make them philosophic lawyers— who demand, and are capable of apprehending, principles, instead of merely cramming their memory with details. And so of all other useful pursuits, mechanical included. Education makes a man a more intelligent shoemaker, if that be his occupation, but not by teaching him how to make shoes; it does so by the mental exercise it gives, and the habits it impresses.[7]

By the late 1860's three originally subsidiary issues had become important. The problems of what should be taught, and why, had become merged in the questions, first, of who should be taught; second, of how they should be taught; and third, of the responsibility of a university toward learning and research. These issues were closely related not so much by internal logic as by the circumstances of the discussion.

Like most English institutions, the older universities were modernized by changing their content and function as much as necessary while preserving so far as possible their form and spirit. And in England the extension of higher education, like the extension of the vote, was generally

supported on grounds of national expediency and social necessity rather
than of abstract right. That parallel is striking (even to the point that the
claim of feminism to a right in the franchise and in the universities was
an exception to the usual type of argument). And in both cases, the first
steps—the Parliamentary Reform Bill of 1832 and the opening of closed
scholarships to competition after 1850—had the incidental effect of dis-
qualifying the lower classes in a few regions where the old anomalies had
favored them.

No school of opinion regarded poverty as having in itself any claim on
the universities. After 1870, it is true, a number of dons in Oxford and
Cambridge undertook the leadership of the university extension movement, but this, though worthy and sincere, was adult education of the
night-school type. It did not involve opening the universities themselves
to the poorer elements of the population. Such an impossible project
would have destroyed academic standards, and, even if it had been practicable, conservatives feared that it would succeed only in making people
unwilling to do manual labor and in overturning the social structure.
Higher education for the working class in places like Owens College got
off to a very slow start and encountered considerable hostility. "Anyone educated in Manchester," said the *Saturday Review* in 1877, "would certainly
be dull and probably vicious."[8] In 1850 the proposal to base scholarships
in Oxford and Cambridge on need was thought illiberal. It would have put
a premium on poverty and closed an avenue of profit to ability. "What the
State and the Church require . . . is not poor men, but good and able men,
whether poor or rich"—so thought the Oxford Commission of 1850.[9]

Although critics of Oxford and Cambridge made much of the point
that the university endowments were a trust to be used for the benefit of
the nation, most radicals had no notion of generally extending higher education beyond the middle classes. And after 1850 university conservatives also accepted and even embraced the inevitable. Freeman inquired:

> Shall the University endeavour to influence the great middle class
> of England? Surely the University hardly fulfills its mission as a
> great national institution, a corporation charged with the guidance
> and nurture of the national intellect, unless it at least attempts to ex-
> tend its benefits to these most important classes. Surely it is hardly
> true to itself if it does not endeavour to gain by all honourable
> means the vast accession of strength which would be conveyed by
> the adhesion of those orders of men who now possess the primary
> political influence in the country.[10]

Radicals regarded colleges as hotbeds of clericalism, privilege, and idleness. They attacked the collegiate organization of the universities and the tutorial method of instruction as vested interests, and the more extreme critics looked to the professorial university of the German or Scottish type as the progressive alternative. Professorial instruction was thought to be demanded by the nature of modern subjects, particularly the sciences, since these required a degree of specialization and a provision for laboratories possible only on a university level. Besides this, the expansion of higher education at new universities would necessarily involve teaching by means of lectures to large numbers of students.

Outside of radical and Scottish circles, however, the idea of professorial universities was never very well received. "For it is truly urged," thought Mark Pattison, "that the collegiate life and domestic discipline are what make Oxford what it is. What the B.A. at present carries away with him is made up of a small modicum of acquired learning, and a peculiar stamp which remains upon him through life, and which constitutes undoubtedly a relative social advantage, whatever its true worth may be."[11] Professors might be perfectly satisfactory instruments for presenting grubby subjects full of facts, but for making the whole man, for sharpening the understanding and elevating the mind, what was required was the intimacy of the tutorial relationship. As a type, moreover, professors were generally supposed to be both learned and Germanic and likely, therefore, to act as instruments of the state and to do research—neither of which was a function of the pre-eminently English tutor.

Illogically enough, democratization also involved the introduction of original research into the universities, though only in an incidental way. Most of the provincial universities started as technical and professional schools, emphasizing the teaching of science and applied science, and the professors were often research scientists. To Englishmen of the mid-century, however, professorial research was primarily a German phenomenon, and its most obvious products were unorthodox views about the Bible. For Pusey, at least, this was the conclusive consideration against the expansion of professorial teaching in England—it was sure to lead to science, research, rationalism, and infidelity.[12]

So far the English discussion of universities had treated them simply as places of undergraduate education. There were, of course, always a few productive scholars at Oxford and Cambridge, but facilities for science were almost nonexistent until the late 1860's, and the importance attached to research in other fields may be measured by the fact that until the 1850's no books might be taken out of the Bodleian Library, which,

nevertheless, closed at three o'clock. The idea that a university is peculiarly a center of research and that a major function is postgraduate training was not widely established until the present century, and before 1870
or so the suggestion was usually advanced only to be refuted.

Newman defined a university as "a place of *teaching* universal *knowledge*. This implies that its object is ... the diffusion and extension of
knowledge rather than the advancement. If its object were scientific and
philosophical discovery, I do not see why a University should have students...."[13] There are, said the *Times* in 1867, two opposing conceptions
of the university: its function is either research or teaching. Most Englishmen take the second view and expect their children to be educated. If
research and the cultivation of learning interfere with teaching, parents
expect these pursuits to be restricted, or else they will cease to send their
sons to the universities.[14]

Until late in the century, liberal opinion on research was indifferent
rather than actively hostile. Mark Pattison, the most radical of the internal
critics of Oxford and Cambridge, agreed with Newman that a university
existed for the sake of the students, not the subjects. He did urge that the
endowments should maintain a professional body of men of science and
learning. It is not, however, to forward the interests of science and learning that they are associated, but in order to increase the efficacy of the institution as a seat of education. Pattison's ideal faculties would simply
master and hand on existing knowledge; they would not be expected to
advance it, and he also agreed with Newman and the *Times* that the characteristics which make a man a good investigator are almost certain to dis-
qualify him as a teacher.[15]

A definite shift in opinion on research occurred around 1870. In 1873
the Devonshire Commission described research as a primary duty of the
university,[16] and it is significant that the first official body to assert this
view was a commission on scientific instruction and the advancement of
science.

The scattered supporters of the German conception of a university did
not attempt to appeal to some ideal of advancing the frontiers of knowledge—such an approach would have awakened no response in English
opinion until considerably later. Instead, it was by a demonstration of the
obvious, practical, concrete utility of science that responsibility for facili-
tating original research was first fixed on the English universities. Anti-
German prejudice had stiffened the opposition to this development. This
was also, however, the period when the English began to feel the first
twinges of uneasiness about their industrial position and their commercial and military security. The few intellectuals like Huxley, Matthew

Arnold, and later the Webbs, who admired Prussian efficiency and German scholarship and who wanted to emulate both in the universities, were decisively reinforced by the general apprehensiveness about German scientific progress and German industrial competition.

Once the breach was made, a strong tradition of original scholarship for its own sake did develop within the walls of English universities. The contributions of Cambridge to modern physics go back to the 1870's. Oxford became a center of philosophy in the same period, and before the turn of the century the professional spirit had permeated the history schools at both Oxford and Cambridge. Other examples might be mentioned—following the Commission of 1877–1882, the new emphasis on specialized scholarship was apparent in all fields in the enlargement of the professorial faculty and the curriculum. Nevertheless, acceptance of responsibility for research was always more qualified than in Germany, and in Oxford and Cambridge the effects were more or less grafted onto the parent stem.

After 1880 the influence of the new ideas was more immediately obvious in the agitation in favor of transforming the University of London into a teaching university. The movement was exceedingly complicated, and its objectives were far from being realized by 1900, but a fairly definite ideal of what the university should be had emerged by then. The potentially enormous number of qualified students in London were to be provided with general, professional, and technical education on the undergraduate level. In addition, university examinations leading to degrees were still to be open to students throughout the empire.

No attempt was to be made to imitate the humane, domestic atmosphere of Oxford and Cambridge. Instead, the university intended to exploit and co-ordinate the unique opportunities for research and study in the metropolis. Faculties were to be headed by specialists actively engaged in advancing their subjects, who would provide formal graduate instruction and seek to attract the ablest students both from Britain itself and from overseas. Eventually London ought to rival Paris and Berlin as the seat of a world university, and in addition to being desirable in itself, the brains and talent brought in would be of great concrete advantage to the industrial, commercial, and social progress of the empire. In a word, the requirements of democratization, of imperialism, of technology, of professional training, and of providing for both teaching and research in the sciences, social sciences, and arts, were all reflected in the project for incorporating a federated university in London.

Like London, the other provincial universities were barely getting under way by 1900, but if their history falls in the twentieth century, their origins belong to the nineteenth. Except for Durham, all of the present civic universities started as university colleges which qualified students for degrees conferred by the external examining authorities of the old University of London or the Victoria University. The oldest of them, Owens College in Manchester, had been founded in 1851 by the will of a Nonconformist businessman, but the enrollment was still only about one hundred in 1870, when it began to grow rapidly. The other university colleges were all opened in the last thirty years of the century.

In general they were either outgrowths of schools of applied science and engineering and owed their origin to the increasing demand for technical education, or else they developed around courses of extension lectures sponsored by Oxford and Cambridge. The Yorkshire Science College in Leeds and Mason College in Birmingham were examples of the former type, the university colleges in Reading and Nottingham of the latter. A few, though not all, of these foundations were at first exclusively scientific and technical. Sir Josiah Mason, a self-made industrialist, had refused to permit his money to support anything except scientific, technical, and commercial instruction. In Manchester, on the other hand, Owens College had provided courses in the liberal arts from the beginning, an example which was followed in most other cities.

Despite incidental differences, however, all the provincial university colleges conformed to a general type and met a similar need. Whether founded by an individual philanthropist or by a civic movement, they were frankly the creations of an industrial and commercial society. Even where work in arts was available, the primary emphasis was vocational and professional training because this was what the majority of their students wanted. After Birmingham became a university, one of its innovations was a faculty of commerce, and it even offered courses in brewing. The teaching at all these institutions was professorial. A number of the professors happened to be research scientists, but the support of research could not be a primary objective of the universities themselves until the present century, when their financial resources, though still rather limited, had grown larger.

And in another way the objectives of the lesser civic universities were at first somewhat more restricted than those of London. All of them were regional institutions primarily concerned with providing educational opportunities for local students, both men and women, most of whom would live at home. In the case of London, this was only one of many functions of a university which was also intended to be national, imperial, and even cosmopolitan in its scope. But when the governing body of Mason College petitioned the crown to charter a university in Birmingham, they did so on the grounds that the College must have control of its own degrees if it was to "model its teaching and educational activity on such lines as will be specially advantageous and useful" to the population, manufactures, and industries of the Midland districts.[17]

At the end of the century, the reconstitution of London and the foundation of provincial universities had not impaired the prestige of Oxford and Cambridge, which, despite enormous changes in the past hundred years, remained in many respects unique. The idea or fact that they had a particular responsibility to the governing class died hard, if indeed it did die. In 1909, Lord Curzon as Chancellor of Oxford warned that a democratic zeal for academic excellence must not squeeze out the pass man. If Oxford, he thought,

> is to continue to deserve the name of a University, it has few more important duties to perform than to give a good general education to the man of birth and means. To convert him into a useful public servant is as honourable a task as it is to convert the artisan into a useful citizen, or the solicitor's son into a good solicitor; and many of the men who in later life have done the greatest justice to Oxford have been those who never took more than a Pass degree.[18]

Collegiate residence retained its importance as an integral part of education, and the influence of the system is apparent in the attempt of the new universities to house students in hostels providing some sort of social life. "English opinion," said one commentator, "shrinks from the homeless condition of the undergraduates in Edinburgh or Berlin."[19] Undergraduate devotion to sport contributed to the esteem in which the older universities were held—as far back as the 1840's a boat race had been the one indication which convinced Alton Locke, Kingsley's Chartist tailor, that, despite everything, there was sound British stuff at Cambridge.

Research degrees had been instituted shortly before 1900, chiefly to attract foreign students, but though there were now many productive scholars at both universities, there was as yet little provision for formal postgraduate instruction. Undergraduates were still the main concern, and in the eyes of apologists—and of critics, too—Oxford and Cambridge were more than universities. Oxford in particular seems to have been in addition almost a way of life if not a state of grace, and an education there was not only a training of the mind, but a social, a moral, a British, even a spiritual experience.

Although this attitude aroused some impatience outside the fold, there were very few intramural reformers who altogether rose above it; and, despite the shake-ups imposed by Royal Commissions and Parliament, the special character of the older universities was treated tenderly. Not for Oxford or Cambridge the earnest, the utilitarian, the somewhat dowdy and out-at-elbows intellectuality of Bloomsbury scholarship, and not for them the professional, trade-school atmosphere, the onwards-and-upwards-with-a-degree mentality of the provincial university. For Oxonians the objective remained the development of the whole man, and even critics would agree that Oxford trained, at least, the whole Oxford man.

Notes and References

1. "English Universities," *Edinburgh Review,* LIX (1831), 484.

2. *House of Commons Sessional Papers* (1922), X [Cmd. 1588], p. 11.

3. Mark Pattison, *Suggestions on Academical Organisation* (Edinburgh, 1868), p. 55.

4. Royal Commission on Cambridge, *Report* (1852), p. 202.

5. J. H. Newman, *The Idea of a University* (London, 1891), p. 109.

6. E. B. Pusey, *Collegiate and Professorial Teaching and Discipline* (Oxford, 1854), pp. 215–216.

7. J. S. Mill, *Inaugural Address Delivered to the University of St. Andrews* (London, 1867), pp. 6–7.

8. Quoted in Edward Fiddes, *Chapters in the History of Owens College and of Manchester University, 1851–1914* (Manchester, 1937), p. 73.

9. Royal Commission on Oxford, *Report* (1852), p. 174.

10. Quoted in J. W. Adamson, *English Education, 1789–1901* (Cambridge, 1930), p. 192.

11. Pattison, *Suggestions,* pp. 76–77.

12. H. P. Liddon, *Life of E. B. Pusey,* 4 vols. (London, 1898), III, 387–389.

13. Newman, *Idea,* p. ix.

14. *Times* (London), March 8, 1867, p. 7, cols. 4–5.

15. Pattison, *Suggestions,* pp. 171–173.

16. *Third Report, House of Commons Sessional Papers* (1873), XXVIII [C. 868], pp. xxix–xxxi, lvi–lx.

17. *House of Commons Sessional Papers* (1900), LXVI, Accounts and Papers, No. 22, p. 1.

18. Lord Curzon, *Principles and Methods of University Reform* (Oxford, 1909), p. 117.

19. A. I. Tillyard, *A History of University Reform* (Cambridge, 1913), p. 308.

3

The Formation of Lamarck's Evolutionary Theory

THE FOLLOWING ESSAY originated as a lecture delivered before the History of Ideas Club in the Johns Hopkins University in 1953. Professors Bentley Glass and Owsei Temkin were then meditating a centennial commemoration of Darwin's *Origin of Species* (1859). *Genesis and Geology* (1951) had come to their attention, and I was greatly honored to be invited to give a talk in a series looking to the publication of a suitable volume on Darwin's forerunners.

As I have mentioned in the opening autobiographical memoir, Arthur O. Lovejoy's classic *The Great Chain of Being* (1936) was one of the few works known to my generation that exemplified the possibility of treating, albeit a bit peripherally, the development of scientific thought as intellectual history. More than that, Lovejoy, a philosopher professionally, founded not only the *Journal of the History of Ideas,* but the very practice of intellectual history as a modern discipline. The privilege of speaking before the famous History of Ideas Club, which Lovejoy also founded, was compounded by his presence. He could not have been kinder or more encouraging to a beginner while at the same time participating actively in the discussion and offering cogent criticism.

It might be supposed that choosing Lamarck as the subject was the first step in shifting from the study of English to French scientific topics. It may indeed have anticipated that change of focus, but not, as well as I can recall, intentionally. It was simply that Lamarck was the most famous of Darwin's forerunners, and I thought I would try to find out what he actually did. As is my wont, I began by reading, not the secondary literature, but all of his books. To my surprise, it turned out that his commitment to what he and the French later called "transformisme" began not with zoology or natural history more largely, but with chemistry. When I then looked through writings about Lamarck, it did not appear that others had noticed that in later times. It may be that few people had ever read Lamarck's early writings, which in modern (though not contemporary) eyes are very peculiar indeed. However that may be, the findings seemed worth publishing, and here they are.

Meanwhile Professor Glass was selecting papers delivered before the History of Ideas Club and combining them with other pertinent memoirs

47

for publication in the long-planned commemoration of the Darwin centennial. My contribution, "Lamarck and Darwin in the history of science," draws on the gist of the paper below in order to strike a contrast between the Lamarckian and Darwinian theories of evolution.[1] Reading it over, I doubt that the state of Darwin scholarship needed me even then, and by now the contrast would be self-evident. Not all modern scholars agree with my sense of Lamarck's proper place in the history of evolutionary thought.[2] But no one, so far as I know, has questioned the basic point of this article, which is the role of Lamarck's theories of chemistry in the genesis of his theory of evolution.

Notes

1. *Forerunners of Darwin: 1745–1859*, .d. Bentley Glass, Owsei Temkin, William L. Strauss, Jr. (Baltimore, The Johns Hopkins University Press, 1959), pp. 265–291.

2. The standard works are Richard W. Burkhardt, *The Spirit of System: Lamarck and Evolutionary Biology* (Cambridge, Mass.: Harvard University Press) 1977, and Pietro Corsi, *The Age of Lamarck: Evolutionary Theories in France, 1790–1830.* (Berkeley: University of California Press), 1988.

THREE

The Formation of Lamarck's Evolutionary Theory[*]

—m—

L amarck occupies an unenviable position in the history of science. He is a truly outstanding figure. But a certain ambiguity hangs about the merit of his achievement, arising from an apparent oscillation of his career between poles of futility and pathos. The futility is that of any victim of a plot, however fortunate the outcome—for the outcome was to turn him into what his generation considered a distinguished scientist, not perhaps against his will, but certainly against his inclination. The pathos is of one who achieved recognition for what he held in small esteem, and never for what he prized.

In Cuvier's éloge, "Ce sont ses observations sur les coquilles et sur les polypiers . . . la sagacité avec laquelle il en a circonscrit et caractérisé les genres . . . la persévérance avec laquelle il en a comparé et distingué les espèces, en a fixé la synonymie, leur a donné des descriptions détaillées et claires," by which Lamarck "s'élevait enfin un monument fait pour durer autant que les objets sur lesquels il repose."[1] This was not the credit Lamarck wanted. He often exhorted his students to aspire beyond the detail and pedantry of nomenclature, way beyond it to a philosophy of nature, and he offered himself to his students and colleagues—indeed, he thrust himself upon them—as natural philosopher to his generation.

And so embarrassing were Lamarck's theoretical ventures, embarrassing not, as he imagined, to the security of some vested orthodoxy, but simply for the light in which they placed their author, that he was never refuted. He was not even dignified by unjust treatment. He was simply enveloped in a conspiracy of silence. For years he put his ideas forward. His *Recherches sur les causes des principaux faits physiques* was written

*Reprinted from *Archives internationales d'histoire des sciences,* 9(37) (Oct.–Dec. 1956), pp. 323–338.

in 1776, when he was 32.[2] His *Système des connaissances positives de l'homme* appeared in 1820. There were many similar volumes in the interval, and in all this time, his lifetime, no scientific judgment was published on these writings.[3] Officially, they did not exist. Unofficially, it was different. "Je sais bien," he observed in an exposition of his views in 1802, "que maintenant peu de personnes prendront intérêt à ce que je vais exposer, et que parmi celles qui parcourront cet écrit la plupart prétendront n'y trouver que des systèmes, que des opinions vagues, nullement fondées sur des connoissances exactes. Elles le diront; elles ne l'écriront pas."[4]

It is doubtful whether Lamarck would have been more gratified by the admiration of those modern successors who have distinguished his evolutionary notions from his other, as they agree unfortunate, philosophical ventures. The purpose of this paper is to consider Lamarck's ideas as a whole, but to do so from the point of view of their formation and import, and not of their validity. In this connection, it will not be necessary to discuss Lamarck as the founder of invertebrate zoology. For it was Lamarck's own feeling that taxonomy was not the most provocative aspect of natural science, and one can hardly fail to admire and sympathize with him as he toiled through the last twenty years of his life into blindness on what was expected of him, on what he did superlatively well, and on what he considered to be work of inferior dignity.

Nor is it clear that Lamarck's theories originated as generalizations of his positive contributions, even though that is what they became. His great *Histoire naturelle des animaux sans vertèbres* appeared from 1815 to 1822—but his evolutionary theory was fully developed in *Philosophie zoologique,* published in 1809, and first adumbrated in his course in 1800. Now, in 1800, he was just at the beginning of his work as a zoologist. Since 1793, when—known to science only as a botanist—he was appointed to the new chair of zoology at the reorganized *Muséum,* his thoughts had been absorbed in writings on chemistry, geology, and meteorology.[5]

He tells us himself that he first advanced his evolutionary theory as a pedagogical device—for he was of course a disciple of Condillac—in order to lead his students' minds back down the path which nature had followed upwards.[6] Consistently enough, he first approached the animal series as a study in degradation, not development. Moreover, what is clear is that the actual taxonomic work of Cuvier and Lamarck was complementary and executed on similar principles, which Lamarck learned primarily from Cuvier.[7] Like some pair of Copernican and Ptolemaic astronomers, they differed not about practice, but only about fundamentals.

The place to begin, then, is with the *Philosophie zoologique,* and at the outset it will be well to be definite about certain points on which interpretations vary. First, the book contains a real theory of evolution: the transmutation of animal species is taken as a fact and presented as a feature of the uniform development of the world. Secondly, Lamarck was a deist of the late Enlightenment, and for him evolution was the accomplishment of an immanent purpose to perfect the creation. Thirdly, there is no hint of the idea of natural selection. Lastly, much of the book is pre-scientific, both because of its finalism and because Lamarck's implicit conception of scientific explanation was Cartesian, not to say Aristotelian. Unless an explanation is causal, it is nothing, and his method is to bring each problem back to some inherent principle or tendency to perfection, carried out by the physical agency of a subtle, imponderable, 18th-century fluid, which is distinguishable only in its effects, but which must exist, for otherwise the effects remain inexplicable.

The organization of the *Philosophie zoologique* offers some excuse for differentiating between the seemingly unscientific parts of the work and the theory of evolution. The book falls into three divisions. Part I treats of natural history, Part II of physiology, and Part III of psychology. But Lamarck was a good *idéologue* and admirer of Cabanis, and the latter two parts discuss the same theme, the physical basis of life and feeling. It will be appropriate, therefore, to consider first Lamarck's views on evolution, and then his more general notions, and to do so in good Lamarckian fashion, tracing them in the inverse order of their development back to some point of common origin.

Stripping the theory of evolution of all color and connotation, it may be summarized quite briefly and baldly. Nature produces all varieties of animals, from the most rudimentary to the most advanced, and does so by progressively differentiating and perfecting their organization. If organic nature were omnipotent, the sequence would be altogether regular. But the tendency to complication is not the only factor at work. The influence of the physical environment causes variations in what the drive towards perfection alone would effect. Changes in the environment lead to changes in needs; changes in needs produce changes in behaviour; changes in behaviour become new habits which, if long-lasting, lead to alterations in particular organs and ultimately in general organization. The environment, therefore, does not act directly. It is a carrot, not a stick. And as a consequence of these principles, Lamarck stated the two laws of the development or decay of organs by use or disuse, and of the inheritance of acquired characteristics.[8]

If attention were confined to certain chapters of Part I of *Philosophie zoologique,* these evolutionary principles might very arguably be taken (as is often done) for the main point of that work—although nowhere does

Lamarck discuss transmutation as if it were itself the "ism." According to Lamarck himself, the book was an elaboration of *Recherches sur l'organisation des corps vivans* (1802). This treatise is devoted to physiology and psychology, and the theory of evolution is stated in a very lengthy preface, the substance of which had formed his introductory lecture for that year. Moreover, the emphasis is different. The main evolutionary principles are all to be found. But he begins with a theory of the *mechanism* of evolution—of which more later—and the argument is rather that species do not exist than that they are mutable. The question of transmutation becomes, if not unreal, at least subsidiary.

What interests Lamarck at this stage is rather the whole tableau of the animal series, and we are to see it not as the chain or ladder, but (if the metaphor may be varied) as the escalator of being. For nature is constantly creating life at the bottom. Moisture and heat and ooze are all she needs. And essential life fluids are ever at work differentiating organs and complicating structures. And there is a perpetual motion of organic matter up the moving staircase of existence, and of its residue spilling back down the other side, the inorganic side. Here, too, Lamarck states laws, but these are different laws. They generalize the facts, not about mutability, but about the whole zoological scale. "La série qui constitue l'échelle animale réside dans la distribution des masses, et non dans celle des individus et des espèces." For there is indeed a linear, regularly gradated series in the scale of nature. "Mais je parle d'une série assez régulièrement graduée dans les masses principales, c'est-à-dire dans les principaux systèmes d'organisation reconnus." The series does *not* reside in species. Preoccupation with species is precisely what has led natural history in a circle. This conception of masses as the systems of organization was very important to Lamarck: "Ainsi, chaque masse distincte a son système particulier d'organes essentiels, et ce sont ces systèmes particuliers qui vont en se dégradant, depuis celui qui présente la plus grande complication, jusqu'à celui qui est le plus simple. Mais chaque organe considéré isolément, ne suit pas une marche aussi régulière dans ses dégradations: il la suit même d'autant moins, qu'il a lui-même moins d'importance."[9]

According to Lamarck's historians, however, the first statement of evolutionary theory occurred two years earlier, on 11 May 1800, in his inaugural lecture for the year 8.[10] As will be explained, we know that in 1797 he still believed in the existence and invariance of animal species. And admiration and astonishment are often expressed that at the age of 57 Lamarck should have conceived an entirely new theory and established thereby his place in the history of science as the founder of evolutionary thought. This lecture was published as a preface to his first extensive zoological treatise, *Système des animaux sans vertèbres* (1800). Once again there will be found, though now in very summary statement, almost all the main positions which were to be occupied in detail in *Philosophie zoologique*. But once again the focus and emphasis are different. Lamarck is here concerned, not to establish the mutability of species, nor even to argue the uniform distribution of organic masses. These points are simply stated. (It was always his way to elaborate in enormous detail in one work a conception which he had simply introduced into its predecessor.) And now these theses are brought in as subordinate propositions, supporting the main contention, which is that the naturalist must make an absolute distinction between living and non-living bodies. This, and not the traditional division into the three kingdoms, must be his starting point as he fares forth to survey the whole panorama of the products of nature.

This insistence is significant for two reasons. On the one hand, it is consistent with Lamarck's preference for classification by dichotomy, which was his method in *Flore française* (1778). But more importantly, this is also the thesis of *Mémoires de physique et d'histoire naturelle,* published in 1797. The *Mémoires* resumed the attack upon contemporary chemistry which as a young man he had launched twenty years before in the *Causes des principaux faits physiques.* That book had been ignored by the *Académie des sciences;* now he set out to read his *Mémoires,* one by one, to the *première classe* of the *Institut.* The central proposition is that all inorganic composites are residues of the life processes of organisms, perpetually repairing the tendency towards decomposition which is the basic process of inorganic nature. In physical nature, there is no tendency for composites to form—there is only the reverse. Nor did Lamarck renounce this view. He recurs to it in passing in *Philosophie zoologique,* where he has simply shifted his attention to the organic branch of natural history. Such importance did he attach to this division, that in the *Causes des faits physiques,* where his primary interest is explaining *inorganic* nature, he says that the phenomena of life are so different that they are hardly to be considered part of nature at all.[11] It was not inconsistent, therefore, when his evolutionary theory described the departures from regularity in the animal scale as consequences of conflict between organic nature and the brute nature of the environment. Quite the contrary—this relationship, at once opposition and dependence, is fundamental to Lamarck's thought, which in this respect is almost dialectical.

Nor is the inconsistency on species other than apparent. It is true that in 1797 Lamarck still believed that animal species are constant.[12] But in a short essay included in *Organisation des Corps Vivants* (1802), he tells us himself how he came to change his views, and it turns out that, far from revolutionizing his outlook on nature, he is simply changing his mind on what he originally saw as a matter of detail.[13] What he did between 1797 and 1800 was to assimilate the question of organic species (or rather of their non-existence) to that of species in general, and of mineral species in particular. For in Lamarck the word "species" had not yet lost its broader connotations. It still has the sense of all the forms into which nature casts her manifold productions in all three kingdoms (or rather in both departments).

Lamarck had long been impressed with the mutability and perpetual decay of the surface of the earth, and for a long time (he writes) he had shared Daubenton's opinion that there are no permanent species among minerals, no real species there at all. The only entities in inorganic nature are the "molécules intégrantes," and the masses which form under the play of circumstance and universal attraction. It is to be remembered that the notion was still very widespread in the eighteenth century that minerals are molded by some plastic force, that they are *bred* in the earth. Lamarck does not state this old view. But he cannot have been unconscious of it, and it is implicit in his chemistry, where he refused to believe that a molecule can be "as old as the world."

Indeed, the analogy with the view which Lamarck came to hold of the living world is striking. In both organic and inorganic nature, the only realities are the individual—the particular animal, the particular molecule—and the system of organization—mammalian quadruped, granitic structure—into which circumstance has fitted it. But the material itself moves along this double chain of systems in the course of world history, from mollusc to man, from limestone (say) to Iceland spar. And this explains, perhaps, Lamarck's pleasure in the concept of masses as links in his chain. It is natural enough to think of the principle of granite in the world in terms of mass; what he does is to think of the principle of mammal in the same fashion.

Once Lamarck began to study zoology, therefore, what he did was to modify his views on animal species, which he had probably accepted without much thought from Linnaeus, to accord with his congenital outlook on the world as flux. He never *did* change his views on species in the larger sense:

> On enseigne à Paris, dans les cours, que la molécule intégrante de chaque sorte de composé est d'une nature invariable, et conséquemment qu'elle est aussi ancienne que la nature. Il y a par conséquent des espèces constantes parmi les minéraux.

Pour moi je déclare que je suis persuadé, et même que je suis con-
vaincu, que la molécule intégrante de toute substance composée
quelconque peut changer de nature, c'est-à-dire qu'elle peut subir
des changements dans le nombre et dans les proportions des principes qui la constituent.

En effet, chaque fois qu'une molécule intégrante, de quelque
composé qu'elle soit, subit quelque changement dans les proportions de ses principes, ceux qui restent combinés après ce changement, forment alors une molécule d'une nature différente de celle
de la première; cela est de toute évidence. La molécule qui a subi
le changement dont je viens de parler, n'a plus nécessairement ni
la même densité, ni la même forme, ni les mêmes qualités particulières qu'elle avoit auparavant . . .

Non seulement les actes des fermentations, des dissolutions,
des calcinations, des combustions, etc. attaquent et détruisent les
masses des corps qui s'y trouvent soumis; mais ils attaquent aussi
les molécules intégrantes de ces mêmes corps; ils dérangent l'état
de combinaison de leurs principes; ils opèrent des dissipations de
certains de ces principes, et souvent des additions de quelques
autres qui n'y existoient pas; en un mot, ils donnent lieu à de nouvelles combinaisons diverses, et conséquemment ces actes chim-
iques exercés, soit par la nature, soit par l'art, changent réellement
la nature des molécules intégrantes des matières qui en subissent
les effets.[14]

Tracing Lamarck's evolutionary theory back through successive muta-
tions, then, one finds oneself in the midst of chemistry. For the main burden of Lamarck's complaint against the Lavoisier school was precisely its
insistence on the fixity of chemical composition. But Lamarck's chemistry
is a looking-glass chemistry, just as Aristotelian mechanics was a looking-
glass mechanics. Nor was there any greater possibility of a meeting of
minds between his chemistry and a real chemistry. For example, he criticized the chemists for holding that addition of one element to another can
form a new compound. This simply supposes that the product was pre-
existent in the reagents. If this is true, nothing has happened. If it is not
true, nothing has been explained. And, Lamarck continues, the proper
business of chemistry is to explain the *cause* of chemical change. To in-
vent a word like "affinity" is to take refuge in a name. Instead the chemist
must show how the active principles involved permeate the bodies and alter their very nature.[15] He offers in fact the old riposte—that to describe

is not to explain. For Lamarck, there were still the four elements and an indefinite number of principles, all capable of existing in various states, depending on how far they were permeated by various subtle fluids which run through all things and which are the physical carriers of effects: color, heat, sound, taste, and odor.

This is not the place to dwell on the details of this archaic chemistry of qualities and principles. It suffices to point out that the first statement of what became Lamarck's evolutionary theory occurs in his assertion of the indefinite variability of chemical composition. And when attention is turned to the second aspect of *Philosophie zoologique,* to the physiology, it emerges that this, too, is ultimately derivative from chemical theories, both in manner and substance. In *Philosophie zoologique* he says of physiology what he says of chemistry in his early books. What is needed is not discovery. All necessary facts are known. He brings to the subject, not research, but the right perspective.[16]

In his 1776 essay on physical causes, Lamarck devotes one section to the activity of living beings. Here he first explains that there can be no chemical fixity because physical nature tends only to decomposition, whereas all composites are originally formed by the activity of organisms. Only life can synthesize.[17] Conversely, in the life processes themselves, growth is the increase in mass resulting from the retention in youth of what is needful from the materials of the environment which the organism passes through its system. As has been noted, Lamarck was always drawn to express problems as situations of equilibrium balancing the activity of life against mass. Aging and death result from the progressive hardening of the soft and pliant organs by the life-long ingestion and digestion of the environment.

In his later works, Lamarck adapted this same principle to provide evolutionary theory with a mechanism. It is erroneous to attribute to Lamarck the explanation that evolution is caused by the giraffe stretching its neck or the serpent slithering under the stone. These are creative responses. The cause is far more fundamental and lies in the very nature of life. "Le propre du mouvement organique"—quoting from the discourse of the year X—"est non seulement de développer l'organisation, mais encore de multiplier les organes et les fonctions à remplir; et . . . en outre, ce mouvement organique tend continuellement à réduire en fonctions particulières à certaines parties, les fonctions qui furent d'abord générales"— reproduction, for example, in simple animals a function of the whole organism, becomes located in certain specialized organs. This comes about because "le propre des fluides dans les parties souples des corps vivants qui les contiennent, est de s'y frayer des routes, des lieux de dépôt et des issues: d'y créer des canaux et par suite des organes divers; d'y varier ces canaux et ces organes, à raison de la diversité soit des mouvements, soit de la nature des fluides qui y donnent lieu."[18]

The mechanism of evolution then, is perfectly analogous to the mechanism of erosion. Eventually, the organism silts up and dies. It is an illustration of the unity and adaptability of Lamarck's ideas, from chemistry through physiology to geology, that he formulated this conception of evolutionary mechanism at the time when he was writing *Hydrogéologie*, a uniformitarian treatise on the development of the earth.[19]

In his essay of 1776, there is an interesting footnote. "Pour expliquer physiquement l'origine et le mécanisme de l'univers," he writes, we need to know three things:

> La première est la cause productrice de la matière munie de toutes les qualités et facultés qui tiennent à son essence. La seconde est celle de l'existence des êtres organiques et de ce qui constitue la vie et l'essence de ces êtres; car la matière avec toutes ses qualités, ne me paroît nullement capable de produire un seul être de cette nature. Enfin la troisième est celle de l'*activité* qui se trouve répandue dans tout l'univers. . . .[20]

It is true that Lamarck was modest. He did not think that philosophical reasoning could ever fully attain to this knowledge. Nevertheless, his career was nothing less than an attempt to "explain the origin and mechanism of the universe," an attempt carried forward with great fidelity, and having dealt with Lamarck's views on the origin and essence of matter and of organism, let us turn briefly to the third problem of his trilogy, the problem of activity.

In all his chemistry, Lamarck attached great importance to the element of fire. He even called his theory the "pyrotic" theory and opposed it as such to the pneumatic theory of the chemists. How gratuitous, he urged, to describe combustion as the combination of the burning body with something in the air. It is not just that no one has ever seen this hypothetical oxygen. The same was true of Lamarck's fluids. The crucial objection is that there are no effects which require it to be posited. Combustion is perfectly well explained as the action of fire, and for the existence of fire we have much independent evidence. It is even observable, in its most violent state in the act of burning, and also in less active states. One can see it, shimmering over a hot stove or a tile roof in the sun of a summer day.[21]

But Lamarck is not simply disagreeing over the most common chemical reaction. For fire is the principle of activity in nature. It is a subtle fluid.

It exists in many states, of which Lamarck undertook a taxonomy.[22] In the fixed state, in coal, wood, or what will burn, fire is the principle of combustion. Conflagration is fire in its state of violent expansion, penetrating all the pores of a burning body and ripping it to shreds. Evaporation occurs when fire in a state of moderate expansion surrounds the molecules of water and bears them upward, so many tiny molecular balloons, to rejoin the clouds, where the specific gravity of the water molecule in its light shell of fire balances that of air. (Lamarck also aspired to found meteorology as a science.) Finally—not to follow fire into all its states—there is a natural state, to which fire seeks to return. And all the phenomena of light and heat, of the action of the sun and atmosphere, are explained by fire in its different states, forever striving to regain that which is natural.

Nor did Lamarck abandon his commitment to fire. It provided him with a physical basis of feeling, and—he seems to have thought—of life itself. From his distinction between organic and inorganic nature, it might be supposed that Lamarck was a vitalist. In fact, the reverse is true. Spontaneous generation is brought about by the stirring of fluids, and Lamarck sometimes infers that this process is quickened by fire—on this point he never quite stated the ultimate mechanism. But as to feeling he is explicit: its physical basis is the nervous fluid, and this is probably the same substance as the electrical fluid, which is only fire in one of its special states.[23] It is not, therefore, too much to say that in the pyrotic theory is to be found the common origin of the three aspects of the *Philosophie zoologique*— the psychology, the physiology, and the theory of evolution.

Moreover, the emergence of Lamarck's evolutionary views from his hatred of Lavoisier's chemistry is not simply a curious adventure of ideas. For chemistry was then the locus of the scientific revolution. But the loss of phlogiston was not what distressed Lamarck. That was a superficial controversy, and the real issues went deeper. They were, indeed, the relics of those that had divided Galileo from his enemies and (at the same time) the harbingers of the opposition that awaited Darwin. For the significance of the revolution inaugurated by Lavoisier in chemistry is that it removed the subject from the humane realm where science explains by seeking out familiar qualities in nature, to the forbidding one where all it offers is abstract and quantitative description of something which is no longer recognizable as nature at all.

Lagrange once suggested that the proper road for chemistry lay in turning it into a kind of algebra,[24] and Lavoisier took his inspiration from the analytical approach of his mathematical colleagues.[25] In the split between natural history and physics which opened as the reflection in science of Jean-Jacques Rousseau's revolt against the Encyclopedic spirit, Lavoisier took up his position with the physicists. And his theory evoked the hostile response that the *honnête homme* often reserves for new steps in abstract understanding—Rousseau's "honnête homme qui ne sait rien, et qui ne s'en estime pas moins." This issue was explicit. Lavoisier paid for his stand. For he died, not just a farmer-general but an embodiment of those elements of arrogance, inhumanity, and incomprehensibility, which the romantic view of science saw in Newtonianism, which the Jacobin view of science saw as the unforgivably vain and aristocratic spirit of the great learned academies and indeed of theoretical science itself.

Lamarck's attack upon Lavoisier, then, is of a piece with Gœthe's *Farbenlehre* and with the scientific undertakings of Marat (whom Lamarck cites with respect) right up to the point that Lamarck, too, hated and resented mathematics.[26] Even the history of his book on physical and chemical causes bears out this pattern. Written in 1776, almost at the moment of formation of the new chemistry, and presented to the Academy in 1780, it was examined and rejected. For the next ten years, Lamarck, though of the Academy himself in his capacity as botanist, moved in the resentful intellectual demi-monde of Paris and formed part of that circle of naturalists, forgotten now, who as youths had occasionally herborized with Rousseau himself, and who in 1789 formed societies which soon joined in a coalition with democratic clubs of artisans, artists, and inventors to bring down the old academies during the Terror.[27] In 1794 he published his book, dedicated to the French people, to whom he addressed an account of the persecutions that had kept it from them until the day of liberty.

This is a set of circumstances which has real historical interest. In the eighteenth century, the preference of romantic thinkers for the sciences of life is as striking as the predilection of rationalistic writers for the physical sciences. Voltaire popularized physics and Rousseau botany; it is no accident that the Jardin des Plantes was the one scientific institution to flourish in the extremist phase of the Revolution, which destroyed all the others. And down to Darwin, who subjected life to nature, the idea of evolution was attractive to most romantic minds.

In the interaction between life and matter which Lamarck saw as the dynamic process of the world, he asserted the primacy of life. He sallied out to meet the enemy on his own ground, in the science of chemistry, into which he introduced life as the sole *constructive* chemical agent, and did so long before the battle had shifted to biology itself. It was a sound instinct, then, that led many neo-Lamarckians to his work, sounder than they knew who stopped with the theory of evolution itself. For not only can it be interpreted as a defense of the study of the nature in which our lives are led against mechanistic science. It originated as that, in a forlorn attempt to save the science of chemistry for the spirit of natural history.

One final remark should be made: in seeking out the origin of Lamarckism, it is not the purpose of this paper to denigrate the scientific value of its theory of evolution, any more than one would impugn the laws of planetary motion because of their emergence from Kepler's Pythagoreanism. Like Kepler, Lamarck drew from his general outlook all but the positive aspect of his theories. As he actually studied zoology, as the panorama of biological evolution came to occupy a progressively larger place in his vision of the world, he improved and sobered his theory, not perhaps out of all recognition, but certainly in the direction of a generalization which, but for Cuvier, might have brought his idea of order into a department of knowledge.

It is extraordinarily interesting to follow the march of his opinions, interesting and (if one may say so) heartening to those who like their faith in the merit of scientific investigation confirmed. For they display an almost Lamarckian evolution toward perfection. In his writings on chemistry, those endless books, the tone is bitter, misunderstood, wronged. In his earliest evolutionary essays, he is still the lonely prophet, reviled by the interests and the men of little vision. But in *Philosophie zoologique*, this air of martyrdom has diminished, though not to vanishing. And what is astonishing is that a man who in his fifties was still striking martyred notes, should in his last years appear to have mellowed and sweetened in his nature. The *Système analytique des connaissances positives*, published as blindness set in, is a book of real serenity. In his contemplation of nature, in his patient and accurate cataloguing, Lamarck found not only materials to substantiate his theory; he seems to have found, too, the consolation which Rousseau only sought.

The pathos of Lamarck, then, is more than the pathos of the unappreciated, the deservedly unappreciated philosopher. It is the pathos of one side of the history of science, of the human mind resisting the implications of its own conquests. For (as the work of M. Koyré has taught us) it is precisely the humane and idealistic elements of science which in its own progress are inevitably exorcised. One cannot but hear the dying echoes in Lamark's work: the striving towards perfection; the organic principle of order over against brute nature; the life process as the organism digesting its environment; the primacy of fire, seeking to return to its own; the world as flux and as becoming. He is perhaps the last scientist who gives them back, these old echoes. Would it be too fanciful to describe him as the Heraclitos of the Enlightenment? In any case, in all but the range of his information, he was clearly closer to the Greeks than he was to Darwin, and this paper might, therefore, be taken as a documentation of the reminiscence of Lamarck's course which Sainte-Beuve imagines for Amaury:

> M. de Lamarck était dès lors comme le dernier représentant de cette grande école de physiciens et observateurs généraux, qui avait régné depuis Thalès et Démocrite jusqu'à Buffon; il se montrait mortellement opposé aux chimistes, aux expérimentateurs et analystes *en petit*, ainsi qu'il les désignait. Sa haine, son hostilité philosophique contre le Déluge, la Création génésiaque et tout ce qui rappelait la théorie chrétienne, n'était pas moindre. Sa conception des choses avait beaucoup de simplicité, de nudité, et beaucoup de tristesse. Il construisait le monde avec le moins d'éléments, le moins de crises, et le plus de durée possible ... De même dans l'ordre organique, une fois admis ce pouvoir mystérieux de la vie aussi petit et aussi élémentaire que possible, il le supposait se développant lui-même, se composant, se confectionnant peu à peu avec le temps; le besoin sourd, la seule habitude dans les milieux divers faisait naître à la longue les organes, contrairement au pouvoir constant de la nature qui les détruisait; car M. de Lamarck séparait la vie d'avec la nature. La nature, à ses yeux, c'était la pierre et la cendre, le granit de la tombe, la mort! La vie n'y intervenait que comme un accident étrange et singulièrement industrieux, une lutte prolongée, avec plus ou moins de succès et d'équilibre çà et là, mais toujours finalement vaincue ... Et puis, dans sa résistance opiniâtre aux systèmes de toutes parts surgissants, aux théories nouvelles de la terre, à cette chimie de Lavoisier qui était une destruction, une révolution aussi, il me rappelait involontairement cette semblable obstination imposante de M. de Couaën dans une autre voie; quand il dénonçait avec amertume la prétendue conspiration générale des savants en vogue contre lui et contre ses travaux, je le voyais vaincu, étouffé, malheureux comme notre ami; il avait eu du moins le temps de se faire illustre.[28]

Notes and References

1. *Mémoires de l'Académie Royale des Sciences de l'Institut de France*, XIII (1835), p. XXVII–XXIX.

2. (2 v., 1794). Lamarck offered the work to the *Académie des sciences* in 1780. See p. VII–XVII for his account of its cold reception and his decision to publish almost twenty years later, now that the Academy had been suppressed.

3. Bibliography in Marcel Landrieu, *Lamarck* (1909), p. 448–475.

4. *Recherches sur l'organisation des corps vivans* (1802), p. 69.

5. A.-P. de Candolle, *Mémoires et souvenirs* (Geneva, 1862), 44.

6. *Philosophie zoologique* (2 v., 1809), I, i.

7. See Henri Daudin, *Cuvier et Lamarck: Les Classes zoologiques* (2 v., 1926).

8. The best summary of Lamarck's theory is in E. Guyénot, *Les Sciences de la vie* (1941), p. 418–439.

9. *Op. cit.*, p. 39–41.

10. Landrieu, p. 297–302.

11. *Op. cit.*, II, 186.

12. An explicit statement will be found in *Causes*, II, 214; and although Lamarck does not discuss the question directly in the 1797 *Mémoires*, he does re-affirm the argument of the passage just cited (see *Mémoires*, p. 270–271).

13. *Op. cit.*, p. 141–156.

14. *Ibid.*, p. 150–152.

15. See, e.g., the discussion in *Mémoires*, p. 7–20, and *Réfutation de la théorie pneumatique* (1796), p. 69–77.

16. *Philosophie zoologique*, I, p. 370; *Mémoires*, 4.

17. *Causes*, II, p. 274–315.

18. *Corps Vivants*, p. 7–9; Cf. Causes, II, p. 184–219.

19. It cannot be too much emphasized that Lamarck saw his own work as a single body of thought. His original plan was to follow his "*physique terrestre*" with the *Hydrogeologie* (1802), a *Météorologie*, and a *Biologie*, and it was material which he had originally reserved for the latter which he drew on for the *Corps Vivants* and ultimately *Philosophie zoologique*. (See *Corps Vivants*, V–VIII). In the event, he never drew his meteorological writings together.

20. *Op. cit.*, II, 26.

21. *Causes*, I, 47–60.

22. This is the subject of most of Volume I of the *Causes;* see, too, tabular classifications of the states of fire in *Réfutation*, 31, 36.

23. See e.g., *Corps Vivants*, p. 163–4, 195.

24. Delambre, "Notice sur la vie ... de Lagrange," *Œuvres de Lagrange*, I (1867), XXXVIII.

25. Maurice Daumas, *Lavoisier, théoricien et expérimentateur* (1955), p. 12–13, 18, 160–161, and for Lavoisier's own (inevitable) failure to make a clean break with the old chemistry, p. 161–178.

26. On this, see the suggestive discussion of G. Bachelard, *La formation de l'esprit scientifique* (1938), p. 226–228. For Lamarck's citations of Marat, see, e.g., *Causes*, I, p. 343–368.

27. This subject will be discussed and documented in a future work. For the moment it will suffice to refer to the Registre ... de la Société Linnéenne, Bibl. Mazarine, MSS 4,441; to the preface to the *Actes de la Société d'Histoire naturelle* (1792); and to the materials bearing on the Point central des arts et métiers, Archives nationales, AD VIII, 40.

28. *Volupté* (1927, ed.), I, pp. 192–194.

PART II

French Science

—ɯ—

4

The Discovery of the Leblanc Process

THE FOLLOWING THREE PAPERS were the firstfruits of the archival research in Paris that the Guggenheim Foundation underwrote in 1954–55. This was my introduction to dusty old documents, most of which no one had ever looked at. Such an experience entails more than the information they contain. That is the main thing, of course, but apart from that there is an immediacy, a sense of presence, about having the yellowed papers right there on the desk. One needs to walk through the streets where these things happened. Not a trace of Leblanc's factory remains, but one needs to take the Métro out to Saint-Denis and find the site where it stood.

Archival research presupposes some notion of what one is looking for. As a sometime student of chemistry and chemical engineering, I was aware of the importance of the Leblanc process for the manufacture of artificial soda throughout much of the nineteenth century and also of the legend of Leblanc as a martyr inventor during the French Revolution. Papers may or may not come to hand in some sort of order, depending on how well the given "fonds" or series has been inventoried and arranged. In this instance they did not. I was able to piece the story together only after having called for every entry under the rubrics "Leblanc" and "soude artificiel." Making sense out of documents that pass under one's eyes at random and chasing down such other characters as Dizé, Shée, Carny—all that adds to the labor and the pleasure of the hunt. Eventually, if luck permits, a credible picture emerges.

The question of context remains. J. R. Partington discusses the basic Leblanc reaction in his magisterial history of chemistry.[1] John Graham Smith places the process in the history of chemical industry. He disagrees with my conjecture that Leblanc came on his method by means of a fallacious analogy with the smelting of iron.[2] The problem for me is the relation of science to technology. What that may have been will appear in the paper that follows.

Notes

1. *A Short History of Chemistry* (London, 1937).
2. *The Origin and Development of the Heavy Chemical Industry in France* (Oxford: Oxford University Press, 1979).

FOUR

The Discovery of the Leblanc Process*

—⁓—

This is the first of two papers addressed to the question of what sort of influence science exerted on industrialization in the eighteenth century. Both are the outcome of studies pursued in France,[1] and they make special reference, therefore, to the shape which this problem assumed in its French development. The second paper will be devoted to certain general considerations. The present paper, by contrast, limits itself to a single instance and offers an account of the discovery by Nicolas Leblanc of the use of limestone in the conversion of salt drawn from sea water into soda for commercial consumption. This approach has several advantages. It is based upon what was done in a specific industry and not upon what was written about industry in the large. There was, of course, no typical discovery. But the questions raised by this one are characteristic. Moreover, the Leblanc discovery was of very great importance. Upon it rested almost the entire alkali industry throughout much of the nineteenth century, during the time when this was the most extensive single chemical industry. And as is often the case with important events in the industrial and scientific past, investigation reveals that the subject has still to be turned from legend into history.

I

The legend of Leblanc is very familiar. It conjures up the vision of a conscientious chemist whose discovery, a classic illustration of science in the industrial revolution, is supposed to have been the due reward of research pursued in penury and self-denial. He founds a flourishing enterprise, but its prosperity is soon to be destroyed by the Committee of Public Safety's

*Reprinted from *Isis* 48, part 2, no. 151 (June 1957), pp. 152–170.

impulsive order to publish the process during the crisis of war production in the Year II.[2] He becomes the innocent victim of the Revolution, his rights sacrificed to the war effort, he himself compromised and excluded from his own factory because the capital had been furnished by the Duc d'Orléans. He emerges the devoted inventor, betrayed by his scientific colleagues and ruined by competitors who took advantage of his patriotism and misfortunes to secure a commanding headstart in exploiting his own process. He dies by his own hand in 1806, the benefactor of his country, broken by its ingratitude, having won for his pains only a pathetic distinction in the martyrology of discovery.

This is a familiar story. It rests on very slender information. T. S. Patterson discussed it without modifying it in a paper published in 1925.[8] But this and all other accounts of Leblanc are derived ultimately from a memoir published in 1884 by his grandson, Auguste Anastasi.[4] Anastasi had been a painter. His little book is a work of family piety. When he wrote it, he had been blind for some years. For Leblanc's personal history he drew upon family papers, which cannot have been very complete, although with due caution the book may still be used for biographical information. For his scientific history, Anastasi relied entirely upon a report to the *Académie des sciences* drawn up in 1856 by J. B. Dumas in adjudication of certain claims to Leblanc's invention originally advanced in 1810 by J. J. Dizé, one of his associates, and later pressed by Dizé's descendants.[5] Materials for a really historical account do exist, however, in contemporary journals, in subseries F-12 at the *Archives nationales,* and in the *Archives de l'institut national de la propriété industrielle,* where are preserved documents relating to all the patents which have been granted under the patent laws of 1791.[6] This last collection does not, indeed, appear to have been much consulted by historians of French science or industry. All told, these sources permit an account of the early history of the Leblanc process, one that would even be definitive if it were not for a single missing piece. Unhappily, the documentation stops just short of the act of invention itself. How did Leblanc decide to try limestone? To that question there can be returned only what seems to be the most plausible suggestion, and to that and to the scientific history of the discovery it will be necessary to return. First it will be convenient to set forth the early industrial history of artificial soda in France.

II

Until the latter part of the eighteenth century, potash (potassium carbonate) or "vegetable alkali," was much the more important of the two "natural" fixed alkalis. But it was obtained from wood ashes, and the supply was reduced, therefore, by the expansion of the metallurgical industries at the very time when demand increased with the growth of the textile trades and the manufacture of soap and of glass. Mineral alkali (sodium carbonate), on the other hand, so-called because of its occurrence as a hydrate in the efflorescence of salt beds in the Egyptian desert, was obtained commercially from the incineration of certain sea-shore plants, which convert sodium chloride from seawater into oxalate, tartrate, or other organic salts, and which on incineration and lixiviation yield soda. The best quality, known as barilla, was produced in Spain, in the vicinity of Alicante.[7] The inelasticity of this natural supply, combined with obvious mercantilist considerations, was what established the incentive to meet the growing demand with a practical method for converting salt from sea-water directly into commercial soda.

Nor was the Leblanc method the first. "La fabrication de la soude avec le sel marin n'est pas un secret aujourd'hui," wrote Tolozan, intendant of the *Bureau du Commerce,* in 1790, over a year before Leblanc received his patent.[8] In France the earliest practicable process was developed by a Benedictine abbé, Père Malherbe, who in 1777 succeeded in converting Glauber's salt (sodium sulfate)—obtained as was common practice by the action of sulfuric acid on marine salt—into soda by use of charcoal and iron scrap. Malherbe's method reduced sodium sulfate to the sulfide by fluxing with charcoal in a reverberatory furnace. He then added scrap iron and secured a mass of ferrous sulfide and caustic soda, formed by the action of oxygen in the fire gases. The residue was cooled and exposed to the air until it crumbled. It was then lixiviated for the carbonate, ordinary mild soda.[9]

Malherbe formed an association with one Athénas, an entrepreneur and chemical artisan, probably of Croisic, who pushed the research further and determined the feasibility of substituting for iron either certain iron ores direct from the mines, or "vitriol martial" (copperas or ferrous sulfate) obtained from the peat bogs of Brittany. This latter variation had the advantage of eliminating the need for sulfuric acid, which was both expensive and likely to be unobtainable whenever the supply of saltpetre was preempted by military requirements. (Sulfuric acid was generally made by burning sulfur and potassium nitrate and dissolving the fumes.) Athénas began by taking fourteen parts of ferrous sulfate, which he heated slowly to a red heat and to which he added ten parts of sodium chloride, well calcined. The mixture was heated further until all hydrogen chloride was liberated and the salt converted into sodium sulfate. At this point charcoal was added, and the heating pushed to fusion. According to Darcet's description, "L'oxide de fer repasse à l'état métallique et se recombine avec le soufre," after which reaction the mass was withdrawn from the furnace and the soda obtained as in the original Malherbe method.[10]

Evidently Malherbe's association with Athénas failed, although by 1794 Athénas was himself exploiting the process. By that time it was also in production in at least two other establishments—at the great glass factory at Muntzthal, and (more importantly) at Javelle, where there was a very considerable factory which, among other enterprises, produced hydrochloric acid for sale to the bleaching industry as a source of bleach, and where Alban, the director, used Malherbe's process to convert his residual sodium sulfate into soda.[11]

Throughout the 1780's the *Bureau du Commerce* under Tolozan's administration had sought to bring about just such developments. It is, nevertheless, doubtful whether those that succeeded did so as a result of this official encouragement. One of the larger producers of artificial soda was Chaptal, who in his factory at Montpellier exploited the reaction of salt and litharge (lead oxide). The method was simplicity itself, but because of the cost of litharge, it was also expensive. A brine solution was poured onto the litharge in great earthen containers, in sufficient amount to form a paste. The mass was stirred and additional brine added from time to time for twenty-four hours. The products were lead oxychloride and caustic soda, which was separated by lixiviation, re-crystallized and allowed to take up carbon dioxide from the atmosphere. The lead oxychloride could be used for pigment, either directly as yellow lead, or, converted into lead sulfate by dilute sulfuric acid, as white lead. This method had been in use for some time in England, where it was worked for yellow lead rather than soda as the principal product.[12]

Chaptal had never made a secret of his procedures, nor asked for government favor. No more had either the management at Javelle, or a bleaching establishment owned by one Ribeaucourt, who for two years had been converting Glauber's salt to soda.[13] On the other hand, although several privileges were granted carrying the exclusive right to exploit particular processes, none was successfully exploited. In every case, the obstacle was economic rather than technical. The *Conseil d'État* granted a privilege only after Macquer or Berthollet, his successor as chemical consultant to the *Bureau du Commerce*, had conducted careful tests. But the government's policy of encouragement moved in contradictory directions.

For example, the prize established in 1783, by the Crown, and administered by the *Académie des sciences*, was never awarded. The process which would have won, and which would as a condition of the contest

have been made public, was the same as that invented simultaneously by
Guyton de Morveau, by which salt was converted to soda by use of quick-
lime. In Guyton's view, this was the method employed by nature herself
in producing an efflorescence of soda on the surface of cement mortar in
certain cellars, and in the dried residue of saline lakes and springs. His directions call for slaking quicklime in water, and adding a saturated brine
solution. The mixture was concentrated to a paste, and left exposed to the
air in some closed and humid place, preferably a cellar. Sodium carbonate
would then form on the surface of the mass. When it was removed another layer would form, and the process would be repeated until the materials were exhausted.

For this process, Guyton received a privilege dated 3 June 1783.[14] But
even the policy of granting inventors exclusive rights to the exploitation
of their own inventions fell victim to conflicting interests. On 23 September 1783, another privilege was granted to one Hollenweger, a protégé of
Meusnier and former artisan at the royal glass factory. His process converted Glauber's salt into the sulfide by incineration with powdered charcoal and then derived soda from the sulfide.[15] Unfortunately, the sources
do not tell us how this was done, and this is a peculiarly regrettable omission, because (as will be seen) Leblanc's inability to carry out this same reaction played a crucial part in the story of his own discovery. Hearing that
another privilege was to be granted, Guyton became alarmed. Macquer,
the member of the *Académie des sciences* whose correspondent Guyton
was, undertook to reassure him to the effect that the Hollenweger process
was altogether different from his own, but the threat to his monopoly and
not to his originality was what deterred Guyton from venturing his capital, and Macquer's letter was the reverse of reassuring.[16]

Later, when widening adoption of Berthollet's bleaching process offered an added incentive to combine in one operation the production
of soda and hydrochloric acid, the *Bureau du Commerce* abandoned its
attempt to set on foot competitive processes in favor of combining the
competing interests. Their hope was to set in motion at least one broadly
based concern. When the Marquis de Bullion came forward in May 1788,
with a request for a privilege, he was put into touch with Guyton, who consented to abandon his rights in consideration of a broader privilege to be
drawn for a merger of their interests.[17] Preparations were made to establish factories in Brittany and Languedoc. But in March 1789, Tolozan received an even more ambitious petition. Géraud de Fontmartin and Jean-
Antoine Carny, who appears to have been the active partner, claimed to
have developed eleven distinct methods of producing soda from salt.
They sought, not only a privilege protecting these discoveries, but per-

mission to establish a diversified works which would also manufacture mineral acids and sal ammoniac competitively, and they proposed to float a stock company.[18] Berthollet found two of the processes to be in fact entirely original.[19] One was apparently identical with Chaptal's litharge method, of which Berthollet may then have been ignorant. The other was more complicated. Organic acids were first prepared by the distillation of wood, beech-wood being the most suitable. Litharge was then dissolved in this "acide pyrolignique," and a saturated brine solution added. The result was a precipitate of lead chloride and a solution of organic salts of sodium which was separated and evaporated and from which sodium carbonate was obtained by incineration and lixiviation of the residue.[20] The *Bureau* notified Guyton that it was impossible to withhold a privilege, and took the initiative in urging all parties to pool their interests.[21] And on 25 October 1789, the *Conseil d'État* approved an Act of Association between Guyton and Bullion on the one hand, and Carny and his partner on the other, with the object of exploiting the advantages of a consolidated privilege drawn in very wide terms indeed. The *Bureau du Commerce* went so far as to consult the convenience of the farmers-general in regions where the salt tax might interfere with the undertaking.[22]

Several facts, then, are clear. By the time of Leblanc's invention, over a dozen laboratory processes for converting salt to soda were known. Of these, at least seven had been tested on a large scale. And artificial soda was being manufactured in at least five separate establishments; but it is true that it succeeded only as a by-product. Nowhere could artificial soda compete as the principal object of manufacture with the importation of so-called natural soda, of which the best quality since time out of mind had been prepared by incineration of the seaside barilla plant,[23] and which France imported from Spain in the amount of several million pounds a year.

III

What is not appreciated, however, is that for a generation the Leblanc discovery did not change this situation. Twenty years later, no one in France had succeeded in exploiting his process profitably. Moreover, there is no reason to suppose that he could have succeeded himself, even if the Committee of Public Safety had not appointed Jean Darcet and his fellow commissioners to examine, evaluate, and publish all the known processes for making artificial soda. They accomplished their mission with admirable clarity in a *Description des divers procédés pour extraire la soude du sel*

marin, published in the Year II (1794), which remains the most important
source of knowledge of the Leblanc process and of all the others.

But Leblanc's famous factory at Saint-Denis never actually became a
going concern. The essence of his discovery is that sodium sulfate is converted to soda by fusion in the presence of charcoal and limestone.[24] He
found this out in the latter half of the year 1789, when he was still a surgeon in the scientific retinue of the Duc d'Orléans. He must have taken his
find to his patron straightway, for a preliminary agreement to form a
company among four associates, the Duke himself, Leblanc, J. J. Dizé, and
Henri Shée, was notarized on 12 February 1790. Shée's interest was very
minor—he had no technical qualifications and served simply as the
Duke's agent. Dizé, a young man of twenty, was the laboratory assistant of
Jean Darcet, Professor of Chemistry at the then *Collège royal.* The Duke
had asked Darcet to verify Leblanc's process. Dizé had been deputized to
run the tests, and it seems probable that he was brought into the association both for his skill in chemical manipulation and because he had
himself invented a method for preparing white lead, which was to be exploited as a sideline.[25]

Successive descriptions of Leblanc's procedures make it possible to follow the development of his process in three well-defined steps from the
laboratory to the industrial stage. When the association was formed, he
had not moved beyond the first step. He carried out the transformation of
sodium sulfate to soda in crucibles; he had yet to find the final proportions
of sulfate, limestone, and charcoal; nor had he made any progress with
the preliminary conversion of sodium chloride to sodium sulfate. His description says only that this is to be accomplished by sulfuric acid, in the
usual fashion.[26]

The association once formed, a location for the factory was chosen near
Saint-Denis, some distance from the abbey, at a place known as Maison-
de-Seine, where access would be easy for barges bringing limestone from
Meudon. Before the end of the year a plant was constructed. Its specifications are described in Leblanc's patent, for which he applied on 15 July
1791. By this time he had perfected a reverberatory furnace for the second
reaction, converting sulfate to soda, and he had come very close to the
most advantageous proportions for his reactants.[27] It should be remembered that Leblanc's invention, strictly speaking, is limited to this second
reaction, which in the parlance of the industry came to be called the black-
ash process. He had also been working on the first reaction, however,
(later called the salt-cake process), and in fact his specification devotes
much more space to describing his procedures for converting sodium
chloride to the sulfate than to the subsequent preparation of soda. But de-

spite all his efforts (or Dizé's—it is, indeed, possible that it was Dizé rather than Leblanc who developed the plant processes), this preliminary reaction was still carried out very clumsily. The apparatus was no more than an oversize laboratory device, nor is it possible that it could have supplied sufficient sulfate to the reverberatory furnace to maintain continuous production of soda.[28] At this point, then, Leblanc could carry out the second step of his process industrially; but the first step remained on a laboratory scale.

The plant visited by the Commissioners of the Committee of Public Safety in February 1794 had been transformed. Potentially, it was a truly industrial establishment. And since accounts of early industrial establishments almost always give an exaggerated impression of their capacity, it may be well to describe briefly the factory and its equipment. Except for a horse-driven mill for pulverizing the raw materials, the entire installation was housed in a building twenty meters long by sixteen wide which was divided into two shops, the first for preparation of sodium sulfate and by-products, the second for manufacture of soda itself. Both steps were now carried out in furnaces similar to the one originally designed for the final reaction. These stood about one and a half meters high and the outside lateral dimensions were two and a half by five. The first shop contained two such furnaces, lined with lead sheets and fitted with inlets for controlled admission of sulfuric acid, and placed side by side for continuous operation. The gases could be discharged into the atmosphere, or alternatively led into a large lead chamber (two-and-a-half by three-and-a-half by six), which served either for collection of hydrochloric acid or for production of sal ammoniac—it was possible to admit either water or the vapors from a large incinerator burning animal matter. The second shop also contained two furnaces, the first for driving off residual acid and drying the sulfate to powder, which then was charged into the second furnace with limestone and charcoal.[29] Leblanc seems to have made little change in this step after 1791.[30] The yield of a single operation was 225 pounds of crude soda, four operations were possible in a fourteen-hour day, and the annual capacity of the plant might therefore theoretically be placed at 275,000 pounds.

In fact, however, the total production of the plant, throughout its entire history, amounted to only 30,000 pounds of soda. We have Leblanc's own testimony that the factory was engaged in "essais en grands" up to July 1793, when the impossibility of procuring sulfuric acid forced him to close down.[31] Nor was it possible to resume operations so long as all stocks of sulfur and saltpetre remained under emergency requisition for the munitions industry. It must be emphasized that the shutdown had nothing

to do with expropriation of the interests of the Duc d'Orléans, now become Philippe Égalité. The Duke was tried and beheaded on 6 November.
But the property was not placed under sequestration by the local authorities until 28 January 1794 (8 Pluviose), the very day (by chance) on which
the Committee of Public Safety adopted its decree on soda and appointed
Darcet and his colleagues commissioners to collect and publish all available information.[32] By that time Leblanc had long since left the idle factory at Saint-Denis—now become Franciade—in the care of Shée and had
gone to Paris, where he had taken a post in the *Agence des poudres* and
was deeply immersed in various political activities.[33]

The decree of 8 Pluviose was essentially a decision to extend the revolutionary effort of war production to the manufacture of alkali, and it was
adopted primarily because of the imperious necessity to expand the output of soda tremendously. Not only was the normal Spanish supply cut
off, but this happened at precisely the moment when soda, in addition
to its usual uses, had to be substituted wherever possible for potash, all
stocks of which and more were required for the emergency manufacture
of saltpetre. Far from having ruined Leblanc by striking down his flourishing enterprise, as is often said, what the decree really did was to draw
technical attention to a small and promising venture which had been in
abeyance for over six months. Nor do the sources bear out the picture of
the callous state crushing out the rights of genius. It may be that in the
exuberance of the Year II the authorities were somewhat grandiose in
their belief that by taking thought they could add a cubit to the industrial
stature of the Republic. But given their responsibilities, the most that can
be fairly charged against them is excessive optimism. In the event, the
publication of the Leblanc process did the war effort no good. But neither
did it do Leblanc any harm. For it must be remembered that the factory
was not Leblanc's property; on the execution of Philippe Égalité, his
share—and he had supplied all the capital—became the property of the
nation. And once Leblanc's process was published, every effort was made
to protect his interests. He was in fact shown unusual consideration.

In the first place, Leblanc never was deprived of his patent. All he lost
was the privilege of keeping it secret, and this was not normally provided
for under the revolutionary patent legislation. Permission to keep any
patented process confidential required a special act of the legislature.
Leblanc never succeeded in obtaining such an act, and the right to guard
the secrecy of his process rested on nothing more solid than a provisional
resolution of the *Comité d'Agriculture et de Commerce*.[34] Moreover, the
decision of the Committee of Public Safety on 8 Pluviose (28 January
1794) to order a widespread revolutionary manufacture of soda, al-

though it had the effect of suspending Leblanc's patent, did not formally abrogate it.[35] On the contrary, in the fall of 1793, Leblanc had taken the precaution of applying for a confirmation "déroyalisé" of his patent "dans les formes républicaines."[36] This was granted him by a decree of the *Conseil exécutif provisoire* on 18 Frimaire (8 December 1793) and forwarded to him on 30 Germinal (19 April 1794)—several months *after* the decree of 8 Pluviose.[37] Since the revolutionary production was envisaged as an emergency measure, there would appear to have been no legal obstacle which, the emergency once past, would have prevented Leblanc from enforcing his rights. Or rather, there was only *one* obstacle. It was of Leblanc's own making—the protection of a patent was conditional on its being exploited, but he made no serious attempt to take advantage of the original intention of the Committee of Public Safety, which was that he and his two colleagues should run their own factory.

Instead, Leblanc's behavior from 1793 until his death alternated between plaintiveness and obstruction. And through the medium of his grandson's memoir, the complaints and claims with which he besieged successive ministers of the interior have become the history of his invention—he was in effect the creator of his own legend.[38] He pursued a two-fold objective—to persuade the state first to resign to him and his associates its rights to Orléans' share in the company, and then to subsidize by way of indemnity the recapitalization of the plant. Not to be satisfied with less, he presumed upon the consideration shown for his interests, particularly by Berthollet and Fourcroy, and frustrated every attempt to reopen his factory. For example, on 29 Ventose an II (19 March 1794), the committee of Public Safety asked Leblanc and Dizé for a report on what steps would enable them to resume production. Two days later the Committee by the hand of Prieur de la Côte d'Or sent the local authorities an order staying their projected sale of the establishment and directing them to protect the factory as an important object of public utility. These measures were taken a full three months before the publication of Darset's *Description* on 2 Messidor (20 June 1794). But up to this time Leblanc's only response was a begging letter on 22 Prairial (10 June) asking an indemnity.[39] Finally, on 26 Messidor (14 July), three weeks after publication of the *Description,* he and Dizé answered the inquiry of the previous March. If their facilities were quadrupled, they wrote, they could produce the 3,000,000 pounds a year asked for by the Committee (a gross overestimate actually). But this would be possible only if the government would undertake to requisition and turn over to them all the sodium sulfate in France. They would also need 100,000 francs.[40]

On 19 Brumaire an III (9 November 1794), Prieur and Fourcroy, both

members of the Committee of Public Safety, visited the factory prepara-
tory to asking Leblanc for more realistic proposals. Once again his answer
was dilatory and impractical—he now wanted not only money but also a
settlement of the title before he could attempt to resume operations.[41]
Meanwhile, a well-known chemical firm, Riffautville, Bernard, et Cie., had
proposed to buy the plant, which was in a state of total neglect and had by
this time deteriorated badly.[42] In referring this proposal for examination
to the *Commission d'Agriculture et des Arts,* the Committee of Public
Safety observed that, since Leblanc, Dizé, and Shée unfortunately did not
seem disposed to resume operations on their own account, this proposal
would have to be considered, but that

> Il faudra aussi dans tous les cas ne pas perdre de vue le citoyen
> Leblanc.... C'est lui qui a fait part au Comité de son procédé, re-
> connu un des meilleurs.... Ainsi en examinant la proposition du
> Citoyen Riffautville & Cie., vous aurez soin d'examiner en même
> tems s'il seroit possible de conserver dans leur entreprise une part
> au Citoyen Leblanc égal à celle qu'il avoit dans son association avec
> d'Orléans et dont il sera nécessaire que vous l'engagiez à vous pré-
> senter l'acte.[43]

Leblanc's response to this attempt to reconcile his interests with those
of the state and the industry was to delay furnishing the commissioners
with a copy of the agreement. When at last he did submit it, he proposed
a different solution. He was now willing to buy out the interest of the state
himself. But the terms of his offer were very vague and when, after further
delay, a member of the commission went to ask him for a surety, he said
that his arrangements with his backers were not completed and that he
would have to renounce his intention of making a bid for the factory.[44] In
the event, the sale to Riffautville also broke down. Whether this outcome
was a result of Leblanc's manoeuvres seems doubtful.[45] But that he was
acting in bad faith seems certain.[46]

IV

It would be idle to follow any further the dreary record of Leblanc's attempts to secure subsidies and title to the factory.[47] In 1801 he finally succeeded in the latter aim. Of course the undertaking failed. That anyone
could have made it profitable is extremely improbable, but in any case the
personal qualities which become evident in this correspondence—a dis-

position to sulk, a certain querulousness, a tendency to nag, a proclivity for blaming misfortunes on the authorities or on the circumstances—are not those of a successful industrialist or entrepreneur, although such characteristics may be, and indeed often have been, compatible with an inventive imagination.[48] Nor is it surprising that the life of one afflicted with the personality here apparent should have ended in suicide.

There is, it is true, one point in the research at which the investigator suspects that Leblanc had really been the victim of skullduggery. For the proposal to collect and publish all soda processes originated, not with Prieur or his scientific advisers—Berthollet, Monge, and others—but with Carny.[49] Carny was very well thought of by Prieur and his colleagues of the Committee of Public Safety. Early in 1793 he had been drawn into war work at the request of Berthollet to help with development of incendiary shells and potassium chlorate gunpowder. Later in 1794, Carny received credit for having devised the expeditious methods used in the revolutionary manufacture of saltpetre and gunpowder.[50] Carny was in fact very adroit, and not only in chemistry. In 1791 he established a manufactory of hydrochloric acid at No. 11, Rue du Harlai-au-Marais. And on 31 January 1792, he himself patented a process for converting salt to soda. What the process was, it is unfortunately impossible to say, for the description has disappeared from the archives.[51] But whatever it was, it is certain that he withheld it from Darcet and his commission in 1794.[52]

Carny next appears in 1796, when like most of the others mobilized during the *levée en masse*, he had left public service and gone back to his chemical industry. And now he came forward with a proposal to establish a new factory for making soda. For this purpose he asked that he be given a loan of 200,000 francs at one percent, and that he also be assigned title to the property next to his—it had belonged to a woman executed as a traitor—to be converted to a workshop. This application was warmly supported by Berthollet, now a member of the *Commission d'Agriculture et des Arts* and by Guyton, Carny's former partner, equally well placed on the Committee of Public Safety. And though Carny had to lower his request to 60,000, that sum was awarded him. One is all prepared, therefore, to discover either that he was using the process he had withheld when all the others were published, or that the whole manoeuvre was a plot for getting access to Leblanc's process.

But the hope of finding a villain for the piece is disappointed. For it turns out that Carny was using neither the Leblanc process nor any of the others invented by the prerevolutionary artisans. Instead, he was at-

tempting to work a method proposed by Darcet and his Commission on
its own initiative, and published at the end of their *Description* of the Year
II. (That the commissioners actually did go beyond the procedures submitted to them by Leblanc and the other inventors is an illustration—one
among many—that military necessities created by the Revolutionary
wars brought about something very like the organized technological development work of modern times.) Ten parts of pyrites and four parts of
common salt were roasted for sixty hours in an iron vessel. Sulfur dioxide
and hydrogen chloride were driven off in quantity, and the residue consisted of a mixture of ferric oxide, ferric chloride, sodium sulfate and salt.
Sodium sulfate was then dissolved out, and the ferric oxide could serve,
with the addition of charcoal, or better yet with coal or peat, to transform
it into soda. Essentially, of course, this differs from the old Malherbe-
Athénas processes only in utilizing iron pyrites instead of iron scrap or
copperas. But despite the availability of pyrites, Carny's venture failed. Inside of a year he too became a well-known claimant at the Ministries of
the Interior and Finance, seeking a remission of interest payments and
further subsidies.[53]

This episode is one of many illustrations that, at the time, the superiority of the Leblanc process was not fully recognized. In 1794 and 1795
its exploitation was in any case precluded by the continuing shortage of
sulfuric acid and the military demand for saltpetre. In one of his abortive
proposals, Leblanc himself suggested converting his plant to the use of
pyrites.[54] But even the Darcet *Description* did not describe the Leblanc
process as intrinsically the best. Its advantage lay in the facilities of his
plant and the accessibility of Meudon. Chemically, the commissioners felt
the use of iron to be preferable, because of its greater affinity for the sulfate.[55] Consistently enough, when Giroud, one of the commissioners, himself submitted a petition for authority to create a soda plant, he intended
to rely on pyrites to work the conversion.[56] Similarly, Bernard, Riffautville
et Cie. proposed to adapt the Leblanc plant to different procedures after
buying it.[57] Finally—not to multiply examples—in 1802, Messrs. Anfrye,
Darcet (fils) et Cie. deposited a description of a new process of their invention in a *pli cacheté* with the *Ier Classe* of the *Institut* (the former
Académie des sciences). They described the Leblanc process as abandoned
even by its inventors.[58]

Nor did it begin to come into its own until the expansion of chemical
industry after 1800 created a growing surplus of sodium sulfate in the operations of several already well-established concerns. Here again Carny
came to the fore. He had left Paris and in 1803 he arranged with the di-

rection of the great salt-works at Dieuzé to convert otherwise unused sodium sulfate to soda.[59] Other producers followed suit. But nowhere was it profitable to divert sodium sulfate from its normal uses to the manufacture of soda. Nor was it until after 1807 that anyone found it economically feasible to carry through *both* steps of the Leblanc process, from salt to soda, and then this was made possible by a special remission of the salt tax and not by some scientific development.[60] The question, therefore, whether Leblanc could have enforced his patent becomes doubly academic—for by 1807 it would have expired in any case. By 1810, production had been notably expanded, in the Paris region by Payen at Grenelle, by Marc, Costel et Cie. at Gentilly, and by Anfrye, Darcet et Cie., who having bought Leblanc's old factory after his death, enlarged it and operated it together with their older establishment at Nanterre; at Soissons by Pajot, Descharmes et Cie.; and by a number of smaller concerns in the regions around Rouen, Montpellier, and Marseilles. But even these manufacturers were unanimous in warning that the new industry could exist only on condition that the prohibition of foreign soda be rigorously maintained.[61] And even then, the quality of artificial soda was inferior to that derived from plants.[62]

V

Twenty-one years after Leblanc's discovery, then, his process was just entering upon the truly industrial phase of its history. To return to the discovery itself, the one circumstance beyond dispute is that Leblanc made it in 1789. According to the statement in his patent, he began his researches in 1784. If so, no record remains of them except a few unimportant papers on the crystallization of alum and other salts. His own testimony is that the idea for his process was suggested to him by reading a discussion of artificial soda which La Métherie, editor of the *Journal de Physique,* published in his *Discours préliminaire* for the year 1789.[63] After dismissing several processes making use of litharge, quick-lime, and iron, all of which he wrongly described as impractical, La Métherie writes in the passage to which Leblanc referred:

> Il y a une manière de faire cette décomposition qui seroit très-sure, mais elle seroit peut-être trop chère. Ce seroit dans des appareils convenables de verser de l'acide vitriolique sur le sel marin; l'acide marin se dégageroit et passeroit dans les ballons, et le résidu seroit du vitriol de natron, ou sel de glauber. On décomposeroit ensuite ce

vitriol de natron en le calcinant avec du charbon. L'acide vitriolique
se dégageroit sous forme d'acide sulfureux, et le natron demeure-
roit pur.[64]

Now, this is most disconcerting. For the reaction here described does not
in fact occur: its product will be not soda, but sodium sulfide, as in the
second step of the old Hollenweger process, which La Métherie does not
seem to have known.

This discrepancy forms the basis of the charge published in 1810 by
Dizé, Leblanc's former associate, contesting Leblanc's authorship of the
invention.[65] In Dizé's account, Leblanc followed La Métherie's suggestion,
mistook sodium sulfide for soda, and rushed all eagerly to his patron, the
Duc d'Orléans, who welcomed the project on condition that it be verified
by Darcet. Too busy to undertake such a chore himself, Darcet commissioned Dizé, his assistant, to test Leblanc's procedures in his private laboratory on the Quai Voltaire. When the tests failed, Leblanc was in despair.
He insisted that the process really had worked in his own laboratory, and
he begged Darcet to delay his report to the Duke and to let Dizé help him
redeem the fiasco by further experiments. Darcet allowed them to work at
the Collège royal, and for three months they labored night and day, using
different methods of incineration, but with no success. In the course of
these efforts, they did sometimes use limestone as a source of carbonic
acid, with which they attempted converting the sodium sulfide to sodium
carbonate in solution. Then by chance, Dizé tried heating this mixture
with a little charcoal in an iron flask, and found that the resulting solution
contained much less sulfide than that from the other tests. Each night he
reported the day's work to Darcet, and it was Darcet's suggestion to try
the same ingredients dry and to heat them to fusion. For the first time,
the test succeeded—to Leblanc's disgust, who stayed away for a day and
then came back to say he knew it all the time.

Revived by Dizé's descendants, this story was rejected out of hand by
J.-B. Dumas in 1856,[66] and the popularity of Leblanc as a martyr dates
from that sentimental period. After the Dumas judgment was announced,
streets began to be named for Leblanc in various industrial cities. Nor
does the weight of evidence confirm Dizé's account. For at each stage of
the early development of the process—in March 1790, in August 1791,
and in February 1794—Darcet served on the examining commission, and
he always gave full credit for the invention to Leblanc.[67] It is unlikely that
he would have done so if he had thought of the idea himself. Nevertheless,
Dizé may probably have played a larger part than is commonly appreciated. For the following sentence appears in the Darcet Description:

L'établissement est déjà tout formé à Franciade; le citoyen Dizé, l'un des co-associés, en a dirige particulièrement la construction: elle est faite de manière qu'il peut servir également à toute espèce d'usages et de procédés de ce genre; c'est une justice que lui rendent ses co-associés.[68]

Dizé claimed the invention, not for himself, but for Darcet, and the picture he evokes of the frantic scenes in the laboratory is sufficiently plausible to suggest that if (as must be concluded) the invention was Leblanc's, he no doubt had his difficulties.[69]

Except to write that the idea came to him upon reading the La Métherie *Discours,* Leblanc himself nowhere explained how he came to try his Glauber's salt and charcoal with limestone. La Métherie never mentions limestone in connection with soda. But he does do so in the preceding section of his essay, and necessarily so, for this section is devoted to the iron industry. Not that these passages contain anything original: of no subject was the reader of French technological literature more frequently reminded than of the superiority of British metallurgical practise. Exhortations to use coke abounded long before 1789—whatever else may be true of science and industry, it appears to be a permanent feature of their relationship that French industrialists should be nagged by French technical writers. La Métherie, nephew to the abbé Rozier, was not himself a scientist. He was a scientific journalist, and the *Discours préliminaire* for 1789 was not a scientific memoir but a reporter's account of a technological junket to England and Scotland, a journey which La Métherie had evidently made during the previous year. He had paid the ritual visit to the Soho Works and expresses the conventional admiration for the industrial interests of English scientists and the scientific attainments of English industrialists. He always refers to Boulton and Watt as "cex deux savants"—although "Monsieur Wedyewood" is only "cet artiste." The edifying example of "milord Dundonas" (Dundonald) is duly brandished before the French nobility. And the Wilkinson foundry is described in considerable detail.[70]

From iron La Métherie turns to the manufacture of soda. And it seems very probable that this juxtaposition is precisely what led to the Leblanc invention. If charcoal and limestone in a furnace separate iron from its ore, might they not somehow have a similar effect on Glauber's salt and liberate the soda? Even terminology would have turned Leblanc's thoughts in this direction: iron is "extracted" from its ore; similarly the problem of securing soda is always described as that of "extracting" it from its base. Moreover, other aspects of the Leblanc process are to be dis-

cerned in germ in La Métherie's report of his English journey. Leblanc's
reverberatory furnaces may be seen as adaptations of the coke ovens used
by Wilkinson and also by Dundonald—indeed La Métherie suggests precisely this comparison between the coke oven and the chemical reverberatory furnace. Finally, the paragraph may be quoted by which La Métherie
makes a transition from his discussion of British iron foundries to the
general subject of the manufacture of soda. Lord Dundonald, he writes (after describing his coke ovens):

> est parvenu à décomposer le sel marin en grand, et à en obtenir sé-
> parément l'acide et alkali; unissant cet acide avec l'alkali ammo-
> niacal qu'il a obtenu du charbon, il en forme du sel ammoniac, et
> par ce procédé, empêche la sortie de tout l'argent qu'on portoit en
> Égypte et ailleurs pour avoir cette substance nècessaire dans beau-
> coup d'arts.[71]

And La Métherie even suggests that an incinerator of animal matter
might take the place of the coke oven as a source of ammoniacal gases.

But even if Leblanc did find his process because of this travel article, it
remains nonetheless a discovery and the one soda process of which that
was true. For the reactions exploited by Malherbe, Guyton, Carny, and the
others were known—the works of Scheele seem to have been the great
repository of facts which they exploited. Only the Leblanc reaction was
truly original. Necessarily, therefore, it was also the one with the least understood mechanism—it is a truism that the scientific comprehension of
a genuine discovery must be less advanced than of some process which
has been studied for a certain time. Leblanc's scientific writings show him
to have been a competent and ordinarily uninspired chemical worker—a
scientist only of the second rank, possibly of the third. Nowhere does he
undertake any analysis of the role of limestone, which is always referred
to simply as the intermediary by means of which soda is extracted from
its base. Nor did anyone else offer a better explanation.

But it could hardly have been otherwise. The composition of soda was
unknown. The composition of limestone was unknown. Nor was this reaction (an exceedingly complicated one) properly described until near the
end of the nineteenth century, some time after the Leblanc process had already begun to go out of use. It is ironical, for example, that in 1856 Dumas, pontifically rendering judgment in favor of Leblanc's authorship of
the invention, should have been mistaken about the reaction itself.[72] As
late as 1895, there was still considerable doubt as to precisely what occurs.[73] This doubt was obviously on a different plane from the fumbling

for affinities in the 1790's. But this circumstance suggests that the later history of the Leblanc process was no better an example of some simple debt of industry to science than was the invention itself. The operation and development of the process belong to the history of industry. What belongs to the history of science is only the increasing understanding of the Leblanc process—the knowledge of reactions and not just of results. Nor did practice and understanding ever really coincide historically. The process flourished, not because it was understood (by the time the theory was perfected, the Leblanc industry was in irretrievable decline); it flourished for a simpler reason: it was the best process there was.

Notes and References

1. I am indebted to the John Simon Guggenheim Foundation for a fellowship (1954–1955) which made possible the studies of which these papers are one result. It is a further privilege to acknowledge my deep obligation, one shared by many American scholars, to the staff of the *Archives nationales* in Paris and particularly to M. Guy Beaujouan, whose knowledge of the revolutionary materials relating to science and industry makes research in them rewarding and whose kindness in placing himself at the disposal of one and all makes acknowledgment a pleasure. I should like, finally, to express my gratitude to M. Falala, director of the *Institut national de la propriété industrielle* and to M. Tresse, secretary-general of the *Conservatoire des arts et métiers*, who most cordially gave me access to materials in their care.

2. Jean Darcet, A. Giroud, C. H. Lelièvre, and Bertrand Pelletier, *Description des divers procédés pour extraire la soude du sel marin* (Paris, an II). Cited hereafter as Darcet, *et al.*, *Description*. An *extrait* was published in *Annales de chimie* (1797, *19*: 58–156), which, however, does not include the plates illustrating the construction and arrangement of Leblanc's furnaces.

3. "Soda, Nicolas Leblanc, and the French Revolution," *Proceedings of the Royal Philosophical Society of Glasgow*, 1925, *53*: 113–128. See too, R. E. Oesper, "Nicolas Leblanc," *Journal of Chemical Education*, 1942, *19*: 567–572 and 1943, *20*: 11–20; Paul Baud, "Les Origines de la grande industrie chimique en France," *Revue historique*, 1934, *174*: 1–18; Desmond Reilly, "Salts, Acids, and Alkalis in the 19th Century," *Isis*, 1951, *42*: 287–296. It would serve no purpose to list additional accounts of Leblanc, for all are unreliable. They differ on many details: the date of the academy's prize, the date of the invention itself, the dates and even the nature of the processes ascribed to Leblanc's predecessors, etc. There is no point in taking issue with each error, however, for these accounts are based on each other rather than upon research in the sources. An exception must be made for the article by Baud, which makes use of some documents at the *Archives nationales*. But he does not cite the essential ones.

4. *Nicolas Leblanc, sa vie, ses travaux, et l'histoire de la soude artificielle* (Paris, 1884).

5. J. B. Dumas, "Rapport relatif à la découverte de la soude artificielle," *Compte rendu des séances de l'académie des sciences*, 1856, *42*: 553–578. The commission for

which Dumas was reporter consisted of Thenard, Chevreul, Pelouze, Regnault, Balard, and Dumas himself.

6. This is not the place to discuss patent legislation in detail. In brief the law of 7 January 1791, as amended by the law of 25 May 1791, remains the basic French patent law. Under its provisions, *brevets* were granted on request to any inventor upon the receipt of properly drawn specifications and a fee of 1500 francs. The term was 15 years. Adoption of this legislation was a victory for the interests of inventors over those who favored encouraging invention by a system of open competition for particular grants, on the model of the English Society of Arts. This was the procedure supported by the *Académie des sciences,* a circumstance which explains much of the hostility to the Academy among skilled artisans. See S. J. de Boufflers, *Rapport ... sur la propriété des auteurs de nouvelles découvertes* (Paris, 1791); *Bibliothèque nationale* [Le²⁹ 1206]; and (for a convenient summary) C. A. Costaz, "Notice sur les brevets d'invention," *Bulletin de la Société d'Encouragement,* 1802, 1: 81–85. The *Institut de la propriété industrielle* is at 26 bis Rue de Leningrad, Paris VIIIᵉ. Its archives are well worth calling to the attention of historians of science and technology. It must be noted, however, that rather than take out a patent, many inventors preferred to apply for one of the grants awarded by the *Bureau de Consultation des Arts et Métiers,* which in a sense took over and enormously expanded the Academy's program of prizes. A full census of French inventive activity in the Revolution requires, therefore, a study of the surviving records of the *Bureau de Consultation,* which are preserved in the office of the Secretariat of the *Conservatoire des Arts et Métiers.* Only extracts were published by Charles Ballot, *Bulletin d'Histoire Economique de la Révolution* (1913), 15–160. See also *Archives Nationales,* F¹⁷ 1136, 1137, 1138, 1307ª.

7. For details, see George Lunge, *A Theoretical and Practical Treatise on the Manufacture of Sulphuric Acid and Alkali* (2nd ed.; 3 vol.; London, 1895), II, pp. 77–80.

8. Memoir dated 12 May 1790, *Archives Nationales,* F¹² 1505, dossier on "Les Sʳˢ Bourgogne and Baudoin." For the work of the *Bureau du Commerce,* see Pierre Bonnassieux and Eugène Lelong, *Conseil de Commerce et Bureau du Commerce 1700–1791, Inventaire Analytique des Procès-Verbaux* (Paris, 1900).

9. Macquer to Tolozan, 24 August 1783, F¹² 2242, dossier on "Hollenweger"; see, too, Report of the *Commission d'Instruction Publique,* 6 Ventose an III, F¹² 1508. Macquer reported favorably on the process on 13 March 1778. A second favorable report was filed on 16 August 1779 by Grignon, Inspector of Manufactures, on further tests made at the request of Tolozan and conducted at Le Croisic. See, too, Darcet *et al., Description,* pp. 16–17.

10. Darcet *et al., Description,* 15–23; Athénas was granted a privilege on 16 April 1782 (Memoir of Tolozan, dated 23 August 1788, F¹² 2242, dossier on "Les Sʳˢ de Bullion, Guyton de Morveau, Fontmartin et Carny").

In the 1850's this process was taken up again by E. Kopp, who attempted to exploit it as an alternative to the Leblanc process, and who proposed to use iron in the form of native ores, ferric oxide. Since the use of either metallic iron or copper as involves a preliminary oxidation, the reactions are essentially the same. In Kopp's view they were as follows:

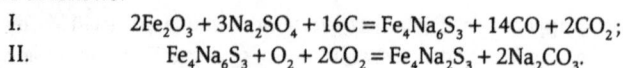

I. $\qquad 2Fe_2O_3 + 3Na_2SO_4 + 16C = Fe_4Na_6S_3 + 14CO + 2CO_2;$

II. $\qquad Fe_4Na_6S_3 + O_2 + 2CO_2 = Fe_4Na_2S_3 + 2Na_2CO_3.$

Finally, $Fe_4Na_2S_3$ was oxidized to recover the Fe_2O_3 and one-third of the original Na_2SO_4.

III. $$Fe_4Na_2S_3 + 7O_2 = 2Fe_2O_2 + Na_2SO_4 + 2SO_2.$$

This process was investigated by Stromeyer, who approved it in practise, but disagreed with the theory advanced by Kopp. In Stromeyer's view, the double sulfide which occurs in the process is $Fe_2Na_2S_3$, and much sulfur is lost as H_2S in the second stage. For a discussion of the problem and references to the literature, see Lunge, III, pp. 222–227.

11. Darcet *et al., Description,* 13–16; on this establishment, see too the notes of Jerome de Lalande, *Bibliothèque publique de Lyon,* Fonds Coste, 1281, f° 129.

12. Memoire by Chaptal, dated 5 Pluviose an II, F^{12} 1508, which gives a detailed description of the process "pratiqué à Montpellier depuis 5 ans." See, too, Darcet *et al., Description,* pp. 23–27.

The reaction was as follows: $2NaCl + 2PbO + H_2O = 2NaOH + Pb_2OCl_2$ (Lunge, III, pp. 178–181).

After which: $2NaOH + H_2CO_3 = Na_2CO_3 + 2H_2O$

13. Darcet *et al., Description,* pp. 33–36.

14. Macquer to Tolozan, 24 August and 3 November 1783, F^{12} 2242, dossier on "Hollenweger"; *Bureau du Commerce, Procès-Verbaux,* p. 433; Darcet *et al., Description,* p. 29.

The essential reaction was:

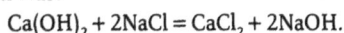

$$Ca(OH)_2 + 2NaCl = CaCl_2 + 2NaOH.$$

See too, Georges Bouchard, *Guyton-Morveau* (Paris, 1938), pp. 103–105.

For the terms of the prize, see Ernest Maindron, *Les Fondations de Prix de l'Académie des Sciences* (Paris, 1881), pp. 39–40.

15. The language in which Macquer distinguishes Hollenweger's process from Guyton's may be of some interest, as an illustration of the level of theory and knowledge which guided a leading chemist in his evaluation of the chemical procedures proposed for government favor. Macquer, who belonged to the generation next older than that of Lavoisier, served as counsel to the *Bureau du Commerce* until 1784, when he was succeeded by Berthollet. Hollenweger, he writes, "commence par appliquer l'acide vitriolique au sel marin, ce qui en dégage l'esprit acide qu'il recueille pour être employé dans les arts aux quels il est propre. Le résidu de cette première opération est un nouveau sel neutre provenant de la combinaison de l'acide vitriolique avec l'alkali marin et connu sous le nom de Sel de Glauber. C'est sur ce Sel de Glauber que le Sr Hollenweger opère ensuite pour dégager et rendre libre l'alkali marin qu'il contient; pour y parvenir il applique de la poudre de charbon à ce Sel. Par cette seconde opération, l'acide vitriolique du Sel de Glauber quitte son alkali pour se combiner avec le principe inflammable du charbon, avec lequel il forme du soufre; mais comme l'alkali a la propriété de se combiner avec le soufre, et de former avec lui un nouveau composé qu'on nomme *hépar* ou *foie de soufre,* il en résulte qu'après cette décomposition du Sel de Glauber, son alkali marin n'est pas encore libre et pur, et qu'il faut avoir recours à une troisième opération qui décompose le foie de soufre et dégage enfin l'alkali marin, lequel alors devient pur et pr[és]ent toutes les propriétés du sel de soude. (Macquer to Tolozan, 3 November 1783, F^{12} 2242).

It is not clear why Macquer does not describe the final step. But what is possible is that it was never in fact carried out. In earlier tests Hollenweger appears to have used the Malherbe process—Macquer says as much in an earlier report (22 July), in which he writes vaguely of several alternative methods for converting Glauber's salt. It may well be that, having been called on for a definite distinction, he is just taking Hollenweger's word for this last step, and that what Hollenweger actually planned to do was get a privilege, after which he would use the Malherbe methods. Such minor chicanery was very common on the part of ambitious artisans and inventors, and both because of this and their secretiveness, it is always very difficult to find out what some process really was and whether it was ever actually practised. Even the official reports are often disappointing—usually stating only that a test succeeded, and not saying what it was. In general, all the records make it appear that inventive activity was pervaded with a curious air compounded of optimism and secrecy, pretense and credulity.

16. Macquer to Tolozan, 3 November 1783, *loc. cit.;* Memoir to *Bureau du Commerce,* 23 August 1788, F¹² 2242, dossier "Les Sieurs de Bullion, Guyton de Morveau, Fontmartin et Carny"; on Guyton's process, see too, G. Bouchard, *Guyton-Morveau* (Paris, 1938), pp. 103–105.

17. Berthollet to Tolozan, 27 May 1788, F¹² 2242, dossier "Les Srs de Bullion. . . ."; this dossier contains also a draft of the privilege (23 August 1788); see too, *Bureau du Commerce, Procès-Verbaux,* pp. 456, 459.

18. Carny's petition was referred to Berthollet by Tolozan on 4 April 1789, F¹² 2242, dossier "Les Srs de Bullion. . . ."; see too, *Bureau du Commerce,* Procès-Verbaux, p. 473.

19. Berthollet's report is dated 13 May 1789, F¹² 2242, *loc. cit.*

20. A number of processes were submitted by Carny and Guyton to Darcet and the other commissioners in 1794; these seem to be the two that had belonged to Carny, whereas the quicklime process had been Guyton's. *Description,* pp. 27–32.

21. Tolozan to Guyton, 18 June 1789, F¹² 2242, *loc cit.;* see, too, Memoir of Tolozan, 12 May 1790, F¹² 1505, dossier "Bourgogne and Baudoin"; and *Bureau du Commerce, Procès-Verbaux,* p. 476.

22. Letters of Tolozan, 10 and 17 September 1789, F¹² 2242, *loc. cit.*

23. See, for example, J. B. van Mons, "Extraits du Journal de . . . M. Kasteleyn," *Annales de Chimie, 1792, 13:* 212–216.

24. The circumstances of the discovery are discussed below, section V.

25. Since the company never made a profit, there is no point in giving details on the terms of the agreement.

Suffice it to say that the capital furnished by the Duke was 200,000 *livres tournois.* Further agreements, regulating the share of each of the partners in the anticipated proceeds, were notarized in Paris after the construction of the factory and the beginning of operations. The first, between Leblanc and Dizé, is dated 15 January 1791 and the second, which was the definitive agreement constituting the company, 27 January 1791. The texts of the agreements of 12 February 1790 and 15 January 1791 and the terms of that of 27 January 1791 are given in Dumas, *loc. cit.* The full text of this last will be found in the *Archives nationales,* F¹² 1508, dossier "La manufacture de soude de Franciade." Why that of 1790 should have been notarized in London, before James Lutherland, notary public, is one of the unsolved mysteries of the subject.

26. The interests of the parties were protected by filing the initial agreement together with descriptions of the two processes attested by Darcet in the office of a notary in Paris, M^{tre} Brichard. This *dépôt* occurred on 27 March 1790. The texts are printed by Dumas, *loc. cit.*

27. The original *brevet* is in the *Archives de l'Institut national de la propriété industrielle,* dossier Leblanc. Though it was approved by the king on 25 September 1791, the effective date was 19 September, when the required documents were filed with the authorities of the *Département de Paris.*

28. Leblanc himself describes it as being "n'autre chose que l'appareil hydropneumatique en grand." The main vessel was made of lead and consisted of a large, fairly deep tray, in the shape of those used by masons for mixing plaster, to which a top could be fitted. It was heated on top of a coal-burning stove, and very complicated arrangements had to be made to keep the lead from melting. (It was the desirability of finding a way to apply the heat from the top which pushed Leblanc in the direction of adapting his reverberatory furnace for use in this reaction as well as in preparation of soda.) The gases were carried off through lead tubes to a ceramic container where HCl might be recovered or NH_4Cl prepared. All the connections had to be disassembled after each operation, and of course reluted on each reassembly.

29. Darcet, *et al., Description,* contains very detailed plans of the plant in the appendix. It should be pointed out that the original project envisaged two further installations which never were constructed—a shop for manufacture of the sulfuric acid used in the first step, and another for purification of the crude soda. (*Ibid.,* p. 11). None of the writers who have described the Leblanc factory as a flourishing establishment, in full production for a period of years, have noticed that even the easily accessible Darcet *Description* refers to it in one passage as "cet atelier naissant" (p. 4).

30. He had modified the proportions very slightly since 1791. His patent calls for 100# of Glauber's salt, 100# of limestone, and 50# of powdered charcoal. In the *Description* of Darcet and his fellow-commissioners, the proportions are given as 100, 100, and 55, respectively. According to a modern authority (J. R. Partington, *The Alkali Industry* [New York, 1925], p. 97), the theoretical proportions are 100, 79.42, and 16.9, and the practical proportions are 100, 100, and 35.5.

31. Leblanc to the Minister of the Interior, 15 Brumaire An II. *Institut de propriété industrielle,* dossier Leblanc.

32. The original draft of this decree, along with the original drafts of a number of others relating to war production, is in the *Archives nationales.* F^{12} 1508.

33. Leblanc to the Committee of Public Safety, 22 Prairial, An II, F^{12} 1508; see, too, Shée to Leblanc, 13 Pluviose An II, Dumas, *loc. cit.,* pp. 566–567, from which it is apparent that LeBlanc was kept informed of the state of affairs at the factory only through the letters of Shée. For LeBlanc's political activities, see Anastasi, 31–33.

34. *Extrait du procès-verbal du Comité d'Agriculture et de Commerce,* 2 September, 1791, *Institut de propriété industrielle,* dossier Leblanc.

35. The original draft of this decree, signed by Carnot and Prieur, is in the *Archives nationales,* F^{12} 1508.

36. Leblanc to Minister of the Interior, 15 Brumaire an 2, *Institut de propriété industrielle,* dossier Leblanc.

37. *Loc. cit.*

38. This correspondence will be found in the *Archives nationales*, F[12] 2243.

39. By a resolution of 19 Fructidor (5 September 1794), this request was met with a grant of 3,000 francs, under the legislation establishing the *Bureau de Consultation*. LeBlanc's patent having been published, it was intended that he should be treated like other inventors. It is true that this impartiality was strictly observed and that, as with other inventors, his grant was never paid. But in any case, this measure too presupposed that he would reopen the factory. *Archives nationales*, F[12] 1508.

40. All the documents mentioned in this paragraph are in the *Archives nationales*, F[12] 1508.

41. *Loc. cit.*, Leblanc to the Committee of Public Safety, 3 Pluviose an III (23 January 1795).

42. The proposal was dated 20 Frimaire an III (20 December 1794). F[12] 1508.

43. Fourcroy (for the Committee of Public Safety) to the *Commission d'Agriculture et des Arts*, 22 Pluviose an III (10 February 1795). *Loc. cit.*

44. *Rapport de la Commission d'Agriculture et des Arts*, 5 Germinal an III (25 March 1795). This report was drawn up by Berthollet, who was a member of the Commission. *Loc. cit.*

45. Fourcroy to *Comité des Finances*, 1 Germinal an III, F[12] 1508. For the complete dossier of this affair, see F[12] 2244.

46. Leblanc's conduct was very exasperating to those who were trying to arrange that he should derive some advantage and profit from his invention. See, for example, a letter written by Fourcroy and Guyton as members of the Committee of Public Safety and dated 19 Brumaire an III (9 November 1794): "Le Comité, fortement occupé de l'établissement des soudières artificielles, voulant décidément connoître jusqu'à quel point il peut compter sur le zèle des citoyens qui ont commencé des établissements en ce genre, t'invite à lui faire sans aucun délai les propositions que tu croiras propres à assurer en grand la fabrication de la soude: tu sens que des objets de cette importance ne peuvent pas être livrés à une languissante indifférence" (F[12] 1508).

47. This correspondence will be found in the *Archives nationales*, F[12] 2243. In the light of the facts, the eventual arbitration of Leblanc's claims by Vauquelin and Deyeux may appear not unjust. It is true that the sums agreed on were never paid, but there is little point in discussing this grievance, which is fully set out in Anastasi, and in the article by Oesper (*loc. cit.*). Leblanc was not alone in suffering from the inability of the French government to honor its commitments.

48. It is also probably significant that in 1793 and 1794 LeBlanc should have pursued a political career with what (if one may judge from the quantity of minor offices he filled) must have been a certain gusto. (See Anastasi, pp. 31–32.) For example, he had a paid post in the *Agence nationale des poudres et salpetres* (from Nivose, an II; he served as Administrateur of the *Département de Paris;* and (to mention only the more important of these offices) he served on the *Commission temporaire des arts*, in which capacity he, together with Berthollet, conducted the inventory of Lavoisier's laboratory (F[17] 1337, dossiers 3 and 5). Nor is there convincing evidence to support Anastasi's statement that he found this a painful duty. Leblanc belonged to the scientific *demimonde* of the 1780's, and it is not inconsistent with what can be gleaned as to his tem-

perament from other sources that he should have been a member of the largely Jacobin *Société des Inventions et Découvertes,* which in 1792 and 1793 formed part of the coalition of artisans' societies that overthrew the *Académie des sciences.* (Fragmentary *procès-verbaux* are in the *Bibliothèque nationale, Manuscrits français, ancien supplement,* 8045. These documents are erroneously catalogued as pertaining to the *Bureau de Consultation*).

49. Original draft of the Decree of 8 Pluviose, F^{12} 1508. This is in Prieur's hand. The draft of an article providing for publication of Carny's process is crossed out, and there is substituted for it the provision that "Tous les citoyens qui ont commencé des établissements ou qui ont obtenu des Brevets d'Invention pour retirer la soude du sel marin sont tenus même dans le cas où ils se proposeraient de donner a cet établissement toute l'extension dont ils sont susceptibles, de faire connoître à la Commission la situation de ces établissements . . ." See, too, Carny to *Commission d'Agriculture et des Arts,* 2 Brumaire an III F^{12} 2244, dossier "Carny; an 3."

50. See, for example, testimonials by Fourcroy and Guyton, 14 Vendemiaire and 5 Brumaire, an 5, F^{12} 2244; see too Camille Richard, *Le Comité de salut public et les fabrications de guerre pendant la terreur* (Paris, 1922), Chapter 12.

51. *Institut de propriété industrielle,* dossier Carny.

52. Darcet *et al., Description,* reports no processes submitted by Carny except those developed before or during his association with Guyton, whereas Carny in his application for a patent specifically says that the process in question differs from all those for which he and Guyton had been granted a privilege (*loc. cit.*).

53. The documents relating to this affair are in F^{12} 1508, dossier "an 3," and F^{12} 2244, dossiers "Carny, an 3" and "Carny, ans 5 and 6." For an account of the pyrites experiments, see Darcet *et al., Description,* pp. 37–43.

54. Report of 14 Messidor an II, F^{12} 1508.

55. Darcet *et al., Description,* p. 44.

56. Petition of Alexandre Giroud, 1 Messidor An II, F^{12} 1508. For other projects, all abortive, see F^{12} 2244.

57. Bernard, Riffautville to Committee of Public Safety, 20 Frimaire An III, F^{12} 2244.

58. *Archives de l'Académie des Sciences,* "Mémoire déposé au Sect le 16 floréal an X" (Plis cachetés N° 12). This process reduced sodium sulfate to the sulfide by means of charcoal and converted the latter to the oxide by means of cupric oxide. It must have been expensive.

59. This enterprise is very well documented in F^{12} 2244, dossier "Carny à Dieuzé." But all of Carny's attempts to persuade the management that it would be profitable to allot him a greater amount of sulfate, or to undertake manufacture of soda as a principal product, were to no avail.

60. Memoir, *Bureau consultatif des Arts et Manufactures,* 24 November 1807, F^{12} 2245.

61. A complete census of the artificial soda industry was undertaken in 1810, by direction of the Minister of the Interior. The prefectorial returns, together with other documents in F^{12} 2245, would permit a complete account of the soda industry from 1806 until 1810.

For an outline of the later development of the alkali industry, see P. Baud, *L'In-dustrie chimique* (Paris, 1932). There is, however, no reason to think that the industrial history of the nineteenth century has been written any more authoritatively than that of the period covered in the present paper.

62. Report of the *Bureau Consultatif des Arts et Manufactures*, 14 June 1810, F[12] 2245; see too Memoir, Chamber of Commerce of Lyon to Minister of Interior, 23 September 1814, *loc. cit.* It is to be noted that the published reports of the juries, which surveyed the state of various industries at the Napoleonic expositions, for example in 1806, are not always trustworthy. For accurate information, it is necessary to go behind them to unpublished sources, among them the reports of the *Bureau consultatif*, which are often less flattering to industry.

63. Leblanc, "Observations sur la manière d'extraire la soude du sulfate de soude," *Bulletin de la Société d'Encouragement* I (1802), p. 170; see too J. C. de La Métherie, "Des manufactures de Soude," *Journal de Physique*, 1809, *69*, 421–28, quoting a memoir of Leblanc to the *Lycée des arts* (1798), which I have been unable to unearth.

64. *Journal de Physique*, 1789: *34*: 44. It is curious that this suggestion is identical with the first step of the Hollenweger process, for which a privilege was issued in 1783 (see above, section II). But no one noticed this at the time—an indication that lack of communication still played a negative part in applied science.

65. J. J. Dizé, "Mémoire historique de la décomposition du sel marin," *Journal de Physique*, 1810, *70*: 291–300. In 1809, La Métherie had called attention to his own part in the discovery by republishing an account by Leblanc (see note 64), and this served as the occasion for Dizé's attack on Leblanc. It is to be emphasized that Leblanc and Dizé agree that the experiments which led to the invention began with the La Métherie article in 1789.

66. *Loc. cit.*

67. This is the only conclusive evidence, however. But it is curious that Dumas attached less importance to the direct testimony of Darcet than to the fact that Dizé erroneously placed these events in the year 1790, from which he concluded that Dizé's story was fabricated out of whole cloth. He dismisses, by an argument too intricate to repeat, the possibility that Dizé might simply have misremembered the year. In fact, however, there is ample evidence (the patent, the La Métherie article) that Darcet *did* examine the process in 1789.

68. Darcet, *Description*, p. 49. It is perhaps worth mentioning that neither Dizé nor Shée collaborated with Leblanc in the effort to reactivate the factory after 1801. In 1810, Dizé was working for Marc, Costel et Cie. in their factory at Gentilly, where he was placed in charge of the manufacture of soda which the firm undertook for the first time in 1809 as a sideline to its main business in mineral acids. (Report of the Prefect of the *Département de la Seine*, F[12] 2245.)

69. It may perhaps be worth while quoting Dizé's estimate of Leblanc as one "n'ayant aucune connaissance du calcul, spéculatif, nullement habitué aux manipulations manufacturières, et naturellement peu activ." (*Loc. cit.*, p. 296). It is a commentary on the state of the literature that in a recent article, R. E. Oesper should describe the systematic experiments by which Leblanc allegedly reached his discovery. Oesper cites no source, but it seems evident that this narrative is a reconstruction of Dizé's account in the light of a modern chemist's knowledge. If so, this amounts to accepting

Dizé's account of the episode while rejecting his claim to the accidental discovery of the process. See Oesper, *loc. cit., 19:* 569–570.

    70. *Journal de Physique,* 1789, *34:* 43, 46–47.

    71. *Ibid.,* p. 43.

    72. According to Partington (pp. 83–85), the reaction occurs in two stages:

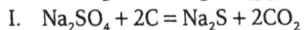

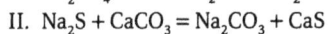

       I. $Na_2SO_4 + 2C = Na_2S + 2CO_2$

      II. $Na_2S + CaCO_3 = Na_2CO_3 + CaS$

And the reduction is carried out by carbon taking oxygen from the sulphate to form carbon dioxide:

$$Na_2SO_4 \rightleftharpoons Na_2S + 2O_2$$
$$2C + 2O_2 \rightleftharpoons 2CO_2$$

    73. Lunge, II, pp. 458–480.

⟿ 5 ⟻

The Natural History of Industry

THE FOLLOWING PAPER is self-explanatory, but perhaps it would be well to say something about the background behind the research in the French archives. While preparing these articles I was commissioned by the publisher of Dover Publications to prepare a reprinted edition of the technical plates from the *Encyclopédie*.[1] It was then borne in upon me, what is too often not noticed by intellectual historians, that this cardinal monument of the Enlightenment is not merely a sly ideological work slipping liberal criticism and ideas into accounts of current practices, institutions, and doctrines. It is also and equally a technological reference work reporting on the most accomplished methods, apparatus, and procedures in the arts and trades and exhibiting trade secrets in the light of day. The purpose was not only technical improvement. It was also in a sense democratic. For in Diderot's eyes the two were related. In the article "Art," he writes

> Let us at last give the artisans their due. The liberal arts have adequately sung their own praises; They must now use their remaining voice to celebrate the mechanical arts. It is for the liberal arts to lift the mechanical arts from the contempt in which prejudice has for so long held them.... Artisans have believed themselves contemptible because people have looked down on them; let us teach them to have a better opinion of themselves; that is the only way to obtain more perfect results from them. We need a man to rise up in the academies and go down to the workshops and gather material to be set out in a book which will persuade artisans to read, philosophers to think on useful lines, and the great to make at least some worthwhile use of their authority and wealth.

The first edition of the *Encyclopédie* consisted of seventeen volumes of text and eleven of plates. Publication of the former was interrupted with volume VII in 1759, when the Council of State ordered suppression of the whole work along with other dangerous writings. There could be no objection to the technical plates, however, and Diderot carried on with them until the remaining ten volumes of text were completed and passed en bloc through the censorship. In effect, therefore, though this may not have been Diderot's intention, the technology carried the ideology.

One correction is in order. The stricture passed on Berthollet in the last two sentences of the first paragraph in section V is mistaken. The essay is indeed better than its expansion into the book there mentioned, but the two together were the starting point of physical chemistry.

Note

1. *A Diderot Pictorial Encyclopedia of Trades and Industry: Manufacturing and Technical Arts in Plates Selected from "Encyclopédie ou Dictionnaire Raisonnée des Arts et Métiers."* 2 vols., New York, 1958: Dover Publications.

FIVE

The Natural History of Industry[*]

—ɯ—

In a previous paper, the problem of science and industrialization was pursued in some detail through an account of the discovery of the Leblanc process. The present paper approaches the question on a broader plane and ventures to offer some general considerations. Both articles are a result of investigations pursued in France, and refer specially, therefore, to the pattern of French scientific and industrial development. This was necessarily affected by certain factors peculiar to French history, notably the ever-growing centralization of cultural development in Paris and its learned bodies and the profound instinct of everyone concerned— scientist and statesman, industrialist and artisan—to expect of a paternalistic state the impetus which their British counterparts drew from private venture and expected only of themselves. But these are only social influences. Neither science nor industrialization has ever been national in scope, and to consider a question related to the industrial revolution—a framework other than British may have the merit of modifying the tendency to see this great event in the exclusive perspective of the Midlands of England and the Lowlands of Scotland.

In histories of 18th-century and imperial France, the assertion is often encountered that science was revolutionizing manufacturing, and Napoleon's encouragement of this process is frequently described as a major reason for the success of his industrial policy until the crisis of 1810. Historians have obviously drawn this judgment from contemporary writings

* Reprinted from *Isis* 48, part 4, no. 154 (December 1957). pp. 398–407. These two papers and the next one have been cited more frequently than any other in this collection. This one was reprinted in A. E. Musson, ed., *Science, Technology, and Economic Growth in the 18th Century* (London: Methuen, 1972). Together with the Leblanc piece it was also reprinted in the Bobbs-Merrill Reprint Series in History of Science (HS 20).

by scientists themselves. That science ought to receive public credit for the unprecedented progress of the arts since the 1770's or 1780's is perhaps the most protean reflection in the technological literature of the time. Of chemistry, for example, Chaptal writes:

> On l'a vue donner de nouvelles méthodes pour le blanchissage des toiles; fabriquer, de toutes pièces, le sel ammoniaque, l'alun et les couperoses; décomposer le sel marin pour en extraire la soude; enrichir la teinture de nouveaux mordans; former le salpêtre et le raffiner, par des procédés plus simples; composer la poudre par des méthodes plus promptes et plus sûres; réduire le tannage des peaux à ses vrais principes, et en abréger l'opération; perfectionner l'extraction et le travail des métaux; simplifier la distillation des vins; rendre les moyens de chauffage plus économiques; établir la combustion de l'huile et l'éclairage de nos habitations sur de nouveaux principes, et nous fournir les moyens de nous élever dans les airs et d'aller consulter la nature à trois ou quatre mille toises audessus de nos têtes.[1]

No one, it is true, specifies exactly how science was changing the face of industry. But knowledge is power; yesterday's banality is today's source material; and on the basis of all this authoritative testimony, the modern historian has very naturally supposed that theoretical science exerted a fructifying and even a causative influence in industrialization. The proposition is inherently persuasive, and there must be something in it.

II

The problem, however, is to know what there is in it. For if the question be approached in detail—as for example in my study of the origin of the Leblanc process—it proves extraordinarily difficult to trace the course of any significant theoretical concept from abstract formulation to actual use in industrial operations. The objection might be raised that such an approach loses sight of the forest for the trees. But in this case no coherent pattern emerges from the widest perspective or the most distant focus. Consider the more significant achievements in basic science—the progress of taxonomy in botany and zoology, the theory of combustion, the foundation of exact crystallography, the discovery of the electric current, the extension of the inverse-square relationship to magnetic and electrostatic forces, the analytical formulation of mechanics, the resolution of the planetary inequalities—and range these accomplishments side by side in

the mind's eye with the crucial points of technological advance in the industrial revolution—deep plowing and crop rotation, the development of
power-driven textile machinery and factory production, the discovery of
coke and its substitution for charcoal in smelting ores, the improvement
of the steam engine by the separate condenser and the sun-and-planet
linkage, the puddling process for the conversion of iron to steel—and it is
immediately evident that no apparent relationship existed between these
two sets of achievements except the vague and uninteresting one that
both occurred in a technical nexus. In a recent book Mr. and Mrs. Clow
have demonstrated with great force that the rapid expansion of chemical
industry must henceforth be numbered among the basic factors in industrialization.[2] But although the chemical industry was no doubt closer to
science than any other, their discussion does not suffer in the slightest
from the fact that its references to the revolution in theoretical chemistry
are fleeting—to have brought this into an account of Scottish industry
would have been to introduce an extraneous element. And this excellent
work does not illuminate the question, therefore, as indeed it was not intended to do; in fact the authors fall into the altogether trifling error of repeating the legend about Watt's invention of the separate condenser having been a practical outcome of Black's formulation of the principle of
latent heat.[3] Ultimately, then, one is tempted to fall back upon a generalization of L. J. Henderson's much quoted dictum to the effect that science
is infinitely more indebted to the steam engine than is the steam engine
to science. But this is too easy a way out, for repeating this famous remark
does not conjure away the sources. It cannot very well be supposed that
men of the eminence of Chaptal, Lavoisier, Cuvier, and many others had
no more in mind, in their frequent references to the utility of science, than
to win for it public esteem, and that no real substance lay behind their belief in its practical contribution to the arts.[4]

Neither does it clarify the issue to turn to economic history. For, on
the one hand, the sources do not exhibit a steady movement forward in
French industry, and on the other hand, except in the case of chemistry,
no correlation is to be discerned between the areas of greatest progress in
industry and in science. More intensive scientific attention was devoted,
for example, to the extractive and metallurgical industries than to any
other. To cite only the most obvious illustrations, Gabriel Jars' splendid
Voyages métallurgiques was published between 1774 and 1781;[5] in 1788,
Berthollet, Vandermonde, and Monge printed a very lucid memoir on iron
and steel;[6] and in 1794, a further treatise by Monge was the most solid
contribution to the technological series commissioned by the Committee
of Public Safety in order to stimulate war production.[7] But what practical

effect all this study had is far from clear. The metal trades proved the least resilient in recovering from the drastic technological setbacks dealt to French industry by the revolutionary disturbances. There is some doubt whether rehabilitation was complete even by 1815.[8] All through this period, inspectors of the *Agence des mines* were very fretful about the reluctance of iron masters to disturb themselves by shifting to coke, and with good reason, for at the end of the Napoleonic wars the only foundry in all France using coke was the furnace established at Le Creusot, which had gone into operation in 1782.[9] In any case, if it be agreed that the central feature of industrialization was the development of factory production, the crucial role must clearly be ascribed to the textile industries, and one searches their history in vain for any trace of scientific influence, except in the bleaching or dyeing of the finished product. In textile manufacturing—and even in metallurgy—French entrepreneurs were shown the way, not by scientific research, but by Englishmen and Scotsmen. John Holker's establishment at Rouen is the most famous example, but judging from the number of traces which his compatriots have left behind them in the commercial sources at the *Archives nationales,* there must have been literally hundreds of British artisans selling eagerly sought skills and ingenuity in France in the latter part of the eighteenth century.[10] Many remained even through the 1790's. Yet despite the ambiguities just discussed, and notwithstanding the economic disarray of the 1790's, this is precisely the decade from which Cuvier, for example, illustrated the benefits conferred by science on the arts.[11] The literature, therefore, exhibits real contradictions, and it is with this in view as the problem to resolve that the remainder of this paper is addressed to the question of what sort of influence science actually did exert in manufacturing.

III

In approaching the problem, it will be convenient to recall the history of the Leblanc discovery, which resists with great firmness being fitted into either of the extreme views that are sometimes expressed about the historical relations between science and invention. It cannot be used as an illustration of the austere doctrine that there was no real connection— Leblanc was closely associated with leading members of the scientific community, and it is a revealing fact that every one of the processes for making soda invented in France between 1775 and 1800 was verified and evaluated by one of a succession of three scientists, Macquer, Berthollet, and Darcet, who among them were informed about every development and exercised, therefore, a measure of control over the field. (This indeed was typical of the relationship, if not between science and the arts, at least between scientists and artisans, which obtained in 18th-century France and which lends to the phrase, "Science governs the arts," a literal sense not always appreciated, one that goes far to explain the rebellion of the artisans' societies against the *Académie des sciences* at the time of the Revolution and the enmity of the popular clubs for science.)

On the other hand, neither does the Leblanc invention bear out the contention that there is no essential difference between science and applied science. Leblanc seems to have found his process, not through some flashing theoretical insight, but by means of a fallacious analogy with the smelting of iron ore. Not only so, but after he had worked it out, neither he nor any of the other artisans interested in alkali production made any attempt to investigate or explain the nature of the reactions involved. They concentrated their efforts—though for a long time with no success—on trying to make money by one method or another—in Leblanc's case by first persuading the government to subsidize him.

Instead, therefore, of substantiating a doctrinaire interpretation of the relations between science and industry, the history of the Leblanc discovery suggests rather that the two main departments of technical activity are distinct but related. It is a relationship which may, perhaps, be observed quite generally throughout the technical history of industrialization—it could equally well be illustrated by a detailed history of any of the innovations credited to chemical science by Chaptal in the passage quoted in the second paragraph of this paper, as well as in many others of those 18th-century inventions which owed anything at all to scientific activity.[12] And there may even be discerned a certain structure in the relationship. For schematically there were two phases through which science moved in becoming effective in industry: first, the exploitation of science by inventors, artisans, and industrialists, and second (often simultaneously), the conscious application of science to practical problems.

IV

The first phase is precisely illustrated in the activities of Carny, Malherbe, Guyton de Morveau and Leblanc's other fore-runners, whose processes for making soda were more scientific and ultimately less successful than his—more scientific in that instead of being discoveries, they were attempts to draw advantage from reactions already known to chemical science. This practical exploitation of scientific knowledge may also be exemplified by the telegraph, the origins of military aviation, the emergency production of saltpeter, and indeed by most of the items which, like these, are cited by Pouchet, Despois, and even Mathiez in their jejune portrayals of the great revolution as favorable to scientific activity.[13] Actually, of course, these accounts simply adopt the Jacobin view which found value in science only insofar as its utility could be demonstrated. In fact, however, the distinction between this sort of thing and science is obvious and elementary. It can even be brought down to the level of persons, where the difference between the scientist and the practical man is basically a matter of temperament.

The point was discussed by Lavoisier, when in his forlorn defense of the *Académie des sciences,* he was trying desperately to preserve the independence of science, of free inquiry, from the Jacobin passion for assimilating it to the useful arts. However useful to industry science may be, wrote Lavoisier—and he thought it very useful indeed—the spirit which moves the scientist is fundamentally unlike that which animates the artisan. The scientist works for love of science and to increase his own reputation. When he makes a discovery, he is eager to publish it, and his object is only to secure his intellectual property in his achievement. The artisan on the other hand, whether in his own research or in using the research of others, is always thinking of his economic advantage. He publicizes only what he cannot keep secret and tells only what he cannot hide. Society benefits both from the disinterested investigation of the *savant* and the interested speculation of the artisan. Confound the two, however, and both will lose the spirit distinctive to them.[14]

Already in the eighteenth century, France was playing Greece to the modern world, and men of learning clearly and instinctively distinguished between the domains of science and practise. Throughout its history, the *Académie des sciences* had had two duties: to advance the sciences and to act as a panel of experts who evaluated the projects of aspirants for the favors offered by the state in its perennial effort to encourage French industry. And the very procedures of the *Académie* as recorded in its *registres* make it evident that disinterested research was the main interest of the foremost members, and that technological *expertise* was at best their corporate duty.[15] To say that they despised the latter would strain the classical analogy. On the contrary, they regarded it as a just and important obligation, if sometimes an onerous and tiresome one; but however important to society, the realm of practise belonged to a different and a lower order of consideration than the realm of theory and abstraction and advancement of the understanding of nature.

In this attitude, French scientists were more severe, perhaps, than their colleagues in other countries and particularly in Great Britain. Accompanying this paper, there is another by Dr. Robert E. Schofield, who offers an

account of his most meticulous researches into the industrial orientation
of science in the Lunar Society circle. And in a large sense Dr. Schofield's
subject would appear to be a further illustration of that mutual stimula-
tion of science, commercial enterprise, and Puritan influence (now soft-
ened into Nonconformity or Unitarianism), successive manifestations of
which so forcibly strike students of British social, intellectual, and scientific history from the seventeenth through the nineteenth centuries. But
there is perhaps an underside to the coin of this famous correlation. For
technical activity is one thing, but power of abstract thought is another;
and it may well be wondered whether a certain vulgarity in this British
utilitarianism—thoroughly evident, for example, in Bacon—was not responsible for the relative poverty of British achievement over the centuries in the abstract reaches of scientific thought; and further whether
the French instinct to separate thought and practise, while giving each its
due, was not by the same token responsible for the formal elegance and
intellectual eminence of French scientific leadership in its great days.

But however that may be in general, it is at least notable that after the
mobilization of talent in the year II, the men who were truly scientists—
Berthollet, Monge, Fourcroy, and Vauquelin, for example—moved back
from war service to science, whereas those whose instincts and interests
lay in production and enterprise—Carny, Deyeux, Pelletier, Chaptal, and
many others—moved outwards in the favorable climate of Thermidor towards the exploitation of whatever opportunities their war service had
suggested to them. It is true that, on the whole, the scientists had served
the state in a higher advisory capacity. They saw the problems and possibilities more clearly and steadily, and they had prestige. But it is an illusion that theoretical science was applied to war production, even in the
terrible emergency of the year II. Science was only exploited. Scientists
were what was applied.

V

Turning then to the second phase, the phase of science consciously applied, it is in general an illusion to suppose that it consisted in any uti-
lization of the latest theory. The literature of chemical manufacturing
makes frequent reference to the guidance offered by chemical theory to
those well versed in it. But the theory in question was the theory of affinities, not the theory of combustion, and the theory of affinities was devoid
of abstract interest. It did not in fact amount to much more than a tentative classification of substances according to their relative chemical activity. If one read only the memoirs on the soda industry, one would suppose

that the great chemist of the century was Richard Kirwan.[16] Lavoisier is never even mentioned. Leblanc and his kind needed knowledge of the chemical properties of substances. They did not need to understand combustion. Similarly in the dyeing industry, the line from Dufay to Berthollet does not pass through Lavoisier. In basic chemistry, on the other hand, neither does the line from Lavoisier to Gay-Lussac and Dalton pass through Berthollet. It was, indeed, the attempt to make it do so, misguided by the current fashion in scientific explanation, which led Berthollet to expand an excellent memoir on mass action into that curious and shapeless book, *Essai de statique chimique*,[17] quite unworthy of his real abilities, in which he tried to fix the elements of chemical science in the circumstances instead of the materials of reactions.

The true mode in which science was actively applied to industry has been obscured by the tendency of modern historical writers to suppose that the framing of basic theories is the main business of science, and that if science was related to industry, it must have been through the medium of advancing theory. They have failed to pay attention to the language of their texts, where the relationship of science to industry is not only clear but so clearly a corollary of the 18th-century conception of scientific explanation that it required no explicit formulation. For the 18th-century scientists do not write of the application of theory. What they say is that science illuminates the arts, that it enlightens the artisans, and that this process honors the century, and in holding this language they are simply considering industry in the light of that pervasive notion of the function of scientific explanation which (after Locke) is to be found in Condillac and Condorcet, in Jussieu and Carnot, in Lavoisier, Lamarck, and Laplace, in the physiocrats, the *idéologues* and the first *École normale*. In this light, science itself is positive knowledge, of course. Its function in the world is essentially an educational one, however, and its mode of procedure is analytical. First, it seeks to discern the essential elements of a complex subject. These once found, it ranges and classifies them according to the logical connections which subsist underneath all the welter of phenomena. Next, it establishes a systematic nomenclature designed to fix the thing in the name, fasten the idea to its object, and cement the memory to nature. In this fashion, the human understanding will be led descriptively towards a rational command over every department of nature by following its inherent order. Scientific explanation, then, consists in resolving a subject into its elements in the objective world in order that it may be reassembled in the mind according to the principles of the associationist psychology. The inspiration was algebra. But the model was botany.

It is consistent, therefore, that when scientists turned to industry, it was to describe the trades, to study the processes, and to classify the principles. In this taxonomic fashion science was indeed applied to industry, and very widely. What were those enormous ventures, the Academy's *Description des arts et métiers,* the *Encyclopédie* itself, the *Encyclopédie méthodique,* if not attempts to lift the arts and trades out of the slough of ignorant tradition and by rational description and classification to find them their rightful place within the great unity of human knowledge? The 18th-century application of science to industry, then, was little more and nothing less than the attempt to develop a natural history of industry. In this sense, the scientific development of an industry is measured, not by the degree to which new theory is used to change it, but by the extent to which science can explain it theoretically. "We are frequently able," writes Berthollet (the quotation is from the first translation of his *Éléments de l'art de la teinture*), "to explain the circumstances of an operation which we owe entirely to a blind practice, improved by the trials of many ages; we separate from it everything superfluous; we simplify what is complicated; and we employ analogy in transferring to one process what has been found useful in another. But there is still a great number of facts which we cannot explain, and which elude all theory: we must then content ourselves with detailing the processes of the art; not attempting idle explanations, but waiting till experiments throw greater light upon the subject."[18] Similarly, the metal industries were not at first much changed by the development of the science of metallurgy; they simply began to be understood. But that processes will be altered for the better if their principles are understood, that artisans will improve their manipulations if they know the reasons for them, are simply illustrations of the 18th-century faith in progress through classification and industrial examples of the 18th-century belief in scientific explanation as a kind of cosmic education.

Accordingly, the revolutionary manufacture of saltpeter consisted essentially in subjecting the French people to mass instruction in a simple technical process. What Monge, Fourcroy, Guyton, Berthollet, and the others did, when they were brought into this enterprise, was to give popular courses.[19] The military crisis of 1793–1794 created the greatest practical incentive ever experienced for the application of science to production in general. The Committee of Public Safety urged it forward with all its fearful authority. And inevitably the project assumed an educational form; science was mobilized in defense of the Republic, and at great speed scientists produced a series of textbooks, instructing practitioners, not so much in new methods, as in the best methods.[20] Finally (not to labor illustrations), it may perhaps be consistent with the nature of the mathe-

matical achievements of Monge in descriptive geometry that he should
have been the one important mathematician who consistently expressed
the utilitarian valuation even of mathematics and that he should also have
been the moving spirit in the foundation of the *École polytechnique,* that
portentous institution.[21]

Nor was the faith in science as the educator of industry simply abstract,
for their relations can be brought down to a question of what people actually did. Chaptal's testimony may again be quoted:

> Mais du moment que la chimie est devenue une science positive;
> surtout lorsqu'on a vu des chimistes à la tête des plus grandes en-
> treprises, et faire prospérer dans leurs mains plusieurs genres d'in-
> dustrie, le mur de séparation est tombé, la porte des ateliers leur a
> été ouverte, on a invoqué leurs lumières; la science et pratique se
> sont éclairées réciproquement, et l'on a marché à grands pas vers la
> perfection.[22]

Chaptal dates this change from the Revolution when the government
"pressé par le besoin, a successivement tiré plusieurs savans de leur cabinet pour les placer dans les ateliers, et la plupart y ont fait des prodiges en
très-peu de temps." And there is ample confirming evidence that a new
generation of scientifically instructed entrepreneurs and managers came
into control of industrial operations during the Revolutionary period, and
that the Revolution saw the culmination and the end of the very real hostility between scientists and artisans.[23] As a result, it was no longer necessary to complain constantly of the obstruction of rational procedures by
the ignorance and traditionalism of the ordinary artisan, always singled
out as the greatest barrier to progress. Popular superstition was the *bête
noire* of the rational writer, and whether he looked to religion or technology, he found it flourishing in ignorance and secrecy. To publicize processes, therefore, to get them out in the light of day, must be the business
of science, and the importance attached to this is evident in the condition
of publicity attaching to the prize programs of the *Académie des sciences*
and the grants of the *Bureau de consultation.* That such was also the policy of the English Society of Arts is an indication that the apparent differences in the relations of science and industry in France and Britain were
more verbal than real.[24] There is no contesting the reality of British industrial leadership in the eighteenth century, but it is conceivable that the
industrial interests of British scientists were more a result than a cause of
this pre-eminence. If they did not relate them to some general outlook,
this circumstance may possibly be taken as another illustration of the

historical law binding the two great western societies which prescribes that the French should formulate what the British only do.[25] In any case, it is, perhaps, not overly fanciful to summarize the problem by contrasting the enterprising, bold manufacturer of the nineteenth century, the engineer, the industrialist, in whatever country, to the Gothic master-craftsman of olden times, protecting his secrets and his mysteries, bending over his cauldron and stirring some traditional receipt, some confidential brew. The application of science to industry takes on real meaning, then, if it is seen, not naively as the alteration of old practises by theoretical concepts, but rather as an intellectual process and a chapter in the history of the Enlightenment. "Au tableau des sciences," writes Condorcet of his century, "doit s'unir celui des arts qui, s'appuyant sur elles, ont pris une marche plus sûre, et ont brisé les chaînes où la routine les avait jusqu'alors retenus."[26]

Notes and References

1. J. A. C. Chaptal, *Chimie appliquée aux arts* (3 vol.; Paris, 1807), I, p. xiv.

2. A. and N. Clow, *The Chemical Revolution* (London, 1952).

3. *Ibid.,* p. 590; cf. Donald Fleming, "Latent Heat and the Invention of the Watt Engine," *Isis,* 1952, *43:* 3–6.

4. Though outdated, the most comprehensive account of technological history is still Charles Ballot, *L'Introduction du machinisme dans l'industrie française* (Paris, 1923). To mention only the most obvious printed materials, Ballot may be supplemented with G. and H. Bourgin, *L'industrie sidérurgique en France au début de la Révolution* (Paris, 1920); Odette Viennet, *Napoléon et l'industrie française* (Paris, 1947); A. L. Dunham, *La Révolution industrielle en France* (Paris, 1953); Henri Sée, *Histoire économique de la France,* Vol. II (Paris, 1951); *Documents relatifs à la vie économique de la Révolution* (5 vol., Paris, 1906–1910), a series which in 1911 became *Bulletin d'histoire économique de la Révolution* (5 vol., 1912–1919); and among older works: J. A. C. Chaptal, *De l'industrie française* (2 vol., Paris, 1819); C. A. Costaz, *Essai sur l'administration de l'agriculture, du commerce, des manufactures, et des subsistances* (Paris, 1818); the *Bulletin de la Société d'Encouragement* (from 1801); the *Journal des Mines* (from 1794); and the *Annales de Chimie* (from 1789).

5. (3 vol.; Lyon, 1774–81). Published posthumously, these memoirs describe the iron industries of Germany, Sweden, Norway, England, and Scotland which Jars had studied between 1757 and 1769 in a series of expeditions originally undertaken at the instance of Trudaine.

6. "Mémoire sur le Fer," *Mémoires de l'Académie royale des sciences* (1786), pp. 132–200.

7. *Description de l'art de fabriquer les canons* (Paris, 1794). Monge begins this treatise with a little disquisition on the composition of the atmosphere and the occurrence of iron as the oxide, but his description of smelting procedures does not, of course,

differ in any essential from that given in the 1786 *Mémoire* (note 6), in which what iron ore is deprived of is its dephlogisticated air (*loc. cit.,* p. 133).

8. Ballot, pp. 494–527.

9. Dunham, p. 78.

10. André Rémond, *John Holker* (Paris, 1946).

11. Georges Cuvier, *Rapport historique sur les progrès des sciences naturelles* (Paris, 1808).

12. See, for example, Cuvier, pp. 345–58, for an enumeration of innovations attributed to scientific progress.

13. G. Pouchet, *Les sciences pendant la terreur* (Paris, 1896); Eugéne Despois, *Le Vandalisme révolutionnaire* (Paris, 1868); Albert Mathiez, "La mobilisation des savants en l'an II," *Revue de Paris* (November–December, 1917), pp. 542–565.

14. Lavoisier to Lakanal, *Oeuvres de Lavoisier,* (6 vol.; Paris, 1864–1893), IV, p. 623.

15. Unfortunately the *procés-verbaux* of the academy remain unpublished. They may be consulted, however, at the *Archives de l'académie des sciences,* at the *Institut de France.*

16. See, for example, the translation of Kirwan's procedure for the analysis of soda, *Annales de chimie,* 1793, *18:* 163–220.

17. (2 vol.; Paris, 1803); Berthollet's adumbration of the law of mass action, "Recherches sur les lois de l'affinité," was published in the *Memoires de la Classe des Sciences . . . de l'Institut,* III, pp. 1–96.

18. *Elements of the Art of Dyeing* (Edinburgh, 1792), p. 17.

19. On the *cours révolutionnaires,* see C. Richard, *Le comité de salut public et les fabrications de guerre sous la Terreur* (Paris, 1922), pp. 469–86; *Procès-Verbaux du comité d'instruction publique* (Paris, 1901), IV, pp. xxi–xxviii; the text of the courses is at the *Archives nationales,* AD VI, 79, pièce 69.

20. For the present purpose, it would burden this note unnecessarily to cite full titles. Suffice it to say that treatises by leading scientists were published on the following trades or processes: iron-working; small-arms; the casting and boring of artillery pieces; the incineration of plants (for potash); the soda industry; the soap industry; tanning; the separation of copper from bell metal; etc. Most of these treatises are gathered together in the *Archives nationales,* AD VIII, 40. But relatively little was attempted in the way of innovation. Perhaps the most interesting exception was the establishment in the year II of the *Atelier de perfectionnement,* an embryonic technological development laboratory, under the direction of Vandermonde, where experiments were made looking toward standardized and rationalized production, including the introduction, where appropriate, of interchangeable parts. But little came of this hopeful venture, which was eventually absorbed in the *Conservatoire des arts et métiers.* For documents concerning this project, see *Archives nationales,* F¹² 233, 234, 1310; F¹⁸ 288.

21. Besides the excellent study of Monge's work by René Taton, *L'oeuvre scientifique de Monge* (Paris, 1951), there is also available a rather inferior recent biography, Paul V. Aubry, *Monge, le savant ami de Napoléon Bonaparte* (Paris, 1954). For the caution with which this must be used, see my review in *Scripta Mathematica,* 1956, *22:* 245–246.

22. *De l'industrie française* (2 vol.; Paris, 1819), II, pp. 38–9.

23. See, for example, C. A. Costaz, *et al.*, "Rapport au Ier Consul," *Bulletin de la Société d'Encouragement,* 1802, 1: 45–48.

24. See Derek Hudson and K. W. Luckhurst, *The Royal Society of Arts* (London, 1954).

25. Far from regarding Dr. Schofield's accompanying paper, "The Industrial Orientation of Science in the Lunar Society of Birmingham," as calling for conclusions different from those which I base on a study of French developments, I think his material calls for a similar conclusion and could be better taken as a confirmation than a refutation of my point. The difference lies no doubt in what we mean by science. I mean abstract understanding of nature. Dr. Schofield seems to mean technical activity. What is the relationship of any of the scientific work alluded to by Dr. Schofield to basic theory? Indirect, at best—in fact this work was precisely the sort which led French writers sympathetic to the utilitarian or Jacobin tradition to describe the Revolution as favorable to science. But perhaps this question of interpretation had better be left to the reader.

26. *Esquisse d'un tableau historique des progrès de l'esprit humain* (Paris, 1795), p. 296.

—✿— 6 —✿—

The Encyclopédie *and the Jacobin Philosophy of Science:*
A Study in Ideas and Consequences

THIS PAPER WAS DELIVERED at an Institute for the History of Science held under the auspices of the Department of History of Science at the University of Wisconsin in the first ten days of September 1957. The initiative had come from the Joint Committee of the Social Science Research Council and the National Research Council under the chairmanship of Richard H. Shryock. Its intent was to devise ways and means to stimulate study of the history of science. One of the means suggested was to sponsor a selective gathering that would be longer and more intensive than the annual two or three days meeting of the History of Science Society, itself a very loose and heterogeneous organization, one might almost say disorganization. The venue was the Department at Madison, the oldest in our field, founded with the appointment of Henry Guerlac in 1941, suspended in 1943 when he accepted the post of in-house historian of Lincoln Laboratory, and resumed in 1947 when the University appointed Bob Stauffer and, as chairman, Marshall Clagett. It thus fell to Clagett to organize the 1957 Institute in consultation with many colleagues. With equal skill, he thereupon edited the proceedings.

The occasion was a delightful one, and also an important one for the nascent discipline, or subdiscipline, of history of science. It succeeded as well as, or perhaps better than, the sponsors could have hoped and had what one may call a crystallizing effect. It brought together in a common undertaking those of us who in a scattered way at our various institutions were attempting to find a footing for the subject. Not all of the participants, some twenty-six in all, even knew each other until then. Besides us fledging historians of science were such senior people as Ernest Nagel in philosophy, Robert K. Merton in sociology, E. J. Dijksterhuis in mechanics and its history, Richard H. Shryock in History of Medicine, J. Walter Wilson, and Conway Zirkle in Biology and its history. The pace was leisurely and the weather beautiful. There was time to develop topics of considerable substance. For ten days we took meals together and came to know one another, to absorb the criticism our papers incurred, and to exchange our experiences in our different institutions.

As for this paper, the central events it concerns are suppression of the

Académie Royale des Sciences in 1793 and the almost simultaneous foundation of the Muséum National d'Histoire Naturelle. I also treat those topics in my recent *Science and Polity in France: the Revolutionary and Napoleonic Years* (2004). There I put less emphasis on the ideological background and go into more detail on the institutional and political circumstances than in what follows. Not that I think the former is wrong. The sections here on Diderot and the *Encyclopédie* do hold up. Were I to write this paper today I would not reach back to Stoic antiquity, otherwise than to suggest parallels rather than to imply lineage. A better title would have been "Jacobin Attitudes to Science" rather than "Philosophy of Science." Nor am I happy about some of the unnecessary asides or the ruminations in the last paragraph. It turns out, moreover, that the statement in the penultimate paragraph, to wit "It is hardly conceivable that anything of the sort could happen today" has proven to be incorrect.

Similar things have happened, and perhaps the most interesting thing about the paper is the change over time in its reputation. It has been frequently cited and also assigned as reading in undergraduate courses at Princeton and elsewhere. Despite what seemed to me the clearest evidence, I failed conspicuously at Madison and afterward in the later 1950s and 1960s to convince more progressively minded colleagues that there could have been any conflict between science and political radicalism. Such a proposition was unthinkable to those whose minds had been touched, however lightly, by Marx. They attributed the undeniable suppression of the Academy early in the Terror rather to its elitist form. When by contrast the paper was read by scholars and students at the same end of the political spectrum amid the counterculture of the early 1970s, they accepted the findings as self-evident and disagreed with me only in thinking that the left was right to distrust exact science.

For my part, however, I have come to think that the question goes deeper than ordinary political differences into regions where the extremes sometimes touch. Others have discerned features of a similar antiscientific mentality in proto-Nazi aspects of the culture of Weimar Germany.[1] In all these contexts, France in the 1790s, Germany in the 1920s, and America in the 1970s, as well as in more partial and more personal instances, a sense of the failure of institutions and of corruption in high places has provoked the diagnosis of a sick society, which appears to have been connected with these manifestations in consequence of the breakdown of legitimate authority. It is certainly a feature of the scientific enterprise that it has always needed to draw upon authority, externally for funds and institutionalization, internally for the maintenance of standards, standards of rigor and discipline, and also for motivation to work.

I do not think that these necessities are inconsistent with a liberal polity. The legitimacy of the authority is the limiting factor. But to the rebellious temperament, authority and discipline are anathema, and we can understand why it finds science uncongenial and oppressive. For such temperaments resent limitations upon personality and will and deny the existence of any boundary separating one aspect of existence from another, whether that boundary separates man from man (hence the get-it-altogether spirit, fraternity and equality, but not necessarily liberty); whether it separates man the maker from what he makes (hence health in craftsmanship and dehumanization by externally powered machinery); whether it separates consciousness from nature (hence the yearning for a world alive and would-be substitution of psychology and biology for physics and mathematics as the ordering sciences); whether, finally, such a boundary separates science from its object (hence nature is to be known through penetration and sympathy, not through analysis and abstraction).

In Blake's graphic demonology, Newton measures and divides.

Note

1. Notably, Paul Forman, "Weimar Culture, Causality, and Quantum Theory, 1918–1927," *Historical Studies in the Physical Sciences* 3 (1971), pp. 1–115. See also, Peter Gay, *Weimar Culture* (New York, 1968); G. Lukács, *Die Zerstörung der Vernunft* (Berlin, 1954).

SIX

The *Encyclopédie* and the Jacobin Philosophy of Science: A Study in Ideas and Consequences*

—ɯɯ—

"La distinction," wrote Diderot of *Jacques le fataliste*, "d'un Monde physique et d'un Monde morale lui semblait vide de sens,"[1] a remark which provides this study with a text. For ultimately the subject emerges as an instance of the tension between science and the aspirations of humanity to participate morally and through consciousness in the cosmic process. These aspirations demand a nature different from that embraced by post-Galilean science, the nature not of the atomists (and much less of the Aristotelians) but of the Stoics. Newton's world offered virtue no purchase, and the Enlightenment saw, in consequence, a curious and important effort to transcend Newtonianism in a modernized Stoic physics through which virtue could, as in ancient times, be drawn from Nature.[2]

For such a physics the model of order is the organism, some unitary emanation of intelligence or will, or else identical with intelligence or will. In the long dialogue which science throughout its history has conducted between the unity of nature and the multiplicity of phenomena, this image is the dialectical opposite of that objective order into which analysis ranges whatever particulate entities it discerns as the term of measurement, whether actual or conceivable. And the renewals of this subjective approach to nature make a tragic theme. Its ruins lie strewn like good intentions all along the ground traversed by science, relics of a perpetual attempt to escape the consequences of western man's most characteristic and successful campaign, which must doom to conquer. So, like any thrust

* Reprinted from Marshall Clagett, ed., *Critical Problems in the History of Science* (Madison, 1959: University of Wisconsin Press), pp. 255–290.

in the face of the inevitable, it induces every nuance of mood from desperation to heroism. At the ugliest, it is sentimental or vulgar hostility to intellect. At the noblest, it inspires Herder's vision of history and Goethe's of nature, the poetry of Wordsworth and the philosophy of Whitehead or of any other who would make a place in science for our qualitative and aesthetic appreciation of nature lying all around us.

I shall pursue the theme in historical compass, considering as an example the attempt to alter the image of nature, first by the science and philosophy epitomized by the project of the *Encyclopédie,* and then by political actions expressive of that philosophy, which culminated in the liquidation of the foremost scientific institution of the world in the French Revolution. Since the subject is controversial it would, perhaps, be well to define the terms of discussion at the outset. The aspect of the Revolution with which I am concerned is Jacobinism, its most active element, in Burke's words, political "fixed air . . . broke loose." For Jacobinism will not here be taken as simply a collection of expedients. On the contrary, I think it the most passionate attempt in western history to realize moral ideals of virtue, of justice, equality, and dignity, in political institutions. It was obviously not a system developed *a priori* like its successor, Marxism. But only in this sense was it an improvisation, and system in it can indeed be discerned *a posteriori* by analysis. One must, therefore, distinguish its philosophy of science, like its political philosophy, in its acts as well as in the words which inspired it. I should say further that with certain reservations, I adopt Professor Talmon's recent theory of Jacobinism as messianic democracy.[3] I also accept M. Belin's claims for utility as an *idée-force* in the Revolution.[4] My thesis is that once affairs were engrossed by Jacobinism, science was bound to incur the enmity of the Republic. This was not because scientists were unpatriotic. On the contrary, they pressed into the service of the State on a scale unequalled until the twentieth century. But in its intrinsic combination of assurance and irrelevance, science all unintentionally stood across the cosmic ideals of the Republic in a posture nonetheless insulting for being unassumed. Only by ceasing to be science, could it give the Republic the nature it needed. And the Republic, true to its inspiration, could ask no less.

Élie Halévy remarks somewhere that whoever would be definite about the evolution of religious opinion must fix attention on the sects and organized churches in which alone it takes a form accessible to historical enquiry. Something of the sort may be the chief interest offered by the fortunes of scientific institutions for the history of scientific opinion, and it is in order to body ideas out into reality that the reader's attention will be returned from time to time to the law of August 8, 1793. This abolished the learned academies of France as incompatible with a republic.[5] It may fairly be described as a measure inimical to science and learning. In the historiography of the Third Republic, however, it became canonical to represent the Academy of Sciences as compromised by the undoubted sins—corporatism, favoritism, futility—of the other academies, to the point that the Convention's innate respect for its work and for science only just failed to save it.[6]

The facts are otherwise. The defenders of the Academy had been careful to disassociate it from the stigma of corporatism. They never disputed the justice of the impeachment when directed against the humanistic academies. Ultimately this strategy would save it at their expense by merging it into the republican educational system as the scholarly apex of the nation.[7] To this end, its friends on the Committee of Public Instruction inserted in the decree abolishing the other academies five clauses conserving a provisional existence to the Academy of Sciences.[8] The tactic miscarried. A single speech by David secured adjournment of those clauses. No voice spoke in their favor. This maneuver, therefore, enables us to estimate the attitude of the Jacobin Convention to the scientific community. As a community it was not to be permitted to exist. The attitude was hostile, then, and the question is, what did it spring from, this hostility?

The suppression of the Academies came in the late summer of defeat, alarm, and treason, in the months when the Jacobin regime was establishing its extraordinary efficacy on a rising curve of idealism, patriotism, and terror. It was the successful culmination of a considerable campaign, waged in pamphlets, journals, and popular clubs which since early in the Revolution had been heaping opprobrium on the Academy, certain of its members, and its influence in society. Three distinct themes recur throughout these writings.[9] To take them in the order of increasing political importance, they first of all express resentment for the new chemistry, directed expressly against the person and influence of Lavoisier, who appeared to his detractors not in the humble guise of chemist, nor simply as a financier, but as the arrogant spokesman and evil genius of science. Secondly, a gentle sentiment appears, which might seem inharmonious in this hostile chorus. Enthusiasm for natural history was unanimous. The paradox, however, is only apparent. Finally, there was a political assertion of the sort which rings of injured interests. Science was undemocratic in principle, not a liberating force of enlightenment, but a stubborn bastion of aristocracy, a tyranny of intellectual pretension stifling civic virtue and true productivity.

These were deep emotions, instinctive responses to an image of science as alien to the common man, but their revolutionary expression was patterned rather than informed. To understand their import, and the passion in them, the question must be taken back into the history of the Enlightenment, back to where it was explicit among the makers of opinion who knew why they felt as they did, who saw the point of the world picture of Newtonian physical science, and who rejected it as uninteresting and unsatisfying. The trail leads to the mid-century crisis of philosophy between the surrender of Cartesian science and the rising of the Romantic phoenix. For through the ashes between them runs the bond of dissatisfaction with a poverty in the Newtonian conception of scientific explanation. To the Cartesians, nature was the seat of rationality, and Newton's laws appeared intellectually trivial. To the Romantics, Nature was the seat of virtue, and Newton's laws were morally unedifying. In both views, nature is congruent with consciousness. The work of Diderot's generation, therefore, had to be to preserve the continuity of man and nature by opening its personality to reality rather than its intellect. For if nature *is* congruent with man, if science is the correspondence, it has to be a continuum, a whole, a "tout," as d'Alembert is made to see in his dream. But it is the whole personality which communicates, and not just the heart, because until Rousseau's revolt there is no question of irrationality.

—⁂—

The first thrust back to this more intimate sense of Nature came from chemistry, in an attempt to deepen the concept of matter, to give back body to what physics deprives of every attribute but surface and dimension. In speaking of physics as concerned with bodies in motion, we tend to forget that the word "body" originally implied organization, internal material organization, to which chemistry might properly address itself. The article on that science in the *Encyclopédie* is extremely significant. It is by Venel, a disciple of the elder Rouelle. He invokes a new Paracelsus, who will make of chemistry the science that understands nature and displaces geometry from that pretension. He will be gifted, this Paracelsus, with the sheerly technical insight to penetrate beyond physics, but he will have a spirit and imagination like that of the pre-Newtonian philosophers. For Venel agrees with Buffon that theirs was a less limited genius, a more extensive philosophy, that they "voyaient mieux la nature telle qu'elle est," because, "une sympathie, une correspondance, n'étaient pour eux qu'un phénomène, tandis que c'est pour nous un paradoxe, dès que nous ne pouvons pas le rapporter à nos prétendues lois de mouvement."

Physics, to pursue the comparison, is superficial. Chemistry is profound. Physics measures the gross characteristics of bodies—surface, shape, position, and motion. Chemistry penetrates their essence. Physics confounds abstract notions with verities of existence. One asks for a fact; it gives one a theorem. The physicist uses rigorous calculation to arrive at those exact theories, which experiments then confirm "à peu pres." The chemist, by contrast, never deludes himself by calculations. He apprehends his theories rather by a "pressentiment expérimental," and his theories are only approximate. But as a reward for his modest humility, in his case it is the fit with nature that is exact.

The question, then, is nothing less than the structure of nature. Mechanics will never bear the chemist into the heart of things, for the texture of reality is not corpuscular. Mass, the superficial aspect of matter which is the object of physics, doubtless is corpuscular. But—and this is the whole point—the essential merit of chemistry is to take the sting out of atomism. For it allows the masses in which atomism resides no ontological interest. So it is that chemistry carries the empirical answer to Newton's unreal abstractions. It is the qualities in things which impress our senses, our windows on reality, and this reality inheres—not in mass, be it repeated—but in the principles which run through the world as activities, as bearers of qualities and causers of perceived effects. The physicist, therefore, who denies existence to entities like yellowness and fire is simply presumptuous. They are not in his field.

It is not for the physicist to study quality, nor for the chemist to study quantity. Even his laboratory operations will be different from those of physicists like Boyle or—to anticipate—Lavoisier. His laboratory will offer no scene of weighing and measuring. What concerns him is the combinations and separations in matter, its state of interpenetration, and masses do not combine or penetrate. They only aggregate. Principles are what combine in perfect mixture, and the chemist, therefore, will catch glimpses of "la vie de la nature" coursing through his laboratory in phenomena which run all through, around, and under mass: in effervescences and distillations, evaporations and condensations, rarefactions and expansions, elasticity, ductility, malleability, and liquidity. The vision is of stretching and blendings in depth, full of that Faustian sense that nature has an inside. It is not alchemy, but like some alchemist, Venel is always moving in the mind's eye from fermentations in his laboratory through digestions in the animal down into the mineral gestations of the earth, whose cosmic womb is the source of unity in nature. The chemical authority Venel most admires and follows is Becher's *Subterranean Chemistry*. But back in his laboratory, your chemist, true to his calling, wields his materials with art. His hands are gentle. It is the physicist who brutally pulverizes, ignites, and destroys. The chemist does not analyze. He divines.

The chemist's world, then, is a palpable continuum—his science is Cartesianism stripped of geometry with its clear ideas. To replenish the Newtonian destitution of nature, it sees down into a world of matter pulsing with activity. In place of universal attraction between particles, Venel has discovered that the fundamental property of matter is universal miscibility. But chemistry is more than intimacy with nature. It has the common touch. It is everybody's science, the poor man's manual metaphysics, whereby that artisan in whose skills true wisdom lies manipulates reality, not in the humiliating abstractions of mathematics, but with his own hands: "La chimie a dans son propre corps la double langue, la populaire et la scientifique." And all this seems harmless enough until suddenly, out of the *Encyclopédie,* there speaks in one startling sentence, the authentic voice of the *sans-culotte:* "Parlez plus bas," the metaphysical theorist is told, "vous feriez rire nos porteurs de charbon, s'ils vous entendoient."[10]

—⁂—

In the *Interprétation de la nature* Diderot calls attention to this science of the chemists and to the work of Benjamin Franklin as examples to be emulated in handling nature with the surety of experimental art.[11] But sentience and organism weave a more grateful veil, and though Diderot drew his conception of material reality from the palpable continuum, he transposed it out of chemistry into the far more plausible terms of natural history, launched it in the flow of time, and incorporated it into a significant *Weltanschauung.* Diderot's prophecy about the decline of mathematics is sometimes taken as a passing slip in a prescient vision of the biological shape of things to come. This seems a misreading. Anticipations in the history of science are curious but by definition almost meaningless. To understand what a man meant, it is always not only more historical, but also more helpful, to look backward rather than forward to Darwin or to Einstein; and to read the connotations of modern biology into the eighteenth century, when the word itself was not yet invented, is to obscure that what distinguished natural history was not just its taxonomic method, but the uncontrollability of its metaphors by its evidence.

Moreover, Diderot's philosophy of science put no confidence in abstract conceptualization of any sort. His rejection of mathematics was fundamental. He objected to its claim to be the true language of science on all grounds, metaphysical, mechanical, and moral. It is not just that geometry idealizes—it falsifies, by depriving bodies of the perceptible qualities in which alone they have existence for an empirical science.[12] The comparison of mathematics to games of chance has sometimes been cited as exemplifying stochastic foresight: "Une partie de jeu peut être considerée comme une suite indeterminée de problèmes à résoudre, d'après des conditions données. Il n'y a point de question de mathématiques à qui la même définition ne puisse convenir." But he goes on: "La *chose* du mathématicien n'a plus d'existence dans la nature que celle du joueur."[13] Chance exists in affairs, not in things.

Even the science of mechanics has been rendered trivial by mistaking mathematics for understanding. To suppose that a body is indifferent to its state of rest or motion is to suppose it purely passive, without activity or force, which is to say without existence. But at the very moment that the physicist annihilates some block of marble with his calipers, it gives him the lie. For it is not still: It is a hive of decay and disintegration, a little world of living forces. Diderot will not even accept on Newton's terms inertia or the inverse square law. To make a universal constant vary with distance is a clear contradiction in terms, and "pesanteur" is no tendency to repose, but to local motion.[14]

The mathematical spirit, in fact, is a blight. Fortunate but rare the mathematician whose own aesthetic sensibilities are not blunted by his subject, which has fallen into aridity and circularity as must any science which ceases to "instruire et plaire." Once idle curiosity is satisfied and novelty wears off, only its power to edify will keep a science living. "Je n'en excepte pas même l'histoire de la nature."[15] But Diderot gives back to the Enlightenment, perhaps from his chemical studies, a more ominous note, which echoes down the whole romantic movement. Mathematics is worse than inhumane. It is arrogant, "orgueilleux." In a sense, no doubt, everyone who has felt himself reach his mathematical frontier, whether at long division or out somewhere beyond the calculus, must know something of the helpless resentment engendered by the hidden beauty of the abstract. But Diderot's own mathematical competence was by no means contemptible, and he fully appreciated its value as an instrument of precision in subordinate matters.[16] His indictment is curious and interesting and not mere petulance. Mathematics is the science by which a finite intelligence purports to plumb the infinite. Now, man aspiring above himself incurs the classic guilt of *hubris,* the Christian guilt of pride, and the prospect of an infinite universe has always disconcerted those who would render science humane. But Diderot was no Pascal to agonize over infinity. We are in the eighteenth century, and he responds with perfect nonchalance. He simply dismisses infinity as uninteresting. Since we shall need some criterion to establish bounds between knowledge and the infinite unknown, why let it be our interests. "Ce sera l'utile qui, dans quelques siècles, donnera des bornes à la physique expérimentale, comme il est sur le point d'en donner à la géometrie."[17] So Diderot restores the mind, in a sense, to a finite cosmos, by wrapping science tight around humanity.

Nor in form are Diderot's writings on nature an artless collection of *aperçus.* To see clear, d'Alembert, the geometer, is put into a dream, almost a delirium, out of which he speaks truths instantly recognized as such, and easily anticipated, by whom?[18] By a doctor, the universal doctor, who sees nature across the perspective of human nature: "Il n'y a aucune différence entre un médecin qui veille et un philosophe qui rêve."[19] And the apparent formlessness of the *Interprétation de la nature* is skillfully adapted to convey the congruence between man and nature. For it is written as a stream of consciousness, a reverie on the Experimental Art, the true route to a Science of Nature, moving out toward three objects: Existence, Qualities, Use[20]—a threefold object, but a single purpose. What is the young man to look for in Diderot's natural philosophy? "Un plus habile t'apprendra a connaître les forces de la nature; il me suffira de t'avoir fait essayer les tiennes."[21]

So he reverses Descartes, who studies himself to know nature. Diderot attends courses on chemistry, he reads Buffon, he studies nature—to know himself.[22] But communication is direct, experiential—it does not lie through mathematics. It lies, instead, through craftsmanship. For Diderot, as for the chemists, truth opens to the common touch, and the importance of right method is that it dispenses the ordinary man from the need for genius. In its pride, genius draws a shroud of obscurity between nature and the people—mathematical in Newton, conceptual in Stahl.[23]

They are in error who say that some truths can never be put "à la portée de tout le monde."[24] Certainly, ordinary people will never see merit in what cannot be proven useful. But in this, they see aright—or rather they are aright to fail to see. For only experiential philosophy is an "innocent study," in that it supposes no prior preparation of the mind.[25] The habit of actually handling materials in dumb, untutored experiments, develops in him who performs the coarsest operations an instinctive *"pressentiment"* which has the character of inspiration. Manual facility gives a power of divination, the ability to "smell out—*subodorer"* how it must be with nature.[26] But how do you know you have this power? How do you know you are right? It is—if the analogy is permissible—like awareness of Grace. It is like Virtue. It is participation in the Truth. You recognize it in yourself, in your own intimacy and more than intimacy, your solidarity with Nature. In such a breast, science and nature are one, the reality of the great organism suffusing for the moment the material consciousness of the little. Not mathematical abstraction from nature then, but moral insight into nature, is the arm of science. And consistently with Diderot's conception of scientific understanding as illumination, he contemns, in an unwonted access of Puritanism, the extravagances, the frivolities, the vanities of those who go down the garden path of abstract reason.[27] Presuming to prescribe as rules for nature his own formulations, the savant in his pride conceals from himself and others that it is not his laws of nature which are simple, but nature herself in her essential unity.

For nature is the combination of her elements and not just an aggregate. Otherwise there is no philosophy: "Sans l'idée de tout, plus de philosophie."[28] And Diderot, therefore, is bound to interest himself in continuity and not in divisibility. "Convenez que la division est incompatible avec l'essence des formes, puisqu'elle les détruit."[29] When he writes of molecules it is of their transience, not their existence. In genetics it is the atomistic as well as the providential implications of *emboîtement* which are unacceptable. For nature knows no limits. The male exists in the female and vice versa. (Hence the curious fascination with hermaphroditism.) Mineral blends into mineral. The qualities of one living species penetrate in some degree the others. Minerals are themselves fused into living matter through the *latus* of the plant which aliments the animal. Individual animals are real eddies of tighter organization, the ultimate but impermanent units, borne along a stream of seminal fluid flowing down through time and out from the matrix womb of nature herself. Even the physicist will do better to devote attention to what endures and spreads—to resonance, for example, to fire and electricity, to sulphurous exhalation, and to the behavior of standing waves. Diderot, too, has a substitute for the universal attraction of corpuscular physicists: It is universal elasticity.[30]

"Tout change; tout passe; il n'y a que le tout qui reste."[31] And Diderot uses two figures to express this unity. The second is the more familiar, the universe as a cosmic polyp, time its life unfolding, space its habitation, gradience its structure, for this embodies the twin ideas of universal sensibility and of evolution.[32] The latter idea Diderot took from Maupertuis and Buffon, but he treats evolution as a consequence of the indivisibility of time, a time which is that of biological subjectivity and in no way dimensional.[33] But although this is consonant with historicism, there is no serious sense in which it foreshadows Darwinism, the success of which, after all, is vindicated by its reducibility in genetics to material atomism. It is, therefore, not this, but rather the first of Diderot's metaphors which is the more significant. In it he evokes the swarm of bees. For the solidarity of the universe is social. On a cosmic scale, it is that community which the social insects know.[34] "Only the bad man lives alone," Rousseau was told,[35] and in social naturalism there is a more prescient concordance between the whole and the parts, the one and the many, than in reversion to an antique hylozoism.

—꩜—

Neither lingering devotion to an archaic chemistry nor growing enthusiasm for natural history—nor both these influences together, for the point is that they were in fact the same thing—would pose a serious threat to the scientific community. That would come only from political forces of sufficient importance. Nevertheless, these intellectual factors played their part in creating an image of science behind which it could scarcely be defended should such forces arise, as in fact they did. For it happened that the Revolution coincided with the sensitive period in the history of chemistry, when that science was the arena of the scientific revolution. Lavoisier's enemies were quite right to single him out as the epitome of science. The new Paracelsus had appeared, but in response to the summons of Lagrange and in what grievous form—betraying hopes with his material algebra.[36]

The word "enemies" is chosen advisedly, and not critics or opponents, for the important discussion was not about phlogiston nor was it held among scientists. Properly understood, phlogiston belongs to the objective history of chemistry itself, and as M. Daumas has shown us, Lavoisier's conception of oxygen represents much less sharp a departure than is traditionally said.[37] The question was rather what nature is like, agitated not as between scientists, but as between scientists and opponents of modern science, who wanted to see nature humanely through subjective perceptions of quality, beauty, and goodness. The point is explicit, for example, in Venel, who very well understood what he was saying: "C'est que la plûpart des qualités des corps que la Physique regarde comme des modes, sont des substances réelles que le chimiste sait en séparer, et qu'il sait ou y remettre, ou porter dans d'autres; tels sont entre autres, la couleur, le principe de l'inflammabilité, de la saveur, de l'odeur, etc."[38] Once the chemical revolution was under way, the issue was equally apparent in the chemical writings of Lamarck, the most considerable in the qualitative vein, and I have argued elsewhere that his emanationist theory of evolution originated as the transfer to, or refuge in, biology of this old sense of nature for which Lavoisier made chemistry uninhabitable.[39] But this archaic chemistry was not a science recorded in books, or taught by treatises. It partook rather of the character of lore, passed on by word of mouth like the trade secrets of the artisan, or taught in the public lecture courses where chemical artisans, the pharmacists, learnt their art. And for them, it was no abstract theory which offended, but the new nomenclature which came as a deliberate injury, reducing honest craftsmen to dependence on the scientist, making a mystery of their livelihoods.[40]

M. Mornet has sufficiently celebrated the vogue of natural history in the Enlightenment that it is hardly necessary to insist further on the theme.[41] Suffice it to point out that if this sentiment, too, is brought to the institutional test in the Revolution, its reality is abundantly manifest. On June 10, 1793, just two months before the *Académie des sciences* was abolished, the staff of the *Jardin des plantes* secured from the Convention a decree vesting the administration in their own hands, establishing twelve professorships in the different branches of natural history, and changing the name to *Muséum national d'histoire naturelle*—this munificent provision at a moment when virtually no chairs existed for teaching higher mathematics or physics. The distinction of the museum as a center of higher education and biological research dates from this reorganization. Its transformation offers, indeed, an epitome of the French Revolution in botanical microcosm: There was the situation of the *Jardin du roi* in 1788, at the death of Buffon, who had run it with a high and feudal hand and who left it bankrupt and with fine resources; there was the staff—Daubenton, Lamarck, Jussieu, Fourcroy, Thouin, Lacepède, and lesser figures—an active and able group suffering (though not seriously) from unjust arrangements (unequal salaries, a variety of obscure and arbitrary tenures) and who knew what they wanted (control of their own affairs); there was the attack on privilege and sinecures, for Buffon was succeeded not by a scientist but by a courtier, one Billarderie, who had secured the reversion by a private arrangement with Buffon; there was the discovery that favoritism does not automatically disappear with Revolution, for the staff got rid of Billarderie only to find the post given, not to one of themselves, but to the moralist and nature writer, Bernardin de Saint-Pierre; there were the projects and pamphlets, the approaches to the Assembly by people whose interests were threatened or who hoped for some advantage, each resting his case on buoyant principles like the equality of naturalists and the general welfare; there was the problem of safeguarding what was good from the old regime—of preventing, for example, the people from picking the flowers now that the flowers belonged to the people; there was, not to go into narrative detail, the ultimate success—rationalization of resources, equalization of employments, the rendering professional of a public institution and its dedication to public service and education.[42]

There is, therefore, no need to disagree with Mornet—except in thinking all this nature study less a vehicle of scientific culture than an escape from it. For it developed a pastoral picture of science, in the manner of Boucher, as of something charming, inoffensive, dilettantish, comforting, even cozy. If disillusionment came, it could only be sharp. In the physical sciences, too, popularization turned out a mixed blessing. It is not sufficiently appreciated how much the cult of science was a cult of marvels, nor how directly the fad of electricity led to the fad of Mesmer. In imagining the layman's idea of the scientist, we are reminded of Condorcet's description of that class of men who have the secrets in their keeping, and whose corporate interest is to obscure reality from common sense.[43] And the Jacobin mentality was not one to resign itself to what seemed to escape its control. But a similar rejection of mystery, a comparable insistence on imposing standards in circumstances which matter, is characteristic of the scientist. It explains, for example, the extreme reaction of the scientific community when confronted with a Velikovsky, or a Robert Chambers, or a Mesmer, its inability to understand, much less tolerate, the layman's pleasure at what in his willful ignorance he takes to be the discomfiture of science at the hands of someone who has done what he would like to do himself, and has broken the rules. Nothing was so damaging to the popular reputation of the scientific community as its maladresse in handling Mesmer. For at first, in refusing to examine his claims, it took the line that what everyone was tremendously interested in was not worth the attention of serious people, which was true but not tactful; and then it created the impression that the issue was between the self-esteem of a few scientists and the hope held out to suffering humanity by one who had the vision and daring to fly in the face of the pedants.[44]

The appeal of popular science, therefore, and of natural history, had the disadvantage of surrounding scientists with a fringe of enthusiasts not very well qualified by temperament or the nature of their interest to participate in the serious work of science. Consider, for example, the experience of the French Linnaean Society, founded in 1787 by Broussonet, under the banner of Linnaeus and in rebellion against Buffon. After a few months, it became known that anyone who hoped for election to the Academy would do well to disassociate himself from this group. The pressure was sufficient—the society collapsed, and the members separated to nurse their grudges. But not for long, for among the rights of man is that of forming voluntary associations, and at a meeting on May 16, 1790, the naturalists reorganized themselves as the *Société d'histoire naturelle.*[45] It soon undertook regular communication with the authorities on matters of natural history, going over the head of the Academy, and it entered into relations with other popular societies—of inventors, artisans, and artists. New members were recruited, of whom no qualifications were required beyond a love of nature. There were arresting figures among the members of the first Linnaean Society: Lamarck, for example, and Dolomieu. Several became deputies in the Revolution: Broussonet himself, Bosc, who sheltered Roland during the Terror, and Ramond, who had been Cagliostro's laboratory assistant and whose writings did much to bring mountain scenery into favor.[46] Perhaps the most influential was Creuzé-Latouche, who, as a member of the Finance Committee of the Constituent Assembly, was the first to raise in the legislature a serious question about the propriety of academies in the new order.[47] Additional notables appeared in the revived society: the abbé Grégoire, Romme, Fourcroy, Hassenfratz, and a number of others who have left names, not in the history of science, but on the rolls of the Jacobins in Paris.

It would hardly be worth going to great lengths to demonstrate that men who came to power in the Year II, their cars attuned to Nature by Diderot and Rousseau, did not understand science, for how could it have been otherwise? And we have a hundred indications that they did not: Thibaudeau's remark that the Jacobin leadership regarded intellect as the enemy of liberty;[48] Collot's instinct that excessive erudition is unsuited to a Republic;[49] Robespierre's attack upon Condorcet's educational proposals as tending toward an intellectual aristocracy and his enthusiasm for the Spartan plan of Le Pelletier.[50] In no serious sense can these men be said to have had any experience of science at all, or any interest in it. The real question, therefore, is what images came to their minds in the few minutes which they must have diverted from saving the Republic to approving the liquidation of the Academy of Science? And put this way, it is obvious that the point is not that they failed to understand science, but that they misunderstood it with a peculiarly damaging moral enthusiasm for nature. For any glimpse they got of science itself could come only with the shock of a betrayal of humane values which, so they had been taught by the scientizing moralists, derive through science from Nature herself.

There is an epitome of how chillingly inconsequential such an encounter might be, of what a gulf could open, in the passage between Laplace and Brissot over the claims of Marat. In 1782, Brissot had published in Marat's behalf a dialogue on academic prejudice. It is based on an actual conversation with Laplace, and he makes his straw geometer, all obstinate in his Newtonian idolatry, say that it is by mathematics that he knows Newton to be right. To which the humane skeptic retorts that Newton's critics too have their calculations: "Que faire dans ce chaos de chiffres? Recourir à la nature," says the skeptic, ending the discussion,

"Voir le fait. . . . J'aime mieux croire mes sens et la nature, que vos volumes de chiffres."[51] And in his memoirs, written in the full tide of disillusion-ment with Marat and in hiding from the guillotine, Brissot admits that Laplace may, after all, have been right. "Mais je ne pouvais supporter qu'il traitât avec insolence et despotisme un physicien parce qu'il ne jouissait pas comme lui de fauteuil. Je suivais depuis trois ans les expériences de Marat, et je croyais qu'on devait quelque estime à un homme qui s'en-sevelissait dans les ténèbres pour reculer les bornes des sciences."[52] In common justice a man ought to receive some return for making over 6,000 experiments, and for Brissot the issue between Marat and science is one to be settled man to man, according to principles of fair play and equality.

—⁂—

Such were the patterns of incomprehension in which statesmen re-sponded to an active campaign mounted against the Academy. It was a campaign founded in real interests, and fired by democratic ideology. To appreciate the ideological setting, it will be well to recur once more to the Encyclopedic movement and its relation to science, about which it is time to propose certain second thoughts. The *Encyclopédie* was, of course, a complex work of many hands. Nevertheless, after all exceptions are made, it is possible to group the contents under the two broad headings of lib-eral ideology and Baconian technology. Diderot's master stroke was to make the latter carry the former—to seem to say that the institutions of the old Europe were absurd and unjust because contrary to nature, not with the voices of Denis Diderot and Jean-Jacques Rousseau, but with the voice of science. Surely, however, it took sleight of hand to invest the util-itarian idea of progress with the high authority of Newton. For when one thinks of it, what had Newton to do with sly or sardonic japing about the Christian religion? How could Newton have inspired the proposition that the route to true knowledge of nature lies through rationalizing the puff-ing of the bellows, the creaking of the water-wheel, the hammering of the blacksmith? In the preface to the *Principia* he says, indeed, precisely the contrary, that he means to give the mathematical, not the mechanical, principles of natural philosophy.

Writers who describe the *Encyclopédie* as the exemplar of Newton's in-fluence on the Enlightenment might further reflect that the philosophic testament with which Diderot accompanied that work was the *Interpré-tation de la nature*.[53] Whoever studies the philosophy of the Enlighten-ment must be struck by its eclecticism and humanitarianism. The interest seldom lies in the originality of the ideas, but in how they are combined in the service of mankind. So it is most strikingly in the work of Diderot. He never drew his image of science from Newton. He drew it from the classics. It is Greek—with one all-important variation. He has abandoned the Greek admiration for vaulting speculation and substituted for it out of Bacon a professed humility (what in Diderot seems even a rather meaching humility) about the human understanding, combined with an equally Christian sentimentalization of honest manual work. He seldom loses an opportunity to strike a contrast between the pathetic and appealing humility of technique and the haughty arrogance of mathematical abstraction.

Certainly—as I have argued in another connection—the *Encyclopédie* represents the education of industry by science.[54] Descriptive science dignifies the arts and trades by bringing them within the ambit of systematic learning, by finding them their rightful place in a great fraternity of human knowledge. Not that the arts and trades were actually languishing in quite the depths of despite from which Diderot and d'Alembert would rescue them, but the *Encyclopédie* did carry out Bacon's injunction when it would have been easier simply to repeat it.

There is, however, another aspect to the problem of science and industry in the *Encyclopédie*, related to the role of that work in the history of populism. Its liberal ideology, couched in innuendo and irony, can never have been other than critical in effect, never have done more than unnerve the aristocracy and titillate the intelligentsia. The people, after all, do not like wit. It was the technology which was truly populist, taking seriously the way people made things and got their livings.[55] Artisans were indeed, in Diderot's words, taught "to have a better opinion of themselves,"[56] an opinion which the politicians learned to respect on the 14th of July, 1789. To the dignification of craftsmanship by science in the *Encyclopédie*, corresponds, therefore, a reciprocal democratization of science which entrained, for a time, a no doubt inevitable cheapening of expectations. For this is the final consequence of that 18th-century humanitarianism which would retrieve from the cold abstractions of classical physics a science warming to man. It assimilates the whole of science to its applications, makes it only the rationalization of technology, and seeks in practice to obey Diderot's injunction to keep man at the center, not only man but everyman.[57] It proposed that dream of a citizen's science which the Revolution turned into actual measures.

—⁂—

The Year II gave artisans their opportunity to reverse the aphorism according to which science governs the arts, and a revolt of technology was one among the many rebellions swelling into the great Revolution. To follow its course is to move down into obscure places among people whose history is hard to come by: craftsmen, engineers, inventors, minor manufacturers—small but solid people in the midst of whom fermented a leaven of cranks and malcontents. Nevertheless, fragmentary traces do remain of the societies which they formed to attain their interests, popular societies of inventors and artisans, of a piece with the famous popular societies of a purely political nature in which transpired the actuality of the French Revolution.

The specific trouble began with the Academy's statutory responsibility for advising the government in administering the encouragements to technological development. These took the form of subsidies, or, more rarely, of monopolies granted to inventive entrepreneurs, and it is not surprising that the Academy's role of referee earned it deep hatred among the artisans whose work it judged. A quotation from a writer in the *Journal du point central des arts et métiers* will give the temper. After dismissing the metaphysical sciences as sophistical, he writes:

Les Arts sont plus surs, et leur bienfaisance est plus certaine! Combien n'étoient-elles donc pas coupables ces formes abusives et tyranniques, qui violant la plus sainte, la première des propriétés, celle de la pensée, celle du génie inventif, ou *perfectionnant,* soumettoient les Priviligiés de la nature, les *Artistes,* à ces Loix gênantes, à ces dures épreuves, à ces censeurs inquiets et durs, dont l'ignorance ou la jalousie inquisitoriale n'avoient pour premier but que le soin d'humilier, ou d'écarter le vrai talent!

Combien n'étoient-elles pas cruelles et vexatoires, ces prétentions éxagerées des corps académiques? Combien n'étoit-il pas revoltant cet empire tyrannique et destructeur de l'industrie, que la richesse donnoit à ces vampires usuraires, à ces frélons despotes, qui toujours prêts à dévorer le miel travaillé par les abeilles, profitoient de leur fortune ou de leur puissance, soit pour s'emparer des Ruches, soit pour réduire les Artistes à des compositions avilissantes et ruineuses et les dépouiller même de l'honneur attaché à leurs travaux, en usurpant leurs inventions; en fatiguant, en rebutant leur zèle, leur courage, et leur constance par les dégoûts de toute espèce; enfin, en les forçant le plus souvent d'abandonner des idées, ou des découvertes très heureuses, soit parce qu'elles contrarioient l'amour propre des premiers privilegiés, soit parce qu'elles nuisoient aux portions d'intérêts données dans les entreprises préexistantes.[58]

Such complaints abound in the archives. But a considerable sampling in the *Archives de l'académie des sciences* of projects submitted to its judgment in the late eighteenth century, yields, in fact, no significant instances of meritorious ideas going unrecognized—much less of suppression or sinister interest at work. The only complaint materially justified was of undue delay, and the work of examination was arduous and exasperating. Reports on technological devices consumed more time than scientific memoirs, and the great majority were valueless. The literature surviving from inventors' claims amply testifies to the intensity, amounting- almost to fever, of inventive activity in the late eighteenth century. But to turn it over is to enter a curious, indeed a feverish, atmosphere, compounded of illusion and delusion, secrecy and deception, avid hope and bitter recrimination. One has the impression that great drafts of publicity and fresh air were needed, but that the altogether understandable impatience of theoretical scientists was probably unhelpful.

Nevertheless, although the charges of academic obstruction of ingenuity are little more than instances of the tendency to blame disappointments on the authorities, this did not lessen disappointment nor solace injured pride. And it is just here, in what concerns self-respect, that the sources are most eloquent and precise. For there emerges the picture of the worthy mechanic, the salt of France, the hero of the *Encyclopédie,* all unversed in polite ways, standing before a committee of scientists with his new machine, on which he has lavished years of labor and pinned his hopes, twisting his hat and trying to answer incomprehensible questions about the laws of statics and dynamics. The artisans insist with very deep feeling on their rights: not only on their right to private property in the fruits of their ideas, but with even greater feeling on their right to be judged by their peers, who look at mechanical problems from their own point of view and not from some high theoretical plane.[59] And on this score, it seems almost certain that these bitter and frequent complaints result from that sort of real injury to human dignity which, suffered by a whole people in its ablest elements, made the great Revolution all it was. The correspondence of foreign scientists visiting Paris just before the Revolution offers confirming evidence of the impression of a certain arrogance which the French scientific community—particularly its mathematicians—might make upon an outsider.[60] Indeed, the Academy itself became conscious of the effect it was creating. Early in 1789, Laplace, impatient at the quantity of chimerical projects, had proposed that every applicant be subjected to a test in geometry before his designs would be considered.[61] In 1791 the Academy did modify its procedures, not however in that direction, but by abandoning explicit approbations or condemnations to limit itself to a simple recommendation.[62]

Early in the Revolution inventive entrepreneurs began overreaching the Academy to submit requests for grants directly to the municipality of Paris and the National Assembly itself. There they were received with sympathy. Projects which had been rejected by the Academy were successfully appealed, and subsidies were voted by the legislature.[63] In 1791 a technological jury, the *Bureau de consultation des arts et métiers*, replaced the Academy in advising the ministry.[64] Half of the thirty-man panel was composed of representatives of the artisans' societies. Delegates from the Academy—Lavoisier, for example, Lagrange, and Laplace—who composed the other half thus found themselves colleagues of men whom they had rebuffed. This body, too, was in principle more lenient than the Academy, though there is no evidence that French technology benefited—perhaps because the state mitigated its excessive generosity by failing in practice to honor many of the grants.

It was, however, in popular societies that the majority of artisans pressed the interests of their trades. Painters and sculptors, it would appear, were the first and the most fiery in the rebellion against the academic regime. Under the leadership of David, the *Commune des arts* demanded absolute freedom of exhibition, and beyond that a liberation of the artistic spirit soaring out into infinite Shelleyan realms of creativity.[65] The mechanical arts assembled in more restrained fashion in the *Société des inventions et découvertes*.[66] There people of substance—men of the calibre of Fortin, Mercklein, Lucotte—stood at the head of the lesser artisans. In general, this body proposed itself as a free or revolutionary association of technology. In particular, it worked as a lobby to secure first passage and then administration of a patent law in the interests of mechanical innovators. The law of 1791, which remains the basic French patent law, was drawn to its specifications. In the view of the artisans, the influence to be overcome was that of the *Académie des sciences*.

Here is the comment of Boufflers, reporter for the committee of the Constituent Assembly which drafted the law. He dismisses the Academy's view that applications for patents should undergo technical scrutiny and rallies to the inventors' desire for a system of patents to be had for the asking. How, he asks, is it possible for any panel to judge fairly of an invention which, by definition, does not yet exist?

Mais les savans eux-mêmes ne sont-ils pas quelquefois accusés d'être partis au procès? Ont-ils toujours été justes envers les inventeurs? Convenons-en: l'étude a peine à croire à l'inspiration, et des hommes accoutumés à tracer les chemins qui mènent à toutes les connaissances, supposent difficilement qu'on puisse y être arrivé à vol d'oiseau.[67]

Finally in 1791 appeared the strident *Point central des arts et métiers,* "composé de tous artistes vrai sans-culottes,"[68] although in an earlier manifesto the constituents are more largely identified as "une classe encore plus directement utile ... celle des manufacturiers et chefs d'attelier."[69] The *Point central* set up a constant clamor for replacement of the academies by a democratic administration of support for the arts. This body was taken seriously by the authorities. Two of its representatives sat on the *Bureau de consultation.* It was encouraged to establish a *Lycée des arts,* a perpetual chatauqua of popular science in the court of the *Palais Royal.*[70] The fraternity of all the arts was its guiding principle, which could not be realized until they were led out from under the despotism of science and academics. A few passages from another of its communications deserve quotation:

> Du haut de leur Montagne nos Législateurs veillent; ils planent; ils guêtent par-tout la malveillance, et si d'une main ils tiennent la foudre toujours prêt á frapper les traîtres, de l'autre ils dispensent les bienfaits. Nous sommes donc assurés qu'il est impossible que l'abandon des Arts échappe à leur vigilance. . . . Laissez aux sociétés libres le soin de reculer, par les perfectionnements, les bornes de nos connaissances.—Que l'industrie pratique réunisse les vrais artistes en assemblées primaires des Arts; et qu'ils choisissent librement des commissions temporaires pour chacune des parties de la nouvelle administration; la liberté fera le reste et les fruits, n'en doutez pas, seront abondants. . . .
>
> Toutes les classifications scientifiques et amphatiques (sic) de nos connaissances, nous les réduisons à six commissions temporaires, et tout ce qu'on nomma *science,* nous le rapportons aux seules connaissances utiles. Enfin, cette nouvelle administration ne coûterait plus d'un million. . . .

Not the sciences, but the arts, will serve as basis of the new education; and the teachers will be not scientists, but artisans:

> En conservant les pères des Arts, en les faisant servir à l'instruction publique, vous occuperez, vous endurcirez cette jeunesse bouill-

ante dont les âmes doivent être préparées avant tout au premier de-
voir du citoyen, celui d'être utile à la patrie.—Voilà les véritables
moeurs républicaines.—Les prêtres hypocritiques disaient: *Sachez
vaincre vos passions,* et ils appelaient cela de la morale.—Le républicain chaud et actif doit dire: *Laissez les passions aux hommes, mais
sachez les diriger.* Ce sont elles qui lui donnent son énergie. Un
homme sans passions n'est qu'un fédéraliste modéré, ou un feuil-
lant hypocrite, incapable de grandes choses. Le véritable Sans-
Culotte, c'est celui qui *travaille.*

Is it fanciful to suggest that this is a view of the human aspect of nature
congruent with that which Venel's chemistry (or Goethe's or Lamarck's)
takes of its material aspect, as a web of activities quickened by sympathy,
vivified by the Stoic *tonos?* In the moral philosophy of Robespierre, it is
the priest who tears the web by throwing that barrier of obscurity between nature and the people for which Diderot blames the *savant.* In any
case, these are passages from the proposal for a new constitution of
science which the *Point central* sent to the Convention on September 26,
1793 along with its congratulations on having delivered to the Academies
"leur extraitmortuaire en bonne forme."[71]

For the fall of the academies on August 8th set in train liquidation of
the entire structure of French science. Rebellion was followed by purge.
On August 10, members of the former Academy resolved to take advantage themselves of the Constitution and to form a "*société libre et frater-
nelle pour l'avancement des sciences*" to carry on their work.[72] Lavoisier
wrote Lakanal for approval of this course, telling him in overwhelming detail of all the valuables and projects entrusted to the Academy: The Academy is trustee of expensive astronomical instruments, and has begun the
construction of new ones; Vicq d'Azyr has undertaken an anatomical treatise, for which 6,000 livres have already been expended; the Academy intends to publish the voyage of Desfontaines along the coast of North
Africa, financed by the nation; Desmarets has been given funds for a mineralogical map of France; money has been awarded to Fourcroy for research on alkalis, to Berthollet for work on dyes, to Coulomb for investigations of magnetism, to Sage for mineralogical experiments, to Haüy
for studies of crystallography; the entire section of chemistry has a grant
for work on the combustion of diamonds; agreements have been made for
publication of many manuscripts; the Academy's own Memoirs are three
years behind, and the work will be lost if not printed. Most important of
all, there is the great metric project. Contracts have been signed with
many artisans, whom the Convention surely will not disappoint. Standard

weights and measures are under construction. The survey of the meridian is in progress.

Lavoisier still could not believe that all this would be simply abandoned.[73] But if the government does, indeed, intend to take over the direction, will it please send precise instructions covering every particular? And for a moment the strategy seemed hopeful. The *Société libre* was authorized on the 14th.[74] But on the 17th the members' papers were placed under seal.[75] By September 1st, Lavoisier had at last despaired. His colleagues, he wrote Lakanal, dared not proceed even on a voluntary basis. To do so would flout "l'opinion dominante du Comité d'instruction publique et de la partic prédominante de l'assemblée." He has little hope of a further report promised by the Committee. He fears it impossible to reconcile the interests of science with the politics of the moment: "Nous sommes dans une situation où il est également dangereux de faire quelque chose et de ne rien faire."[76]

For the popular authorities proved very vigilant. In October a committee of agronomists was appointed to advise on the threatened crop failure. It disbanded after three meetings on being warned that it was about to be denounced at the Jacobins as an attempt to revive the *Société d'agriculture*.[77] At the Observatory the fourth Cassini was ousted as Director by four young men whom he had appointed as student assistants. They stood on the equality of scientists, brought off their *coup* by means of their influence in the *Section Saint-Jacques,* and controlled the Observatory for a year and a half. One was later executed for his part in the Baboeuf conspiracy. The others relapsed into obscurity after discovering an allegedly republican comet.[78]

As for the metric project, it seemed for a time that the government would see the work through. Two teams were in the field running the survey, Delambre from Dunkerque and Méchain from Barcelona. Precise determinations of units were under way in Paris. Administrative preparations were being concerted for the shift from *toise* to meter.[79] On September 11 a Temporary Commission was formed of those members of the Academy who had exercised its responsibility until the suppression. Prieur de la Côte-d'Or, emerging as the member of the Committee of Public Safety responsible for technology and war production, sat as representative of the regime. Like many graduates of Mézières, Prieur was an engineer who had indulged vague scientific ambitions with no success.[80] He may well have felt ill at ease in the company of men like Lavoisier and Laplace. On December 23, 1793, they, together with Borda, Coulomb, Brisson, and Delambre, were removed from the Commission by a decree of the Committee of Public Safety. This document was drawn in Prieur's handwriting. The ground was that in the interests of public spirit, the government must assign missions only to men "digne de confiance par leurs vertus républicaines et leur haine pour les rois."[81] The chairmanship was placed in the nerveless hands of Lagrange. Prieur became absorbed in other problems. The work on the metric system was abandoned.

With the scientific institutions of the old regime overthrown or in patriotic hands, the popular societies launched a movement to replace them with a new, republican organization of science. Each measured itself for the mantle of the Academy. The *Société d'histoire naturelle* proposed a definite constitution to the *Société d'inventions et découvertes*.[82] One clause provided that should a discovery occur of which the utility could not be shown, the freedom of science could be preserved by refraining from publication. A century later Berthelot represented the *Société philomatique* as the shelter in which the sciences took refuge.[83] The surviving papers of that group do not bear him out.[84] We may, indeed, catch a glimpse of a week in the life of Jacobin science in the *Lycée des arts*. It opened its courses in the wooden circus which then stood in the *Palais Royal* on April 7, 1793, before a great assemblage. The president, Fourcroy, flanked on the dais by four members of the Convention, delivered an address. His subject was utility as the object of science. Behind him could be seen a relief plan of a new canal for the city of Paris. Hébert then took the floor, and in an impromptu speech urged that the sanctuary of the arts be invested with the spirit of liberty. He further undertook to have Fourcroy's address printed at the expense of the Commune. A report was read on the culture of silk worms. Prizes were awarded for achievements in the mechanical and the agreeable arts. The session closed with a hymn to Apollo. The courses themselves followed nightly: On Monday, natural history, taught by Millin, who later became librarian of the *Bibliothèque nationale;* on Tuesday, amphibians, by Brongniart; on Wednesday, mineralogy by Tonnelier; on Thursday, vegetable physiology, by Fourcroy; on Friday, technology, by Hassenfratz; on Saturday, physiology, by Sue. The physical sciences had their due. In an evening spectacle open to ladies, citizen Val "a fait des tours de Physique amusante."[85]

Throughout the entire period, the spokesmen for science conducted its defense with neither dignity nor skill. They gave away the case for science as inquiry by themselves professing a vulgarly utilitarian valuation. One searches the sources for a single defense of science as a simple intellectual good, as a seeking for truth about nature, and one searches in vain. Instead, the Academy addressed Baconian platitudes about science and the arts to statesmen who were being told by the artisans themselves that the Academy was stifling creative industry, and who had been taught by the *Encyclopédie* that it is the artisan who knows not the theorist. They drew up memoirs about the advantage and glory of the metric system for Deputies whose constituents were in the act of arresting Delambre on the not unreasonable suspicion that all his transepts and alidades and clambering about in steeples could only be the equipment and deportment of a spy.

Lavoisier was a very great scientist and a very bad politician. He was probably the one responsible for establishing the *Bureau de consultation* in order to divest the Academy of the increasingly vulnerable responsibility for technological supervision. Yet the Academy could not quite bear to let go, after all, for the initial constitution of the *Bureau* was so contrived that undoubted *savants* sat as representatives of the arts and trades, a Gallic maneuver which insured a firm majority of *savants*, deceived no one, and only served to exacerbate the suspicions of the *sociétés libres* for the academy, "don't les ramifications cachées et souterraines resistent encore aujourd'hui avec une force incalculable à ce qu'on arrache le tronc."[86] A plan of education was drawn in the name of the *Bureau de consultation* by Lavoisier, which preserved for science an independent organization in which the Academy might take refuge.[87] It is a fair surmise that this was an attempt to forestall the *Point central* in its favorite project of assimilating science to the arts. Lavoisier spared no efforts to draw off the menaces of that body. He and other academicians even joined it in sponsoring the *Lycée des arts*.[88] But Lavoisier was hopelessly out of his element. He drew up a report, for example, to the membership on the assaying of saltpetre. Unfortunately, so the journal informs us, its reading had to be abbreviated for fear of fatiguing the attention of the public by fixing it too lengthily on abstract subjects, although Lavoisier's resumé of the work in hand in the Academy was accepted courteously enough as "semblable au suc precieux que l'abeille va ramasser sur toutes les fleurs pour le deposer dans le ruche."[89]

As the Academy's days ran out, Lavoisier desperately wrote speeches to put in the mouth of Lakanal and ran about the streets trying to find his self-styled champion to send him to the Convention in time for some crucial vote. At the very moment of the Academy's dissolution, he warned that if France abused the devotion of her scientists, she would deprive herself of their services.[90] Nothing of the sort happened, of course. Reflecting on the behavior of the scientific community under the Revolutionary and Napoleonic regimes, one is tempted to apply to science the Schumpeter thesis about the *bourgeoisie's* needing a master,[91] or to recall Malraux's recent remark that an effective regime, by definition, finds its technicians.[92] As if to refute Lavoisier's warning, scientists were mobilized.

They served brilliantly in the famous effort of war production celebrated in every text book.

France was not deprived of their services. Only science was.

—⦿—

This paper is a study of the consequences—not, be it emphasized, the "results," but the consequences—in actual events of the French Revolution of leading ideas of the Enlightenment. From it certain conclusions suggest themselves, which like the terms of the discussion, may best be stated categorically and explicitly, for they, too, will no doubt be controversial:

1. That the hostility of the revolutionary ideals of human nature to abstract physical science was not a passing irritation with the inhumane arrogance of mathematicians. It was profound and utter, rooted right down in the nature of the ideal itself. The Jacobins in their year of exaltation did not stop with their attempt to change human nature. They—I do not mean only the leaders, but especially the rank and file—proposed to substitute for the image of nature with which science confronts humanity a different one, one sympathetic to the ordinary man. Indeed, if the analogy with Marxism holds, the Jacobins ought to have had a philosophy of science, and out of their acts it is indeed possible to construct one, none the less real for being tacit. Theirs would be a science which, as to its technological aspect, would be a docile servant of humanity, and as to its conceptual, a simple extension of consciousness to nature, the seat of virtue, attainable by any instructed citizen through good will and moral insight: "Fut artiste et savant qui voulut," in Cassini's summary of the Revolution, "Heureuse liberté!"[93]

2. That the attack on the Academies in the Revolution was the political expression of a moral revolt against Newtonian science reaching back into the Enlightenment. But it traces back ultimately not, as I first expected (and, indeed, as I said in the earlier version of this paper) to Rousseau, but to Diderot. This is to dispute neither Rousseau's emotional hold over the revolutionary left, nor the attachment to him of the naturalists of the *Société linnéene,* many of whom had actually botanized in his company. But his resentments were sporadic, unsystematic, and essentially trivial expressions of that petulance which loves nature and hates science. His actual writings on nature are surprisingly uninteresting and innocuous. Rousseau's chemical commentaries were sensible enough. His botany was a consoling hobby, making no pretense to science. On the question of the Linnaean system, hostility to which is the touchstone of romantic and metamorphic (or Stoicizing) tendencies in taxonomy as opposition to atomism is in physics, he was on the side of sobriety, ana-

lytical accuracy, and objective understanding. It would be ironical if his providentialism was what kept him in the objective camp, though that is possible, for providentialism permits the mind to sit loosely to necessity. But however that may be, Rousseau would at most diminish the importance of science. He would only attack it on occasion. He would not presume to alter its structure.[94]

For this reason, behind Rousseau, it was the influence of Diderot which was profound, reaching into the heart of science to turn it into moral philosophy. His was no feminine dislike of precision, no soulful sense of God in nature, but a philosophy of necessitarian and Spinozistic organism, a system according to which science could be, not just denounced, but altered. We must, therefore, review our interpretations of the *Encyclopédie*. In its theory of matter, in its enthusiasm for natural history, and in its sentimental humanitarianism about humility and truth in technology, it presaged in each essential element the events which engulfed the scientific community in the Revolution. Enlightenment Baconianism set the pattern in which those events transpired, and we must take this aspect of the Encyclopedic movement of thought, therefore, not as the expression of a developing scientific culture, but as strictly inimical to scientific culture in effect as in intent.

3. That this explains the curious *distance* which every student of the eighteenth century must feel between the work of the scientists themselves and the liberal ideology attributed to its influence. Practicing scientists did not participate in this revived Stoicism (unless one includes in their company Buffon and the early Lamarck). D'Alembert, Condorcet, and Lavoisier saw the importance of science in the light which the associationist psychology threw on the genesis of ideas. In their view, man is what he makes of his experience, and scientific explanation functions as a kind of cosmic education. Leading from Condillac to the *idéologues*, this is the tradition which after Thermidor exerted the constructive influence of science in the Revolution—in the creation of the *Institut de France,* the *École polytechnique,* the *École normale,* etc.[95] But in view of its issuance in the religion of Saint-Simon and the authoritarianism of Comte, it is an open question whether this was a happier reading of science.

4. That, therefore, one has to go all the way back to Voltaire to find the *philosophe* who correctly understood the implications of Newtonian science for liberalism. For if I read him aright,[96] what drew Voltaire to Newton was precisely Newton's revelation of the absolute irrelevance of physics. This is what deprives dogma of any claim to draw authority from nature. After Newton, thought is really free, not just of the censor, but of the far more damaging tyranny of metaphysics. But science had only this

one liberation to effect, and that, rather than the death of Madame du Châtelet, is why, once Voltaire has understood physics, he goes on about his business, which is not to explain nature, but to play on a world of men a critical intelligence. Nor is it inconsistent that science and theology should have co-existed loosely and easily enough in the breasts of Voltaire and before him Newton. The warfare between science and religion is relatively trivial: It is the conflict between science and any naturalistic moral philosophy which is profound—there is no co-existence in the breast of the thorough scientizing moralist, of a Diderot, a Comte, or a Marx. If science does not give them the nature they want, they will change it so that it does—or says it does.

5. That the interesting thing about the Jacobin philosophy of science is not that it was futile. The attempt to revive the Stoics' nature as the seat of civic virtue was, of course, bound to fail. Nature is not like that, and the interesting thing historically is rather that it happened at all. For it suggests how shallow was the penetration of culture by science at the end of the century which is always taken as the great century for that influence. Ultimately, the analogy with Marxism flags if it does not altogether fail. It is hardly conceivable that anything of the sort could happen today. The Lysenko controversy never went so deep. And, of course, it did not really matter to science that it happened then. It mattered only to scientists. One feature of the impersonality of science is the way it does survive untouched all such persecutions and all the divagations of thought and institutions which it inspires. In moments of discouragement, therefore, one is tempted to fear that the history of the influence of science on culture is bound to be the history of an unavoidable misunderstanding, in which what changes is the way in which the import of science is misunderstood.

For men do not seem to be content to accept science for what it is intellectually: A great creation of the human mind, an inquiry about how nature works, the results of which are admirable and interesting in themselves, but empty of lessons, or promises, or comforts. They are bound to rummage about in it for liberal ideologies or social Darwinisms. They are bound to look for reassurance about free will in the unpredictability of the electron. Nor are they willing to accept as better evidence of what they seek their failure to find it there. So, perhaps, the question for a relevant sociology of science becomes this: What is the effect on men of living in a society whose most dynamic and characteristic activity moves ever further from their comprehension? Were the Jacobins (or the Marxists) right after all, and does it in practice come back to the fruits of technology? And perhaps, therefore, this paper is to be taken as a lengthy exemplification of the moral drawn for the author by a friend with whom he discussed it and who is a logician—one of those alarming friends who would embrace (and extinguish?) all one's interests in a single equation of linguistic analysis. It is impossible, he said, to move from a declarative to a normative statement, from an "is" to an "ought."

If you try, you only lose the "is."

References

1. Denis Diderot, *Oeuvres romanesques,* ed. Henri Bénac (Paris, 1951), p. 670.

2. On Stoic Physics, see S. Sambursky, *The Physical World of the Greeks* (London, 1956), and on its relation to ethics, E. Bréhier, *Chrysippe et l'ancien stoïcisme* (Paris, 1951). For influence of Stoic ideas in the sixteenth century, see Léontine Zanta, *La renaissance du stoïcisme* (Paris, 1914); and for a discussion of neo-Stoicism in 17th-century political philosophy, Gerhard Oestreich, "Justus Lipsius als Theoretiker des neuzeitlichen Machstaates," *Historische Zeitschrift,* 161 (1956), 31–78.

3. J. L. Talmon, *The Rise of Totalitarian Democracy* (Boston 1952).

4. Jean Belin, *La Logique d'une idée-force: L'idée d'utilité sociale et la révolution française, 1798–1792* (Paris, 1939).

5. Printed documents in *Procès-verbaux du comité d'instruction publique,* ed., J. Guillaume (1894), II, 240–60; *Archives parlementaires de 1789 à 1860, fondé par MM. J. Mavidal et E. Laurent, Première série (1787 à 1799)* (82 vols.; Paris, 1867–1911).

6. For example Guillaume's introduction to the volume cited in note 5, pp. lxii–lxxii.

7. On the instructions of the Constituent Assembly, adopted on 20 August, 1790, all academies were to draw up within one month revised statutes to bring their regimes into conformity with the new constitutional order (*Arch. parl.* XVIII, 173–74; *Archives nationales,* AD VIII, 11, pièce 1.) The Academy of Sciences conscientiously had anticipated this requirement, and began debating its own reform on 18 November, 1789, at the initiative of the Duc de la Rochefoucauld. On 21 August, 1790 it resolved to accelerate the discussions and hold four extra sessions a week, no doubt to conform with the will of the Constituent. On 13 September the new *règlement* was completed and adopted, one week under the deadline. In 1954 a copy of it was found by Professor Henry Guerlac in the *Archives de l'académie des sciences,* where too may be consulted the *procès-verbaux* of the meetings.

For a typical defense of the Academy by Lavoisier, see his "Observations sur l'académie des sciences," *Oeuvres de Lavoisier,* ed. J.-B. Dumas, (1868), IV, 616–23, where he points out (p. 618) that arts and letters do not need academies, and for the plan of education of the *Bureau de consultation, ibid.,* 650–68.

8. From Lavoisier's letters to Lakanal of 17 & 18 July, 1793 (*Oeuvres de Lavoisier,* IV 615, 623), it is evident that it was known at least by that date that measures were gathering against the academics. Indeed, the *registres* of the Academy are so slight for all of 1793 that the life of the body must have been increasingly paralysed as political uncertainty grew. But the measure itself was framed in great haste. Two days be-

fore it was adopted, Grégoire, who drew and presented it for the *Comité d'instruction publique,* wrote frantically to the Ministry to ask for information on what academies actually existed—Grégoire to Garat, 6 August, 1793, *Archives nationales,* F[17] 1097, dossier 1, pièce 1.

9. For typical diatribes against scientists and academies, see Marat, *Les charlatans modernes* (1791); Vadier, *Le Montagnard Vadier à M. Caritat* ... (1793), *Bibliothèque nat'le,* Fol. Lb[41].3196; Chabanon, *Adresse à l'Assemblée nationale* ... (1791), British Museum, FR 450, No. 3; *Suppression de toutes les académies* ... BM, FR 450, No. 4; *Suppression des Académies, Archives nationales,* D38 II 19.

10. The preceding five paragraphs paraphrase and quote from the article "Chimie" in the *Encyclopédie.*

11. Diderot, *Oeuvres philosophiques,* ed. Paul Vernière (Paris, 1956), "De l'interprétation de la nature" (xli), p. 217. For convenience, I cite in parentheses the paragraph numbers by which Diderot divided this essay.

12. *Ibid.,* (ii), pp. 178–79.

13. *Ibid.,* (iii), pp. 179–80. For the profound dissimilarity between Diderot's conception of probability, and that of modern physics, notice that he compares "ce que le sort met d'incertitude" in games with "ce que l'abstraction met d'inexactitude" in mathematical science.

14. "Principes philosophiques sur la matière et le mouvement," in *Oeuvres philosophiques,* pp. 393–400.

15. "Interprétation" (v) in *Oeuvres philosophiques,* pp. 181–82.

16. On his *Mémoires sur differens sujets de mathématiques* (1749), see L. G. Krakeur and R. L. Krueger, "The Mathematical Writings of Diderot," *Isis,* XXXIII (1941), 219–32.

17. "Interprétation" (vi), in *Oeuvres philosophiques,* pp. 182–84.

18. "Le Rêve de d'Alembert," in *Oeuvres philosophiques,* 285–371.

19. *Ibid.,* p. 293.

20. "Interprétation" (xxiv), in *Oeuvres philosophiques,* p. 193.

21. *Ibid.,* prefatory apostrophe, p. 175.

22. *Ibid.* (vii and viii), pp. 184–85.

23. *Ibid.* (xl), p. 215.

24. *Ibid.* (xix), p. 191.

25. *Ibid.* (xxvi), p. 194.

26. *Ibid.* (xxx), pp. 196–97.

27. *Ibid.* (xxxi), pp. 197–98.

28. *Ibid.* (xi), p. 186.

29. "*Entretien entre d'Alembert et Diderot,*" in *Oeuvres philosophiques,* p. 277.

30. This is the tenor of the first five "conjectures," "Interprétation" (xxxii–xxxvi), in *Oeuvres philosophiques,* pp. 198–211. See, too, "Rêve," in *Oeuvres philosophiques,* pp. 263–64 and 289–90.

31. "Rêve" *Oeuvres philosophiques,* pp. 299–300.

32. *Ibid.,* pp. 296–303.

33. "Interprétation" (lviii), *Oeuvres philosophiques,* pp. 239–44. See, too, (1), pp. 224–30.

34. "Rêve" *Oeuvres philosophiques,* pp. 291–95.

35. By the way of the *fils naturel.* See Arthur Wilson's *Diderot* (New York, 1957), p. 255.

36. Delambre, "Notice sur la vie . . . de Lagrange," *Oeuvres de Lagrange,* (1867), I, xxxviii.

37. Maurice Daumas, *Lavoisier, théoricien et expérimentateur* (Paris, 1955), 157–78. Though it is only fair to say that M. Daumas does *not* treat the issue as one between objective concepts and qualitative principles, chemistry and anti-chemistry. He regards phlogiston as itself a principle, and oxygen as partaking of this 18th-century quality.

38. *Encyclopédie,* article "Chimie."

39. C. C. Gillispie, "The formation of Lamarck's Evolutionary Theory," *Archives internationales d'histoire des sciences* (Oct.–Dec. 1956), 323–38; see also my "Lamarck and Darwin in the History of Science," in Bentley Glass, ed. *The Forerunners of Darwin* (Baltimore, 1959).

40. See, for example, the serial attack on the new chemistry by J.-F. de Machy in the Masonic *Tribut de la Société nationale des neuf soeurs,* I–IV (1790–91). Machy was a significant figure among the enemies of the Academy, to which he was refused admission about the time of Lavoisier's election. He was a popular private teacher, a leader of the pharmacists, author of the volumes on distillation of the *Encyclopédie méthodique,* and one of the royal censors who refused a license to *Annales de chimie.* See L.-G. Toraude, *J.-F. de Machy* (Paris, 1907).

41. Daniel Mornet, *Les sciences de la nature eu France un XVIII^a siècle* (Paris, 1911).

42 The essential documents were published by E.-T. Hamy, *Les derniers jours du jardin du roi et la fondation du muséum d'histoire naturelle* (Paris, 1893). See, too, *Procès-verbaux du comité d'instruction publique,* I, 479–87 and *passim.* Most of the archives of the Museum have been transferred to the *Archives nationales,* where they occupy sub-series AJ–15.

43. *Esquisse d'un tableau historique des progrès de l'esprit humain* (Paris, 1795), 27–28: "J'entends cette séparation de l'espèce humaine en deux portions; l'une destinée à enseigner, l'autre faite pour croire; l'une cachant orgueilleusement ce qu'elle se vante de savoir, l'autre recevant avec respect ce qu'on daigne lui révéler; l'une voulant s'élever au-dessus de la raison, et l'autre renonçant humblement à la sienne, et se rabaissant au-dessous de l'humanité, en reconnoisant dans d'autres hommes des prérogatives supérieures à leur commune nature."

44. The famous report on animal magnetism by Bailly in the name of a commission composed, in addition to himself, of Franklin, Le Roy, Bory, and Lavoisier, is in the *Histoire de l'Académie royale des sciences, année 1784* (Paris, 1787), pp. 6–15.

45. On the *Société linnéene* and its successor, the *Société d'histoire naturelle,* see its *Procès-verbaux, Bibliothèque mazarine,* MSS. 4, 441; together with fragmentary papers at the *Muséum d'histoire naturelle,* MSS. 298, 299, 300, 1998; and at the *Bibliothèque nationale,* MSS. FR, NA, 2760, f° 162–163 and 2762, f° 60. See, too, *Adresse des naturalistes,* BN Le²⁰. 826. Publications were the *Actes de la Société d'histoire naturelle,* of which one folio volume appeared in 1792 (BN [Inv S 1333] and Academy of Natural

Sciences, Philadelphia) and *Mémoires,* of which one volume appeared in 1799 (BN [Inv S 4649]). These are not to be confused with the *Journal d'histoire naturelle* (2 vols.; 1792), edited by Lamarck and others, which was bitterly hostile to Linnaean methods (BN S 11705 and S 11704). This, too, may be found at the Academy of Natural Sciences in Philadelphia—a collection, by the way, which is too little exploited by historians of science.

46. For Broussonet, Bosc, and Ramond, see documents in *Bibliothèque de l'Institut de France, Fonds Cuvier,* 186, 157, and 154. Bosc and Ramond have left autobiographies in manuscript. For the documents bearing on Broussonet's unsuccessful attempt to escape from France during the Terror, see *Archives nationales,* F⁷ 4619. See, too, the relevant *éloges,* Georges Cuvier, *Recueil des éloges* (3 vols.; Strasbourg, 1819–27).

47. *Arch. parl.* XVIII (session of 16 August, 1790), p. 91. Creuzé's detailed objection to the Academy was submitted as an annex to the report of the session of 20 August, 1790, *ibid.,* pp. 182–84.

48. A.-C. Thibaudeau, *Mémoires sur la Convention et le Directoire* (2 vols.; Paris, 1824), "Parmi les chefs révolutionnaires d'alors, il y en avait qui regardaient les lumières comme des ennemis de la liberté, et la science comme une aristocratie; ils avaient leurs raisons pour cela. Si leur règne eût été plus long, ou s'ils l'eussent osé, ils eussent fait brûler les bibliothèques, égorgé les savans, et replongé le monde dans les ténèbres. Ils répétaient contre les sciences les sophismes éloquens de quelques écrivains humoristes; elles étaient, disaient-ils, la source de toutes les erreurs, de tous les vices, de tous les maux de l'humanité: les plus grands hommes s'étaient formés d'euxmêmes, et non dans les universités et les académies. Ces déclamations flattaient la multitude; les ignorans étaient ennemis des lumières pour la même raison que les pauvres le sont des richesses." Thibaudeau served on the *Comité d'instruction publique.*

49. Paul Dupuy, ed., *Centenaire de l'École normale supérieure* (Paris, 1895), pp. 67–68. Dupuy reminds us that the *Comité d'instruction publique* was allowed little independence by Robespierre (p. 35).

50. C. Hippeau, *L'Instruction publique en France pendant la Révolution, débuts législatifs* (Paris, 1883), pp. 65, 156.

51. J.-P. Brissot, *De la vérité* (Neuchatel, 1782), p. 335.

52. Brissot, *Mémoires* (4 vols.; Paris, 1830–32), I, 199.

53. As has been remarked by Paul Vernière in his edition of Diderot, *Oeuvres philosophiques,* p. 168.

54. Gillispie, "The Natural History of Industry," *Isis,* XLVIII (December, 1957).

55. See my introduction to *A Diderot Encyclopedia of Trades and Industry: Manufacturing and the Technical Arts in Plates from L'Encyclopédie* (2 vols; New York, 1959).

56. Diderot, *Encyclopédie,* article "Art."

57. Diderot, *ibid.,* article *"Encyclopédie."*

58. N° 1, 4 Sept. 1791, *B. N.* [8° Lc². 6381.

59. See, for example, *Adresse du Point central des arts et métiers,* 16 Oct., 1791, *Archives nationales,* AD VIII, 29; *Journal des sciences, arts, et métiers,* 22 Jan., 1792 and 29 Jan., 1792, *B. N. V.* 42735.

60. For example, the letters (in French) of Lexell and Marivetz to Euler, in 1781 and 1782, I. I. Liubimenko, ed., *Uchenaia Korrespondentsiia Akademii Nauk, XVIII veka,*

Vol. II of Akademiia Nauk Soiuza Sovetskikh Sotsialisticheskikh Respublik, *Trudy Arkhiva*, ed. D. S. Rozhdestvensky (Moscow and Leningrad, 1937).

61. *Procès-verbaux de l'Académie des sciences*, 13 June 1789.

62. *Ibid.*, 3 April, 1791.

63. Compare, for example, in documents at the *Archives nationales*, F[17] 1136, F[17] 1137, AD VIII, 42 (for a typical sampling) the treatment of many *projets* already rejected by the Academy, in its *Procès-verbaux.*

64. Extracts from the *Procès-verbaux* of the *Bureau de consultation* were published by Charles Ballot, *Bulletin d'histoire économique de la Révolution* (1913). Three of the four volumes (the fourth has been lost) of the full minutes are preserved in the office of the Secretariat of the *Conservatoire des arts et métiers.* Some relevant documents were published in the *Oeuvres de Lavoisier,* and others in *Procès-Verbaux du Comité d'instruction publique.*

65. *Archives nationales*, AD VIII, 43; F[17] 1097, dossier 4; F[17] 1310, dossier 14; F[17] 1350, dossier 2; *Journal de la société populaire . . . des arts* (B. N., V. 42711).

66. *Archives nationales*, C 686, pièce 2; F[7] 4239, dossier 2; and *Bibliothèque nationale*, MSS. FR, ancien supplément français, 8045, which *cote* is erroneously ascribed by the catalogue to the *Procès-verbaux* of the *Bureau de consultation.* In fact, the *procès-verbaux* are fragmentary remains of those of the *Société des inventions et découvertes.*

67. S.-J. de Boufflers, *Rapport . . . sur la propriété des auteurs de nouvelles découvertes et inventions* (30 Dec. 1790), B. N., Le[20] 1206, p. 12.

68. *Archives nationales*, AD VIII, 40, T. I, pièce 18, *Point central . . . à la Convention nationale* (26 Sept. 1793).

69. *Archives nationales*, AD VIII, 29, *Adresse du point central,* 16 Oct. 1791, p. 11.

70. *Archives nationales*, AD VIII, 29, *Établissement d'une École athénienne, sous le nom de Lycée des Arts et Métiers.* Another version, B. N., Rz3007 & Rz3008.

71. *Archives nationales*, AD VIII, 40, T. I., pièce 18, *Point Central . . . à la Convention nationale.* The proposed constitution itself had been drawn up in March 1792 and sent to the Legislative Assembly with a "projet de décret." It was "redigé par la Société du Point Central des Arts et Métiers, en présence de MM. les Commissaires des Sociétés des Inventions et Découvertes et de la Commune des Arts" (B. N., Inv. Rz3001).

72. Lavoisier to Lakanal, 10 & 11 August 1793, *Procès-verbaux du comité d'instruction publique,* II, 314–17.

73. Lavoisier to Delambre, 8 August 1793: "Je ne sais si j'ai encore le droit de vous appeler mon confrère," he writes, assuring Delambre, however, that some way will be found to carry on the metric survey. This letter is not yet published, I believe. M. René Fric, who is editing Lavoisier's *Correspondence* (Paris, 1955—) very kindly and cordially allowed me to read through the copies of the letters which he has prepared for publication.

74. *Procès-verbaux du Comité d'instruction publique* (14 August 1793), II, 319.

75. Lavoisier to *Comité d'instruction publique,* 17 August 1793, *ibid.,* II, 320.

76. *Ibid.,* II, 331–32.

77. The chairman was Grégoire. *Bibliothèque du Muséum d'histoire naturelle,* MSS. 312 (1).

78. J.-D. Cassini, *Mémoires pour servir à l'histoire des sciences et à celle de l'Obser-*

vatoire Royal de Paris (Paris, 1810); J.-F.-S. Devic, *Histoire de la vie et des travaux scientifiques et littéraires de J.-D. Cassini IV* (Clermont, 1851); C. Wolf, *Histoire de l-'Observatoire de Paris* (Paris, 1902); and for documents illustrative of the history of the observatory under the Terror, *Archives nationales*, F^{17} 1065^A, dossiers 3 & 4; *Procès-verbaux du Comité d'instruction publique*, II, 217–27.

79. J.-B. Delambre, *Grandeur et figure de la terre*, ed. G. Bigourdan (Paris, 1912); G. Bigourdan, *Le système métrique des poids et mesures* (Paris, 1901); Adrien Fabre, *Les origines du système métrique* (Paris, 1931); *Oeuvres de Lavoisier*, VI, 660–712; *Procès-verbaux de l'Académie des sciences*, passim.

80. Georges Bouchard, *Prieur de la Côte-d'Or* (Paris, 1946).

81. F.-A. Aulard (ed.), *Recueil des actes du Comité de salut public*, IX (1895), p. 600.

82. *Bibliothèque du Muséum d'histoire naturelle*, MSS. 299. Article 2 states: "Le gouvernement ne doit encourager les sciences que sous le rapport des arts, et toutes les fois qu'on demande de l'argent au peuple pour cet encouragement, il faut qu'il en puisse saisir facilement le but utile, autrement il croirait le produit . . . employé à satisfaire une vaine curiosité."

83. M. Berthelot, "Notice sur les origines et sur l'histoire de la Société philomatique," *Mémoires . . . a l'occasion du centenaire de sa fondation* (Paris, 1888), pp. i–xvii.

84. Its papers are conserved, though in a state of extreme disorder, at the *Bibliothèque de l'Université de Paris*, where in 1955 they were shelved in the hallway outside the office of the *Conservateur*. Whatever it became later, the *Société philomatique* in the 1790's remained the resort of young men and of mediocrities.

85. *Journal du Lycée des arts*, 15 April 1793 (B. N.: V. 28667).

86. *Journal des sciences, arts et métiers*, pp. 4–5 (B. N., V. 42735).

87. *Oeuvres de Lavoisier*, IV, 649–68.

88. In June, 1793, Lavoisier presided at the session of the *Lycée* (*Journal du Lycée des arts*, 8 July 1793).

89. *Journal du Lycée des arts*, 14 & 29 June 1793.

90. Lavoisier to Comité d'instruction publique, 10 August 1793, *Procès-verbaux du Comité d'instruction publique*, II, 316–17, and "Observations sur l'Académie des sciences," *Oeuvres de Lavoisier*, IV, 616–23.

91. Joseph Schumpeter, *Capitalism, Socialism, and Democracy* (New York, 1942).

92. Interviewed in *L'Express* (29 Jan. 1955).

93. J.-D. Cassini, *Mémoires*, p. 93.

94. For Rousseau's botany, see *Lettres élémentaires sur la botanique* in *Oeuvres posthumes* (Geneva, 1782), and Albert Jansen, *Rousseau als Botaniker* (Berlin, 1885). For his chemistry, *Les institution chymiques*, ed. Maurice Gautier, *Annales de la Société Jean-Jacques Rousseau*, XII–XIII (Geneva, 1918–21). For a discussion of his attitude to science, see F. C. Green, *Rousseau and the Idea of Progress*, The Zaharoff Lecture for 1950 (Oxford, 1950).

95. C. C. Gillispie, "Science in the French Revolution," *Behavioral Science*, IV (January, 1959), 67–73.

96. Particularly the *Eléments de la philosophie de Newton* (1741) where the scorn expressed for metaphysics is that on which classical physics in practice acted, and *Voltaire's Correspondence*, ed. Theodore Besterman (Geneva, 1953–1965), which is indispensable for his state of mind about physics in the 1730s.

Note

This essay is an elaboration of the first of two public lectures on "Science and the French Revolution" which I had the honor of delivering in Oxford University in April, 1955, and the draft of which I sent early in July, 1957, to my critics, Professors Henry Hill and Henry Guerlac, as a preliminary guide to my interpretations. Between then and the convening of the Institute in September, I developed the material into its present form. Upon my arrival in Madison, Professor Guerlac kindly showed me the typescript of a paper, "The Anatomy of Vandalism," which he had given before the History of Science Society, meeting in Washington, D. C. in December, 1954, at a time when I was working in Paris on a Guggenheim Fellowship (of which this essay is one result). Neither of us had wished to publish these papers, of which, therefore, we remained in ignorance, and although we have been aware of the convergence of our somewhat differing interests on the period of the French Revolution, we were impressed at how closely the two papers agreed, even to the point that both of us hit upon the figure of a miniature French Revolution in the *Jardin du roi*. The gist of my own lecture occupies a certain portion of the essay published herewith. Our presentations were most strikingly parallel on the differential treatment accorded the *Jardin* and the *Académie des sciences*. Professor Guerlac went more deeply than I into the anti-scientific influence of Rousseau and Bernardin de St.-Pierre, whereas I developed the political activities of the voluntary societies of naturalists, artisans, and inventors. I set this out since I should not wish it thought that the hostility of the Revolution to physical science and the Academy contrasted to its enthusiasm for natural history and the *Jardin* are my discoveries. Professor Guerlac expressed and communicated them before I did.

Since then I have not significantly revised my views on the events, but I have seen them from a new angle of importance as a result of three circumstances: First, an investigation into the origin of Lamarck's evolutionary theory (see above, n. 49); second, an interest in Stoic physics aroused by Professor S. Sambursky's *The Physical World of the Greeks* (London, 1956); and third, a study of the *Encyclopédie* of Diderot and d'Alembert in preparation for my edition of the technical plates (see note 55). This is the origin of the analysis which I now give the subject, treating the Jacobin attack on the Academy as the political expression of a half-Stoic, half-Baconian attempt to substitute organismic concepts and technology for Newtonian theoretical physics at the center of science. Professor Guerlac was not altogether persuaded by this analysis. He expresses some dissent and some elaboration in his critical remarks in the published proceedings of the Madison conference, (above p. 110, note).

7

Palisot de Beauvois on the Americans

As the second endnote to the text explains, this essay was written for a collection commissioned in honor of Jacques Roger (1920–1990) on the occasion of his retirement from the University of Paris. Roger was a good friend and a fine scholar, one of the few historians of science who gravitated to the discipline from the side of literature rather than of science. He found, he said, a satisfying solidity in the subject that was lacking in literary history and criticism. He became a specialist in the history of evolution. His finest book is a splendid biography of Buffon, a naturalist and writer whose legacy pertains in equal parts to natural history and to literature.[1]

Having taught, lectured, and visited here many times in the 1970s and 1980s, Jacques Roger knew the United States well. His judgment of this country was very different from that of Palisot de Beauvois, a naturalist who lived in Philadelphia as a refugee from 1793 to 1798, and who was elected to membership in the American Philosophical Society. The latter's fragmentary essay on the character of our nascent republic and its inhabitants is reproduced in what follows, preceded by an account of his adventurous life.

Note

1. *Buffon, un philosophe au Jardin du Roi* (Paris: Fayard, 1989).

SEVEN

Palisot de Beauvois on the Americans*

—⚏—

A t only one moment in American history has there been a considerable migration of French citizens to the United States. It occurred, paradoxically, at a time when diplomatic strains brought the two governments close to a state of war, the issues having been what the French perceived as a tilt toward England in American neutrality legislation, and the Federalist Party as interference in American politics on the part of agents of the French Republic. Nevertheless, during Washington's second term and early in the Adams administration, from 1793 through 1798, the most numerous foreign contingent on American soil was French.[1] It is well known that certain famous persons sought shelter here, more or less briefly, from vicissitudes of revolutionary politics: Brissot (in 1788–89), Talleyrand, Chateaubriand, the future Louis-Philippe, the duc de La Rochefoucauld-Liancourt, the comte de Noailles, the marquis and marquise de La Tour du Pin. It is less often remembered that among the white population of Saint-Domingue, planters, artisans, merchants, shopkeepers, indeed all classes, the majority of those who escaped massacre in the slave risings, in which the Republic of Haiti had its origin, took what passage they could get for Savannah, Charleston, Wilmington, Norfolk, Philadelphia, New York, or Boston. The total figure may have been as high as 25,000. A number settled in the Hudson Valley. Several houses they built may still be seen in New Paltz. The largest fraction, perhaps a third, settled in Philadelphia, "Noah's Ark," as the comte de Moré called the federal capital. French was heard almost as much as English in the section from Front to Fourth streets between Arch and Spruce. The

*Reprited from *Proceedings of the American Philosophical Society* 136 (March 1992), pp. 33–50. A French translation appears in *Nature, Histoire, Société: Essais en hommage à Jacques Roger*, ed. Claude Blanckaert, Jean-Louis Fischer, Rosalyne Rey (Paris: Klincksieck, 1995), pp. 371–389.

A

SCIENTIFIC AND DESCRIPTIVE

CATALOGUE

OF

PEALE'S MUSEUM,

By C. W. PEALE, Member of the American Philofophical Society, and A. M. F.
J. BEAUVOIS, Member of the Society of Arts and Sciences of St. Domingo; of
the American Philofophical Society; and correspondent to the Museum of Natural
History at Paris.

Same in French. Miscell.

Vol. 26

NATURE

" The Book of Nature open,
" . . . . explore the wond'rous Work,
" . . . . an Inftitute
" Of Laws eternal, whofe unaltered page
" No time can change, no copier can corrupt."

PHILADELPHIA:

PRINTED BY SAMUEL H. SMITH, N°. 118 CHESNUT-STREET.

M.DCC.XCVI.

Pennsylvania Gazette ran classified advertisements in both languages. Printing of a catalogue of Peale's Museum began with a simultaneous English and French version. The title page of the latter is shown following page 144.

The present occasion seems an appropriate one to publish a set of observations on the Americans by the author of Peale's Catalogue, Ambroise-Marie-François-Joseph PALISOT, baron de Beauvois, to give his full style.[2] The very execution of the catalogue is itself an invitation to cultural comparisons. Charles Willson Peale intended his collection of stuffed animals and other natural objects to edify an uninstructed public, which paid admission.[3] The collaboration with Palisot would invest this gallery of curiosities with the attributes of science by assigning the items exhibited their place in a framework of formal natural history. We do not know who took the initiative, but it would probably have been Palisot, who badly needed employment, rather than Peale, a poly-artist but no polymath. He had begun collecting in 1786 for the Museum that he opened in a wing of his own house. In 1794 he leased the public rooms of Philosophical Hall, the seat of the American Philosophical Society. It would seem reasonable to surmise that Palisot seized the opportunity of the new installation to open a subscription for a *Scientific and Descriptive Catalogue,* to give it the American title.

Intended to be elementary, the catalogue starts with a statement of the Linnaean criteria for assigning productions of nature to the animal, vegetable, or mineral kingdoms according to the mode of growth, the capacity to live, and the capacity to move. Three distinct tabulations are given for distributing animals into classes. The first depends on external or anatomical characters and the second on other, mainly physiological characters. For the third, Palisot adopts Daubenton's dichotomous method of successive divisions by means of arbitrary discriminants chosen for convenience. All three systems range creatures of the animal kingdom in the same eight classes of quadrupeds, cetaceans, birds, lizards or oviparous quadrupeds, snakes, fish, insects, and worms. Quadrupeds are the subject of the first chapter. The subdivision shows that, though Palisot the botanist could and did guide himself by the natural system of Jussieu, Palisot the zoologist was necessarily closer to Buffon than to modern taxonomy. The orders are six: *Les Primats, les Bêtes Féroces, les Dormeurs, les Brutes, les Ruminans, les Bellues,* which the American edition gives as "Primates, Wild Beasts, Dourmouse, Bruta, Ruminants, and Belluae." The primates in turn comprise four genera, Man, Monkeys, Lemurs, and Bats, and Palisot cites Linnaeus's justification for including man and bats in the same order, the principal point of comparison being the male sexual organ.

Entering the Museum, French edition of the catalogue in hand, a visitor would need to bridge a cultural gulf as wide as the Atlantic Ocean between these high-flown taxonomic principles and the homespun specimens to which they were applied. Exhibited under the heading *Homo Sapiens Américanus* were skeletons of a Wabash brave and his squaw. Since the only distinctive feature was a pair of fused vertebrae in the man, the visitor would, so Palisot suggested, be pleased to know the story of the couple in order to form an idea of the customs of this people. The American version begins, all laconically, "At the time of the American war with Great Britain, a part of this nation of Indians joined the American army." Palisot embroiders the background thus: "A portion of this nation united itself in order to fight alongside the American Army when the American colonists shook off the yoke of oppression and, taking up arms, repulsed the English, who wanted to keep them in slavery and servitude." Of the man and woman whose skeletons were on display, he continues, "These Indians, united by seemly bonds of love, had a child, the fruit of their mutual tenderness." When the mother died, American troops cared for the child. It disappeared after the father also succumbed, and the soldiers were shocked to come upon its body buried between his knees. "Who would have taught him to hunt?" asked the elders of the tribe, explaining why they had sent the little boy to join his parents.

A Notice to the Reader explains that the authors had intended an *Introductory Essay* on natural history, "of all the sciences the one that is most useful to man, and, we do not hesitate to urge the point, the one it is most important for him to know." The work was becoming more voluminous than they had expected, however, what with Peale's adding daily to the collection and the detail required to make the catalogue useful. Rather than disaffect the subscribers, to whom the authors had announced a 500-page work, they decided to defer the generalities and press ahead with the descriptions. In a letter of January 1796, Palisot tells Jussieu that the catalogue contains 100 quadrupeds, 1,000 to 1,200 birds, some reptiles, some insects, some fruit specimens, and few minerals.[4] Only the first installment was ever printed, however. The French version breaks off in the middle of article 37 on the *Congonar,* and the American version a bit further on in Article 53. At the end of the last page, Peale was in the midst of an account of the birth process of a female opossum in the Museum. The reader, and indeed the baby animals, are left dangling. Thereafter, the Museum expanded anyway, without the benefit of a catalogue.

It was Palisot's fate, perhaps because his curiosity was stronger than his judgment, that he ended by publishing only fragments of everything he undertook. Palisot was born in Arras on 2 July 1752, younger son in an old family of the judicial nobility of Artois. He was educated at the Collège d'Harcourt in Paris and from the age of 18 he threw himself into the study of botany, having been introduced to the science by a professor of natural history at Lille called Lestiboudois. He was also a passionate reader of travelers' tales and devoured Niebuhr on Arabia. Brooding on the death of Forskahl, the young Palisot determined to finish the journey his hero had begun by crossing Africa from the Red Sea to Senegal. Never mind that the government failed to sponsor him. Never mind that he had made a marriage—of convenience. Never mind that first his father, then his older brother died, leaving him responsible for the family property and fortune. One Jean-François Landolphe, a captain in the slave trade conducted by the house of Brillantais-Marion out of Saint-Malo, had formed the project of establishing a French trading post in the territory of Owara, a dependency of the kingdom of Bénin, in the Gulf of Guinea. Landolphe needed a horticulturalist. Never mind, finally, that the country was on the wrong side of Africa. Palisot would make it a jumping-off place. He leaped at the opportunity, purchased equipment, hired assistants, all at his own expense. Further, he attached to himself one Bondakou, a putative son of the king of Owara, who had briefly fascinated fashionable Paris as a black man about town in the mid 1780s.

The expedition sailed from Rochefort on 17 July 1786, requiring four extremely difficult months to reach the mouth of the Formose river. Mosquitos, crocodiles, malaria, yellow fever, equatorial heat, tropical deluges: Bondakou reverted to his own people, and everything else went awry. Palisot's brother-in-law and his two research assistants succumbed to yellow fever. Of the 300 Frenchmen who had set off to man the trading post, 250 died during the fifteen months that Palisot passed in Africa. He was more or less ill most of the time himself. Nevertheless, he managed to explore Bénin and Owara, collecting plants, seeds, and insects, reaching the capitals of both kingdoms, and witnessing ceremonies featuring human sacrifice and cannibalistic feasting. All the while, he persisted in his determination to set forth across the Continent, if only he could find one bearer to accompany him. He never did. Instead, he grew weaker and weaker until Landolphe decided that the only way to save him from his obsession was to constrain him to depart. The sole transportation available was a slave ship bound for Saint-Domingue. Palisot had managed to consign certain items from Owara and Bénin to A. L. de Jussieu at the Royal Botanical Garden in Paris. The bulk of his collections were left in the care of Landolphe, however, and destroyed when English sailors pillaged the installation in 1791.

The voyage required five months. Of the 250 Africans in the cargo, 180 died of consumption or smallpox. Bodies went overboard almost every day. The captain, incompetent and suspicious, thought that Palisot was an agent of the owners, spying on him. Only the concern of the ship's baker kept him alive. On landing he was riddled with scurvy and covered with an appalling rash. An uncle, the baron de La Valletière, was an important figure in the colony and governor of Môle Saint-Nicolas. In his fine residence Palisot recovered his health, whereupon he settled into the life of the colony, the richest in the world in these, its last days. He joined the Royal Society of Science and Arts of Cap-Français (the designation accorded the Cercle des Philadelphes by the crown in 1784) and, long since a correspondent of the Academy of Science in Paris, Palisot set himself the task of compiling a thorough natural history of the island.[5] In August 1791, he was elected to the Colonial Assembly, which had to deal, first, with demands of the mulattos, and then with the first rebellion of the slaves, who rose against their owners on August 22. In October the Assembly dispatched Palisot to Philadelphia to bespeak the good offices of the French Minister, Ternant, in securing aid for the beleaguered colony from the American government. There he was elected to the American Philosophical Society, improved his English, and busied himself raising money and supplies for Cap Français. Recalled in June 1793 by the Commissioners whom the French Republic had in the meantime sent out to replace the royal administrators in charge of Saint-Domingue, Palisot arrived on the twenty-fourth, three days after rebels had torched the entire city. Bodies lay everywhere. Embers smoldered. The victors celebrated amid the ruins. Palisot's house, his library, and collections were in ashes. Jailed by the revolutionary leaders, he was saved by a mulatto woman who had remained loyal to his uncle and who managed to have him deported on an American frigate instead of guillotined with others among the old establishment.

An English corsair intercepted the ship on the high seas. Its crew commandeered what personal possessions the refugees had been able to save. In early August 1793, Palisot arrived again in Philadelphia, this time with the clothes on his back, a small trunk, ten francs, and his freemason's diploma. There he found himself ineligible for assistance from the French legation for the reason that in Paris his name had meanwhile been placed on the list of emigrés. He would thus have been subject to arrest and execution should he have attempted to return to France.

Among his associates in the Philosophical Society, Caspar Wistar was the one who lent Palisot a hand and, presumably, money. He took lodgings at 199 Sassafras (now Race) Street and supported himself by giving

French lessons and playing the bassoon in a music hall. Though possessed of important and lucrative properties, he had received no funds from home since departing Paris in 1786. The archives of the Academy of Science contain copies of the correspondence he addressed, mainly to Jussieu, from Philadelphia. In the early letters, desperate for money and persuaded that his wife and his agent were at best irresponsible and at worst dishonest, he importuned his older colleague and mentor to assume oversight of the administration of his estates. In that file is a series of certificates, renewed periodically, evidence of how long was the arm of the French bureaucracy, how invariant its procedures throughout all changes of regime. For example:

> I, the undersigned, Jean-Baptiste Petry, Consul of the French Republic in Philadelphia, State of Pennsylvania, on the attestation of the following citizens of the United States, David NASY, physician and resident of that city, James VANUXEM, merchant and resident of the same, and Pierre R. DU PONCEAU, merchant and resident of the same, whom I declare that I know well,
>
> Certify that Citizen Ambroise Marie François Joseph Beauvois, native of Arras, appeared before me today, that he has resided in this city from 5 August 1793 (Old Style) until the present without interruption.
>
> Characteristics of Citizen BEAUVOIS
>
> Age 38 years, height 5 feet 2½ inches, black hair and eyebrows, black eyes, acquiline nose, small mouth, indented chin, round forehead, round face.
>
> Done at Philadelphia this 30th day of Brumaire Third Year of the French Republic One and Indivisible, this Certificate having been signed by the Petitioner and myself, the aforesaid Consul, in the presence of the Witnesses.
>
> Beauvois, Nasy, James Vanuxem, Peter E. Duponceau, Petry, and Beauvarlet.

Not for lack of a documented residence was Palisot going to be denied his rights to his property, unjustly sequestered, and remittance of his income.

The opportunity to work on Peale's Catalogue had permitted him to devote himself entirely to natural history.[6] In May 1795 he was able to explore western Pennsylvania, and on 2 July he wrote Jussieu proposing that the Museum of Natural History in Paris commission him to do a thorough survey of both the natural history and population of America.

His situation had further improved with the arrival in June 1795 of P. A. Adet, first Minister Plenipotentiary since the removal of Genet in the autumn of 1793.

A chemist in his spare time and a physician by training, Adet was sympathetic to science. In the spring of 1796 he undertook to defray Palisot's expenses for an expedition to the Carolinas and Georgia. If not in seventh heaven, Palisot was at least happy to be employed usefully in the service of the Republic. His tone about the United States ("this country about which so many good things have been falsely said," he had earlier written) now changed for the better, as it also did about "my dear wife," whom he now talks of bringing out to join him. Word had reached him in Charleston that he was free to return to France, but he decided that he could not abandon his work in the middle of an expedition.[7] He met André Michaux in Charleston and enjoyed life there and in Savannah, although he was a little disappointed in the low country of Georgia and South Carolina. Palisot's command of English may be judged from a letter to Caspar Wistar reporting that the soil was nothing but sand a foot or so deep. Beneath that was

> very light mine of iron which seems to me very near that call'd by Bomare *pierre de Périgord.* That mine is not interesting and emploied only in buildings for the foundations. . . . I suppose I shall be more happy among the mountains where it is said never a man of knowledge has been and where are very rich minerals of different kind.[8]

Palisot did indeed travel to the region of the Creeks and Cherokees in the western part of South Carolina, and spent some weeks among them observing the institutions, the agriculture, the medicinal practices, and the role of gender. Back in Philadelphia, he planned a more extensive journey through Kentucky and Tennessee to the Mississippi and beyond into the land of the "Akansas." Alas, the stresses of the twelve years since he had left France caught up with him in the Blue Ridge mountains of Virginia. An incapacitating hemorrhoidal attack forced him to recognize that he was no longer young and to renounce the journey he had begun.[9]

He stopped in Richmond on the way to Philadelphia, where he would finally embark for France. There he wrote his farewell (in French) to Thomas Jefferson, giving an account of ". . . the final trip, which I have just completed . . . during which I was compensated for the difficulties and obstacles I encountered by the spectacle, always enchanting for me, of the beautiful sights of every sort that nature placed before my eyes, ever eager to behold them."[10] A final blow awaited his departure. The ship transporting his American collections went down off Halifax. Of the three important collections he had made, in West Africa, in Saint-Domingue, and in the United States, the only items that ever reached Paris were the relatively few specimens he had sent to Jussieu from Africa and the seeds from America he had enclosed with correspondence to Jussieu and Thouin. Most of them failed to germinate.

In all his travels, his grand design was to compile a comprehensive natural history of the country he was exploring complemented by an account of its inhabitants. Back in Paris by the end of 1798, he divorced his wife, re-married, and took lodgings in the Marais at 511, rue du Parc. In 1796 Jussieu and Thouin had arranged his nomination as non-resident associate in The Institute of France. On 17 November 1806, he was elected to full membership in the section of botany and thereafter participated regularly in the proceedings. For a time he gave a course of lectures on natural history at the *Athenée des Étrangers.* He never again traveled farther than to a country retreat near Paris at Plessis-Piquet, settled down to accomplish what he could of his grand design, and died on 21 January 1820.[11]

The African materials that Jussieu had preserved throughout Palisot's twelve years of absence formed the basis of the two-volume folio, *Flore d'Oware et de Bénin en Afrique* and of most of a companion folio *Insectes recueillis en Afrique et en Amérique, dans les royaumes d'Oware et Bénin, à Saint-Domingue et dans les États-Unis pendant les années 1786–1797.* Both of these works were printed off and put on sale in fascicles of six plates each beginning in 1805 (although the title page of *Flore* bears the date 1804) and continuing irregularly until 1821, the year following Palisot's death. A note in the preface to *Insectes* says that, though sold separately, it forms part of *Flore* and that "the two are intimately linked with the Account of My Travels which is in press." No such book ever appeared, although the descriptions of plants in *Flore* contain many anecdotes of the circumstances in which they had been collected. *Flore* contains 20 fascicles and the *Insectes* 15 of a promised 30. Because of the mode of publication, it is extremely rare to find a complete set of either title. The two are a bibliographer's nightmare. For example, bound into the incomplete second volume of the copy of the *Flore* that Palisot presented to the American Philosophical Society is one fascicle of the *Insectes.*[12]

The complete *Flore,* when it can be found, contains 120 plates, each illustrating a genus, accompanied by a discussion of the plant and of its classification. A number of the plants had indeed been unknown to European botanists, but it does not appear that either the *Flore* or the *Insectes* contributed anything intrinsic to the development of systematics in botany or entomology. The drawings are well done. Those in the first fascicle were executed by Mirbel, director of the Empress Josephine's garden at Malmaison, and the remainder by Sophie Luigné, known for her illustrations in the *Annales* of the Museum. For each genus Palisot gives the Linnaean name, followed by the family in both the Jussieu and Ventenat systems. The generic characters are described in both French and Latin. A paragraph or two of informal discussion usually follows. Commenting on the annual cycle of *Myrianthus,* for example, Palisot excuses himself for using the old style calendar names in the tropics: "How can the words *frimaire, nivôse* be employed with respect to a country in which no one has ever seen either frost or snow?" Occasionally he will urge that the characters of a given plant require creation of a new genus, or modification of an accepted one. He is proudest of a bush that Jussieu agrees is transitional between the *Passiflores* and the *Cucurbitacées,* and pleased that the Emperor has graciously consented to his naming it *Napoleona Imperialis.*[13]

Other writings were related, not to his travels, but to a theoretical *idée fixe.* The Linnaean term cryptogamia—"hidden nuptials"—for the lower plants referred to the difficulty of distinguishing their organs of fructification. Johann Hedwig, a botanist working in Leipzig in the 1770s and 1780s, had established that, in the case of mosses, the male organs are the tiny antheridia (not as was previously thought the spores) producing gametes that fertilize the archegonia. Palisot had repeatedly attacked this finding. From his earliest youth as a botanist, he insisted on attributing fecundation to the fertilization of what he believed to be tiny seeds by the greenish dust brushed over them by the hairs surrounding the urn structures. He never persuaded anyone of the validity of this proposition. With original people it sometimes happens, observed Cuvier in his Éloge, "that an initial idea in which they took pleasure thereafter guides them in all their research and even in the formation of their systems. It reproduces itself in many forms throughout their works and, for lack of experiments and facts, they show great ingenuity in framing hypotheses to support it. That is what happened to Monsieur de Beauvois."[14]

In 1805 Palisot published *Prodrome des cinquième et sixième familles de l'AETHÉOGAMIE: Les Mousses et les Lycopodes.* He had long since completed a classification of these two families of the Linnaean cryptogamia (the other five being algae, mushrooms, lichens, hepatica, and ferns). His reason for publishing now is his disagreement with Hedwig's method, which was signaled by the main word in the title. He would replace "cryptogamie," he explained, with "Aethéogamie," coined from the Greek for *insolitae nuptiae* (extraordinary nuptials). The term failed to catch on, and in 1814 Palisot returned to the charge in a paper read before the Institute of France on 27 June.[15] He had copies bound in with a final work, on grasses, *Essai d'une nouvelle Agrostographie, ou nouveaux Genres de Graminées*. A modern specialist has judged this publication, an annotated atlas, with some severity, considering it "from the standpoint of the nomenclature of grasses, a very important work, its importance being due principally to its numerous errors, less so because of its scientific value."[16] Evidently, the author of that stricture did, nevertheless, find it worthwhile to collate Palisot's terminology with current usage over a century later.

Palisot has been credited, not altogether plausibly, with authorship of an anonymous red Indian pastoral, *Odéhari, Histoire américaine; contenant une peinture fidelle des moeurs des habitans de l'intérieur de l'Amérique septentrionale*. The title page is dated thermidor an IX (1801) and also bears the epigraph "Odéhari is the older sister of Atala." The theme does indeed recall Chateaubriand, though the style does not.[17] Neither theme nor style, however, is characteristic of Palisot's writings on the customs of peoples he had dwelt among. His views on race were extremely dire, for one thing.[18] True, the Creeks and Cherokees come off better than the people of Bénin in his accounts of their respective practices.[19] His descriptions are nothing if not graphic, and Palisot did join the Society of Observers of Man in Paris. But these writings belong to the prehistory of anthropology rather than to its early stages. He treats the societies he describes as object lessons in the appalling consequences of ignorance, savagery, fanaticism, superstition, and despotism.

As for the Americans, Palisot's reactions are those of the Frenchman rather than the naturalist. They may be compared to the observations of Volney, equally disaffected from the Jacobin republic at home, equally disenchanted to discover that the Americans, despite the debt of independence they owed to France, obstinately persisted in the ways they had learned with their mother tongue. It would appear that Palisot projected a work on the scale of Volney's *Tableau du Climat et du Sol des États-Unis*. (which itself remained unfinished).[20] His papers contain sheets of jottings on the physical geography of the United States. The paragraphs that follow do not constitute a coherent essay. Palisot evidently made several starts on the work he imagined, and re-wrote certain passages in a number of drafts. I have taken the liberty of piecing them together as coherently as may be, indicating by asterisks the breaks between the sequences. The asides which appear here as footnotes are written in the margins. His handwriting was crabbed at best. Where I am doubtful of having deciphered a word correctly, I have put it in square brackets.

On the American People

The Americans—Is this a term that suits the whole mass of people who inhabit the territory of the United States? Can this general denomination be applied to a congeries consisting of some natives of the country and of collections of immigrants who have constantly followed each other and who still do, either through voluntary emigration or through group emigration of the Irish and Germans organized by profiteers?[21] Can the term, in short, apply to this bizarre mixture of men from every nation, some of whom do finally establish themselves in the United States, others of whom abandon the country after having made a fortune there, or after having spent and consumed their small assets, but all of whom carry with them their respective prejudices, their taste, their values, their habits, and the vices rather than the virtues of the places where they were born and brought up?

The consequence of so mixed a population is that nothing which can properly be called a national character exists among the inhabitants of such a state, or, to express myself in a less abrupt and more correct manner, that this character is not yet generally evident, nor dominant. That naturally gives rise to the following questions, that I shall not undertake to decide, but concerning which let us consider the import of the end of the administration of George I [George Washington] followed by the election of John I [John Adams].

As for the rapid growth of population in the United States (accelerated as much by emigration from all the nations of Europe as by the number of births, which exceeds that of deaths), will it prove a solid and lasting benefit to the country and to its inhabitants?

Will not that compromise, one day, their tranquillity, the liberty they enjoy and the flourishing appearance they have taken on since the French Revolution, and specially since that of the colonies?

Will not, finally, this republic end up, as it has begun, by being influenced to an exasperating degree by the English government, which, sooner or later, when its commercial interests will no longer be jeopardized, will find the way to avenge itself, not so much for the damage caused it by the insurrection of the United States, for this separation has proved to bring with it advantages of a more real sort, but for the humiliation of having lost and of having been forced to negotiate and to grant peace to those it called rebels. This reflection, to which the well known pretentions of as proud a nation as England naturally give rise, should give pause to the true American. It should lead him to keep up his guard constantly, to beware of a nation as proud as it is full of resentment and eager to avenge itself; but, blinded by language; by the customs, the tastes, the habits of behavior, which are the same, in which he was brought up, and which he has imbibed with his mother's milk, so to say; attached by preference to the English products he has always used, and not knowing any others—all these anglo-dispositions draw him insensibly, without his noticing, towards that nation, whence it results that it is preponderant in the United States. It exerts an influence that only time, experience, and an enlightened administration can attenuate.

Whatever degree of enlightenment a people may have reached, and if we are to believe the session of Congress in 1796, that of the United States is dazzling; however enlightened it may be, it is not always proof against that domination by the secret influence which derives from similarity of language, customs, tastes, and habits that it has taken from another nation, especially when descended from it. The only means of dissipating certain of the prejudices in which a whole people may be plunged is a wise, enlightened, and thoughtful philosophy. Such a change, much to be desired for the happiness of the United States, can occur only with the aid of time, of experience, and of an enlightenment they still lack. When I speak of lacking enlightenment, I am far from arguing that there are no enlightened people in the United States. If I were to make such a statement, and were guilty of such unfairness, my pen would be stopped in its tracks by the reputations of Franklin, of Rittenhouse, of Jefferson, of Madison—I leave aside John Adams himself—and of an infinity of others whom it would be too lengthy to enumerate. But I am considering the nation as a whole. I see there a young nation, a new nation, which proclaims the highest standards, which is already more enlightened than any people has ever been at the same infant age. But the total fund of its enlightenment is and can only be relative to its political seniority, and I think that it would be as ridiculous to pretend that the United States is as enlightened as other countries as it is false and unjust to argue, as the abbé Raynal has done, that America had not yet produced any man of genius.

This false assertion by the abbé Raynal has provoked a fair and elegant reply by Mr. Jefferson in which he enumerates certain great men whom America has produced. The only thing missing to complement it is his own name.

The Anglo-Americans have consistently been judged with the most extreme enthusiasm since their Revolution and separation from the mother country. The bold and courageous initiative of the colonists, their firmness, their unwavering conduct, and their bravery in defending the liberty they were determined to win—all this was no doubt bound to excite and sustain admiration. Added to these grounds for enthusiasm were the accounts of certain informed persons who were on the scene. These reports were followed by enthusiastic books, and sympathetic sentiments reached their peak. Since that time the United States is pictured as the ideal country. Americans are the best balanced of men. Endowed with all the virtues, all the vices are alien to them, and they enjoy the sweetest and purest happiness under heaven, which men have ever and vainly sought and which has descended upon the regenerate Anglo-Americans. Such was the opinion generally held of that nation where everyone yearned to be, and where the emigrant Scioto and [Clinch River] companies were formed. All who set out for these destinations were in transports ahead of time over the happiness and riches that awaited them. None has returned, after having consumed the little he still possessed, without cursing those who so wrongfully deceived them.

Such is almost always the effect of enthusiasm. It rarely accords with reason. The product of impulse, of thoughtless exaltation, enthusiasm and the expectations it normally gives rise to, always fall far short of, or much exceed, reality. The one and the other, when they are not simply shielding error, are to the mind and understanding what the magnifying lens of a microscope is to physics and to the eye of an observer. It is through such lenses that the Americans have been seen by almost all the observers who have attempted to depict them. Viewed beside the portraits drawn by the author of Letters of an American Farmer, the men whom he has there painted to appear so wise and so happy, are to our eyes nothing more than real shepherds, gross, ignorant, and insignificant compared to the elegant, intelligent, and happy shepherds of the poets of antiquity. If the portraits on which the author of New Travels in the United States himself lavishes such praise are also set alongside the truth, you will recognize nothing but exaggeration and ecstasy, and all his colossal and superhuman figures will appear to be very ordinary. The excessive praise lavished upon the Americans, these inflated and unnatural fables, have already begun and will increasingly prove altogether prejudicial to them. One gets to know them better every day. Every day these sensible and impartial men, devoid of all pretention, are coming more fully into their own and can in no way sustain the parallel between what they are and what they have been depicted to be.

—∞—

If I wish to offer my opinion today on this nation, it is not that I believe myself to be more enlightened than those who have preceded me. It is possible that I may [be partly in error]. It's not as they really are that I claim to present them (the Americans), but as they have appeared to me. If I have made mistakes, it's in good faith, but I think, and am vain enough to believe, that I am not mistaken since I know that not all the French Ministers have written like Brissot and that the three most recent, less politicized, have unmasked them more honestly.

In advancing the proposition that up till now the Anglo-Americans have been judged and described too favorably, as something other than they are, I do not at all wish to go to the opposite extreme. If they do not deserve the ecstatic admiration born of all they stand for in principle, and which continues to be fashionable, for at the time fashion was everything, neither do I think it would be fair to them to refuse to recognize the qualities they possess.

—∞—

I have already said, I have even begun to prove, that the American people do not and could not exhibit a truly national character. What I still have to say will complete the demonstration that they are almost completely English.

There is no need to observe how difficult, not to say impossible, it is to portray a people who as a whole have no character of their own and who have borrowed everything. I shall, however, attempt to represent them as I have seen them. The inhabitant of the cities must be distinguished from the inhabitant of the country, and among the inhabitants of cities, the rich must be distinguished from the poor. The outlook is not at all the same in these different conditions. Thus I shall often have occasion to portray the American of one sort as very different from one of another. The inhabitants of South Carolina, for example, of Georgia, and even of North Carolina are good, gentle, honest, hospitable, and of very good faith compared to those of Virginia, Maryland, Pennsylvania, &c., as far as

New York. I do not speak of those farther north, whom I have not
seen.

—⚹—

From what I have just said, it will be evident that distinctions are to
be made among the people of the United States, and it seems important to me to define them. I divide them, then, into three classes,
as follows:

The true American

The americanized immigrant.

The English, German, and French merchants, to whom I add the
emigrants from Ireland, Germany, and all other countries.

THE TRUE AMERICAN[22]

I designate under this denomination all those who, born in America, have imbibed principles and prejudices different from those
that their fathers brought into the country, although the latter cannot have been entirely dissipated in the first generation.

In this class, I distinguish, 1st: the independent American, more
or less prosperous, and free of all ambition.

2nd: The American who is ambitious, either for wealth, or for
prestigious positions at the disposal of the executive power, such as
commercial representatives, judges, ministers, &c.

The first sort are not numerous in the cities and are to be met
with there only among the less wealthy class, people dependent in
one way or another on the rich, the merchants, and the high officials. The inhabitants of the country, on the contrary, are independent. They are proprietors of a plantation, either as owners, or as
tenants (who are fewer in number), free and holding none but
sound opinions. These last are deeply attached to their country.
They, who fought the English directly, whose plantations were
pillaged and burned, whose wives and children were persecuted,
detest the English profoundly and are strongly attached to their fa-
therland, to the government which maintains its independence,
and to the French whom they know, without avowing it, to be the
supporters and defenders of their liberty.

The second sort, who live in the cities and urban regions, inces-
santly preoccupied with their wealth or status, indulge themselves
more eagerly and freely in the pursuit of pleasure and in their prej-

udices. They are all English and exploit their domination of the poor and the workers to augment their own situation.

Such is the result of my observations in the course of traveling a distance of over 500 leagues throughout the country, both along the Coast and in the interior. The country dweller, who has so to say no need of English products and no interest in trade with that nation, rather than with any other country, such a one relies for food only on the grain he sows and harvests and the pig he raises and fattens. His clothing consists basically of shirts and hose made of the cotton he sows and harvests, which the family spins and makes for their own use. In the north they are dressed in clothes of flax and linen, equally the product of their own hands. Their shoes are made of the hides of their cattle and calves, which they supply to the shoemaker. They pay for the fabrication in kind, rarely in money. All that remains are their waistcoat, suit, and overcoat, which they are obliged to buy. But it matters little to them which country the material comes from provided the price suits them. The women are almost always entirely dressed in cotton cloth which is woven in the region and which they pay for by the exchange of grain, cotton, &c.

It is not the same with the americanized.

THE AMERICANIZED

In this class of men living in North America, I include emigrants from all nations who, after living in the country for a time, have made more or less of a fortune and established themselves there, though without having entirely lost or renounced their national prejudice. I consider in this class only those who continue to be in commerce and who are resident in the cities, and not those who, content with having achieved a comfortable competence, have retired to the country where they live in peace and tranquillity.

This 2nd class of men, dwelling in cities, have become attached to the federal government by reason of their associations with Americans, but are always inclined to favor their native country and language. The Irish, Scots, and English are much more numerous than the French and Germans combined. It follows that the English party and system draw more support from them in proportion to the importance they attach to trade and an alliance with England. What is more, the Germans in the cities also favor the English.

THE FOREIGNERS

I include in this class only foreigners who are permanent residents, without adding in the French, German, and other emigrants who have come in search of temporary asylum. But this number of English and Scots, dispersed in all the cities of the United States, those of the interior as well as the ports, set up trading houses, and engage in all sorts of speculations and ventures, and then after a certain number of years return home with the profits. This class of men is the most dangerous for the United States and, furthermore, is the one which lends support and gives preponderance to the influence that England exerts there. In order to give an idea of the number of foreigners of this sort, I shall cite the single small city of Augusta. It contains five or six commercial houses, almost all consisting of merchants of whom more than 19 in twenty are Irish, Scotch, or English. There is, however, in this city, as almost everywhere in Georgia, a strong party of patriots, but in important cases, where the interests of England are weighed against those of France, I doubt that it is really the stronger, even though it may appear so. The foreigners do not come into the open. They act secretly and ruthlessly.

—🙢—

There breaks off Palisot's draft of a classification of Americans. After his return to Paris he sent inscribed copies of his major works and offprints of a number of his articles to the American Philosophical Society.[23] He also remained in touch, intermittently at least, with certain of his former associates. On 14 July 1802 Charles Willson Peale wrote acknowledging receipt of a case of preserved birds for his Museum from the *Muséum d'Histoire Naturelle* in Paris. His sons Rembrandt and Rubens, Peale continues, were on their way to Europe with the skeleton of the mammoth recently discovered in New York State. They planned to exhibit it in a number of European cities and had been promised they should make a handsome fortune. Peale hopes they will also improve their knowledge, and they are looking forward to renewing the family friendship with Palisot. The Museum grows apace, and is to be moved to the State House (Independence Hall) that autumn. He feared that the printing of the catalogue on which they had worked together never would be finished. He had too much else to do. "My young family are fast progressing to manhood, and Mrs. Peale is grown more fleshy. . . . I hope you have recovered your Estate," Peale concluded, "and enjoy every comfort which your favorite country so eminently possesses."[24]

Notes and References

1. Frances S. Childs, *French Refugee Life in the United States, 1790–1800* (Baltimore: The Johns Hopkins University Press, 1940), with an extensive bibliography. See also an unpublished senior thesis at Princeton University, Marc de Lapérouse, "The Philosophes in the Young Republic: a Study of French Emigrés in the United States, 1790–1798 (1977)," in which there is a section on Palisot de Beauvois in Philadelphia, pp. 118–132.

2. This article was composed at the request of French colleagues for inclusion in a collection of essays in honor of Jacques Roger on the occasion of his retirement from the University of Paris in December 1989. Professor Roger, historian of both literature and science, was a frequent visitor to the United States. He lectured at many American universities, and for the last five years of his career spent one semester annually at the University of Virginia. Jacques Roger died suddenly in March 1990, a few months after publication of his fine biography of Buffon. The original version of the texts I have here translated is to appear in what will now be a commemorative volume to be published by Klinckseick.

Palisot's principal biographer was Arsenne Thiébaut de Berneaud, *Éloge historique de A. M. F. J. Palisot de Beauvois* (Arras, 1821), prepared for the Société Royale pour l'Encouragement des Sciences, des Lettres, et des Arts d'Arras, and published in its *Mémoires* 4 (1821), 49–116, as well as separately. There is also an Éloge by Cuvier, *Mémoires de l'Académie des Sciences de l'Institut de France* 4 (1824), cccxviii–ccclxvi. Materials on which it is based are in the Fonds Cuvier, Institut de France. A considerable file of Palisot's manuscripts is in the Bibliothèque de l'État in Mons. They are catalogued in Paul Faider, *Catalogue des manuscrits de la Bibliothèque Publique de la Ville de Mons* (Gand, 1931), cotes 1060–1066. The director has kindly given permission for publication of the passages printed here. They have been transcribed from a microfilm copy in the library of Princeton University. There is a duplicate in the Library of the American Philosophical Society.

3. Charles C. Sellers, *Mr. Peale's Museum* (New York, 1982); also Toby A. Appel, "Science, Popular Culture, and Profits: Peale's Philadelphia Museum," *Journal of the Society for the Bibliography of Natural History* 9, Pt. 4 (April, 1980): 619–634.

4. Palisot de Beauvois to A. L. de Jussieu, 24 nivôse an 4. Dossier Palisot de Beauvois, Archives de l'Académie des Sciences.

5. On the subject of natural history in Saint-Domingue, I have had the benefit of consulting a work by James E. McLellan III, *Science and Colonialism in the Old Regime: the Case of French Saint-Domingue,* published by Johns Hopkins University Press.

6. Palisot de Beauvois to Jussieu, 21 nivose, an 4 (11 January 1796). Archives de l'Académie des Sciences, dossier Palisot de Beauvois.

7. Palisot to (probably) Thouin, 16 pluviose, an V (4 February 1797), Archives de l'Académie des Sciences, Dossier de Palisot de Beauvois.

8. Palisot de Beauvois to Wistar, 20 May 1796, American Philosophical Society MSS. B/W76.

9. Palisot de Beauvois to Jussieu, 27 frimaire an 6 (17 December 1797), Archives de l'Académie des Sciences, dossier Palisot de Beauvois.

10. Palisot de Beauvois to Thomas Jefferson, 25 April 1798, American Philosophical Society, MSS. B/W76.

11. E. D. Merrill gives a summary and bibliography of his scientific publications, "Palisot de Beauvois as an Overlooked American Botanist," *Proceedings of the American Philosophical Society 76* (1936): 899–920.

12. E. D. Merrill gives an inventory of the contents of the *Flore*, Ibid., 914–920, and Francis J. Griffin, "A note on Palisot de Beauvois' *Insectes recueillis en Afrique et Amérique*," *Annals and Magazine of Natural History* 10 (1932): 10, 585–588; and "A Further Note on Palisot de Beauvois . . ." *Journal of the Society of Bibliography of Natural History* 1, Pt. 4, (1937): 121–122.

13. *Flore d'Oware et de Bénin*, 2, 29–32, Pl. LXXVIII.

14. Cuvier, Mémoires de l'Académie, cccxxiii.

15. "Lettre de M. Palisot de Beauvois à M. Delamétherie . . . suivie de quelques Réflexions sur les organes de la fructification des mousses," *Journal de Physique 76* (1814), 5–15.

16. Cornelia D. Niles, "A Bibliographic Study of Beauvois' Agrostrographie," Smithsonian Institution, *Contributions from the U.S. National Herbarium 24* (1931).

17. Gilbert Chinard came upon a copy of this forgotten work in the Newberry Library in Chicago in 1911. He discussed the comparison with Chateaubriand in his *L'Exotisme américain dans l'oeuvre de Chateaubriand* (Paris, 1918). Victor Giraud, meanwhile, found an earlier version printed in a 1795 collection called *Veillées américaines*, the preface of which is signed with the initials PB. That in turn led Paul Hazard to make the attribution to Palisot de Beauvois. Chinard remained skeptical and sets forth his reservations at length in the introduction to an edition of *Odéhari* reprinted in 1950 (Paris: Raymond Clevreuil).

18. A long polemic, *Réfutation d'un écrit anonyme intitulé: Resumé et témoignage touchant la traite des nègres, adressé aux différentes puissances de la chrétienté* (Paris, 1814) argues against the precipitous abolition of the slave trade. Palisot considered that Africans were better off as slaves than as subjects of the despotic potentates in West Africa.

19. "Notice sur le peuple de Bénin," lue à la séance publique de l'Institut du 15 nivôse an ix (1801), *Décade philosophique* 30 fructidor an ix, 513–518; "Extrait d'un Voyage chez les Creecks et chez les Chérokées," lu à l'Institut dans la séance publique du 15 messidor an ix (1801), *Décade philosophique* 20 messidor an ix, 94–103. Chinard has reprinted the latter in his 1950 edition of *Odéhari*, 221–230. There is a draft of a piece on the people of Owara among Palisot's papers (n. 2 above).

20. Paris, 1803.

21. "Certain merchants, for whom any means of making money[a] are good and legitimate provided they are sure, have evidently seized on this new branch of trade, which the labor shortage in the United States opens for them, and which England has condoned, as much in order to populate her colonies as to rid herself of the poor, the idle, and those incapable of paying the high taxes that the needs of government impose on the people. Consequently, they travel to Ireland carrying American products which they sell at long term, then they assemble poor people whose debts they pay. Some are families who like the idea of a change and hope to make a fortune in a far country always pictured in rosy hues. Some are bad characters of both sexes delighted

to find a way to escape the shame they have brought on themselves by their conduct in their own country, &c., &c., &c. The return cargos are composed of this collection of the wretched and the unfortunate, with whom America is being populated. It then offers a curious spectacle to see the dissipated rake, the fanatical Quaker, the possessed Methodist, the rigorous Presbyterian, the intolerant Catholic, &c., boarding these vessels to seek among the cargo a young and pretty servant whom he wants to make his mistress. The others go along with mortgaging the liberty of these unfortunate creatures, obliged to bind themselves to certain associations in order to be able to pay the villain who sells them for their passage. Thus it is that in the much vaunted land of liberty, the arrogant hypocrite publicly frees his slaves (always a calculated act) while bargaining away for the time being the liberty of whites.

(a) God forbid that I should impugn all trade and traders. No one is more aware than I of the necessity of trade, and no one feels more respect for honest merchants. But in every hundred, how many who are honest and respectable can be called to mind? I have known a few such in every country, but the number is so small, that they are the exception.

22. The American has been brought up to share English prejudices against the French. So long as he remains English, so long as he does not wish to be American, so long as he does not rid himself of the prejudices of taste &c. that he takes from England, so long as has not acquired enough knowledge to know how to recognize and appreciate the French at their true value; so long as, like the English, he does not [renounce] superiority over everyone else; and so long as he thinks that, what he does not know, nobody knows, he will still be influenced by England without his being aware of it.

23. For example, a report of a very nice experiment in vegetable physiology, "Notice sur une nouvelle expérience relative à l'écorce des arbres," Lue à . . . l'Institut le 5 août 1811, *Journal de Physique* 73 (1811), I, 209–212.

24. Charles Willson Peale Letterbook, Library of the American Philosophical Society, B/P31–3.

PART III

General

—m—

8

Science and the Literary Imagination:
Voltaire and Goethe

THIS ESSAY WAS COMMISSIONED in the later 1960s by a letter from Anthony Thorlby, one of the editors of the collection *Literature and Western Civilization* cited in the note below. The other, and I gathered senior, editor was David Daiches. I no longer have the file of correspondence, but recall that one or both of them had come upon *The Edge of Objectivity* (1960) and had liked the chapter on Science and the Enlightenment. The essay that follows is an enlargement on the treatment of Voltaire and Goethe, two of the three principal writers—the third is Diderot—who figure in that chapter with respect to their engagement with the science of their time.

My impression was that this essay was not what they expected and that they did not like it much. At least Thorlby, who conducted the correspondence from their side, never said they did. Looking through the collection now, I suspect that they had in mind something more like Marjorie Nicolson's well-known *Newton Demands the Muse*[1] or the elegant chapter that David Daiches contributed to the next volume, "Literature and Science in 19th-Century England."[2]

In fact he there ranges right back to Newton and draws upon an encyclopedic knowledge and tempered judgment first of poets and then of such prose writers as Huxley, Arnold, Butler, and Shaw. For the poets he exhibits the poetic inspiration and for the pundits their cultural response to the findings of contemporary science, principally those of Newton and Darwin. He does see, as do I, a hostility to science or at best a wariness setting in with the romantic movement in the late eighteenth century. His interest is in the poetic imagery itself, whether in Pope or Blake, and in the moral attitudes, whether of Huxley or Arnold, and not at all in the cogency of the readings of science and its findings, either favorable or adverse. In brief, Daiches's concern is with the literature, not the science. I was and am insufficiently versed in the entire literature of the Enlightenment to write such a piece, and my interest was both in the writers and the cogency of their readings of science.

At all events, I have never seen a citation to my chapter, and only one person has ever mentioned it. I doubt that the collection as a whole succeeded very well. The volumes have been checked out of the Princeton University Library only once or twice in the last thirty years. It appeared on the verge of the transition between literary scholarship and literary theory, and would have been ignored as pedantic in the latter phase. Readers might do well to turn to it now. Among the contributors are eminent scholars, and I find a sampling full of interest.

Notes

1. Princeton: Princeton University Press, 1936.
2. Vol. 3, part 2, pp. 441–460.

EIGHT

Science and the Literary Imagination: Voltaire and Goethe*

—ω—

I

It has become a convention of the history of ideas that, following the triumph of Newtonian physics in the seventeenth century, thought, letters, and science came together in a common movement of culture that distinguished the eighteenth century, the Age of Enlightenment. Did not Voltaire's *Élémens de la philosophie de Neuton* establish Newton in the French-reading continental world, and the foremost French writer thereby participate in the work of science? Did not Goethe, the greatest of German writers, augment the knowledge of anatomy with an actual discovery, enrich the study of botany with profound morphological insights, and anticipate the complement that the psychology of perception would bring to the science of colour only in our own day? The argument of the present essay will be that on the whole they did not, and that deep and important though the interactions of science and the literary imagination have been and are, their actuality is not to be seized by taking too seriously the forays that writers felt freer in the eighteenth century than in more recent times to make into scientific subjects, nor even through exhibiting the deployment of Newtonian imagery in the poetry and prose of their proper writings, but rather through considering their scientific sensibility as a function of their purposes in literature itself.

---

* Chapter 6 of *The Modern World,* vol. 3, part 1 of the collection *Literature and Western Civilization,* ed. David Daiches and Anthony Thorlby, (4 vols. London: Aldus, 1972), pp. 167–194.

II

It will be best to begin with each of our authors by being explicit about what may be attributed to him in point of science. Voltaire (1694–1778) published the first edition of his major scientific effort, the *Élémens de la philosophie de Neuton,* in 1738.[1] Let us reverse the usual procedure of historians and review the content before the circumstances. A reading of that work in conjunction with Newton's *Principia mathematica* (1687)— for most readers an inspection of the latter would be a more realistic undertaking and would suffice—will show that, whatever else Voltaire conveyed to the public, it was not knowledge of classical mechanics, either in its specific problems or in its mathematical mode. The account of Newton's optical work, the second great branch of his physics, is somewhat more adequate because the *Opticks* (1704) is not mathematical; but that does not qualify the judgment that the purpose of Voltaire's book can only have been quite other than professional. Not that Voltaire misrepresented his vein or misled his public: the title specifies philosophy rather than natural philosophy, and he promised the reader merely a verbal description of the laws of nature that Newton had discovered. In doing so he must needs proceed as would a statesman who, in framing policy, relies on reports of facts that, lacking technical proficiency, he could never have worked up for himself. The reader who might need a physicist's account is referred to the manuals of 'sGravesande, Keill, Musschenbroek, and Pemberton, on which Voltaire himself relied more directly than he was equipped to do on Newton.[2]

A reader incautious enough to take his idea of Newton's own achievements directly from Voltaire's *Élémens* would form a peculiar and inaccurate view of their order and proportions. The *Élémens* is organized in three parts. Part I contains 10 chapters in 70 pages and is largely concerned with the metaphysical and theological issues that occupied the correspondence between Leibniz and Samuel Clarke, the published texts of which bulked larger in Voltaire's own reading and study than did the *Principia.*[3] To these matters Newton himself, in his properly scientific writings, had devoted the five pages required for the Preliminary Scholium on space and time in the first edition of the *Principia,* to which in the further two editions he added the concluding General Scholium on the relation of God to nature, and in the later editions of the *Opticks* certain remarks in the Queries, notably in the last of them, say another 20 pages altogether. (In this comparison we are concerned with Newton's public writings, not his private thoughts.) Part II of the *Élémens* is about the theory of light and colours. In 14 chapters and 141 pages it summarizes the results of the experimental investigations that Newton had set forth in the *Opticks,* a closely written book of experimental science of over 300 pages. Part III of the *Élémens* deals in 15 chapters and 140 pages with attraction, gravity, and cosmology, the subject-matter of the *Principia,* one of the most relentlessly mathematical treatises ever written, and one that contained a great many other topics.

The disproportions in Voltaire's coverage are even greater than would appear in such an overall summary. In Part II only four chapters—IX, X, XII, and XIII—explain Newton's actual work. The rest is a popularization of 17th-century optics in general, into which Voltaire imported certain views on perception dear to the sensationalist psychology following Locke. He made great play with the famous operation that a surgeon called Cheselden had performed at Chelsea Hospital in 1728 to restore the sight of a boy 14 years old whose eyes had been blinded from birth by cataracts. The lad was observed closely as he learned to use his eyesight. It appeared that the visual instinct is not innate, that at first he associated his vision with touch in a way that might have been taken as reminiscent of Descartes' model of the blind man tap, tap, tapping his way to perception with a stick or rigid body, which image had introduced the mechanistic model of vision as a sensation produced by pressure in a medium.[4] Voltaire's purpose being anti-Cartesian, he did not point out the analogy. The technical level of Voltaire's account is indicated in his exposition of the law of refraction. The reader is told that there is a constant proportion between the sines of the angles of incidence and of refraction, but is spared an explanation of what a sine actually is, because that would surpass the mathematical demands to be placed upon him.

Newton's optical discoveries consisted, broadly speaking, of two sorts of phenomena in the production of colours. The first take up Book I of the *Opticks* and had been the subject of the first paper he addressed to the Royal Society in 1672.[5] The experiments exhibited the dissociation of white light into colours of the spectral band in consequence of the differing degrees of refrangibility of the rays when light was passed through a prism of glass. The second set of phenomena occupy Book II of the *Opticks* and had been the subject of Newton's second, more speculative and controversial paper of 1675.[6] The observations exhibited what in later physics are called interference effects: colours of the sort perceived in soap bubbles, or in flakes of mica, or in Newton's experiments on the ring-shaped segment of air created by pressing a plane surface of optical glass against a slightly convex one. In Newton's own view the former experiments revealed that white light, far from being simple and elementary as was universally supposed, is in fact a mixture compounded of the several colours, whereas the latter phenomena were interesting primarily for the insight they afforded into the structure of matter. It is not clear that Voltaire fully seized the distinction that had been in Newton's mind. Whether he did nor not, his account of the latter experiments is much more adequate to the subject than the former, or indeed than any other aspect of Newton's physics, perhaps because the evidence for the interaction of light and matter involved the one among Newton's views that held the greatest appeal for Voltaire: the principle of attraction.

From the transition that Voltaire employed in moving from Part II to Part III of the *Élémens,* from colours to universal gravity, the reader would be led to suppose that the natural course scientifically was to extend the principle of attraction, this "new power," from optics to cosmology. In Newton's own physics the direction had been the reverse. In the *Principia* he had attempted to deduce the law of refraction from the principle of attraction.[7] Although the effects of gravity were indeed formulated in terms of a force of attraction, the two concepts were not simply interchangeable as Voltaire reported them to be. The most serious shortcoming of the *Élémens* taken as an account of Newton's actual work, however, is that it conveys virtually nothing of the science of mechanics. Newton began the *Principia* with the definitions of density, mass, inertia, and central forces, followed by the laws of motion. That is what the book is about:

> I offer this work as the mathematical principles of philosophy, for the whole burden of philosophy seems to consist in this—from the phenomena of motions to investigate the forces of nature, and then from these forces to demonstrate the other phenomena. . . . [8]

Its subject is mechanics, and Voltaire nowhere attempted to explain the laws of motion. Newton had taken care to instruct the reader about what was minimally necessary: the Definitions and Laws of Motion, the first three sections of Book I about motion in conic sections under central forces, and Book III on the System of the World.[9] Voltaire gave a qualitative sketch of the latter, but of the former only a précis of Proposition I.

It is merely a definition of proficiency, of course, and not a denigration of Voltaire, to observe that no layman could have understood the demonstrations of the *Principia,* for that kind of command entails not only the innate ability but, even more, the regular practice of posing and working similar and further problems. In the years between 1734 and 1738 or 1739 when—installed in the Château of Cirey in the far reaches of Champagne with the companion of his scientific years, Émilie, marquise du

Châtelet—Voltaire was grappling with the study of physics, they did not fail to enlist by correspondence the assistance of the professional scientists, notably Alexis Clairaut and Pierre de Maupertuis, who could reply to questions with professional authority. It is to be doubted that Voltaire ever actually worked his own way through many propositions in the *Principia,* or thought through its structure of axioms and assumptions. Whatever is omitted from his account is left out by design, however, and not through oversight, for it is clear that Madame du Châtelet, at once colleague and mistress, did come to know the full contents of the *Principia* in the course of preparing a French translation. Her rendering, published in 1759, ten years after her death, is acknowledged by Newton scholars to be superior in fidelity to any translation that has yet appeared in any language including English.[10] It is not possible to translate a piece of writing without reading it word by word and mastering the meaning at least sentence by sentence. Any historian of science who has himself done his duty by Newton's own text is bound to record in passing his respect for everyone else among the very small company, past or present, who have really read the *Principia,* a task from which Madame du Châtelet's devotion to their mutual enterprise of comprehension almost certainly dispensed Voltaire.

Differences developed between Voltaire and his mistress over the true measure of force, whether it is to be expressed as a function of momentum, as the strict Newtonians insisted, or of *vis-viva* (later, kinetic energy), as Leibniz had held. It indicates the greater flexibility of her grasp and the rigidity of Voltaire's *parti pris* against Leibniz that she could see the point on both sides. Their approach to Newton was a bookish one, obviously, but it was in accordance with Voltaire's views about physics that it should rest upon experiments. They installed a laboratory at Cirey, not to verify Newton, but to try their hands. They chose a favourite subject among amateurs of science in the eighteenth century, the nature of fire. Voltaire affected to consider, and perhaps persuaded himself, that their experiments revealed it to be a substance.[11] There must be hundreds, if not thousands, of such memoirs, published and unpublished, that survive from the attempts of enthusiasts and cranks to show that it was still possible in the eighteenth century to take a short cut from curiosity or ambition to discovery without passing through the discipline of science. If occasionally that did still happen, it was in natural history, not in physics or chemistry. The memoirs by Voltaire and Madame du Châtelet on light and fire made no detectable impression on those sciences. That those on fire were published by the Académie des Sciences in a volume of memoirs submitted to a prize contest announced for 1738 argues rather for the accessibility of that body to influence than for Voltaire's scientific talents having been those of a physicist *manqué.*[12]

Let us turn now from Voltaire's discussions of physics, in which it is clear that he was no physicist and only rarely posed as such, to his reason for writing on physics, in which it is equally clear that he saw the point of what it was that science held for humanity. The actual text of the opening part of Voltaire's *Élémens* will be surprising to those who would make the Enlightenment in its relation to science an affair of materialism and scepticism as understood in the nineteenth century. The book opens with a paraphrase of Newton's view that God is not merely creator but also lord of nature, and Voltaire quoted Newton's remark in the General Scholium to the effect that we say "My God" but not "My Eternity" or "My Infinity" because these attributes of the deity bear no proportion to the finite, created beings that we are. Now why did Voltaire begin like that, with the personal existence of God? For that is where Newton ended, in a passage added to the second and third editions of the *Principia* for the purpose of rebutting Leibniz's imputations of a tendency to infidelity inhering in Newtonian philosophy. There is no reason to doubt that Newton believed exactly what he wrote, for he was a devout man and had recorded similar views (without publishing them), when still a young man, in a private memoir against Cartesian identification of nature with God. There is no reason to believe either that Voltaire failed to understand what Newton meant, for he was an intelligent man, or that he accepted it in Newton's sense, for he was the reverse of devout, however consistent a deist.[13] Indeed, the gloss about the infinite bearing no proportion to man the finite is not Newton's. Voltaire may most probably have had it from Pascal, whose abdication of his own powers, and betrayal of them to the enemy, Voltaire otherwise deplored.

Was it tongue in cheek that Voltaire thus opened a book on science with the existence of God, intending an argument for the authorities and a certain irony for the readers? Maybe so, though the tone is not particularly ironic, and perhaps it would be more prudent to try the hypothesis that in a way Voltaire meant what he said. For even if he did not mean it quite as Newton did, yet he still needed Newton's declaration of the existence of God in order to assert what he really cared about, which was the freedom ostensibly of God but really of intelligence.

In the following chapters Voltaire turned the exposition against Leibniz, the enemy in metaphysics as Descartes had been in physics, and all the more urgently to be confounded, therefore. The subject was Newton's view that space and time are attributes of God, who, having been free to constitute them as he pleased, was not to be constrained by Leibniz's principle of sufficient reason, in service to which things had to be some certain way rather than another. It followed from Newtonian philosophy (according to Voltaire) that God is a being infinitely free, who has conferred a portion of that liberty on man in the form of free will. As for the problem of what becomes of liberty in the presence of God's foreknowledge of events, Voltaire avoided it as a labyrinth into which Newton had been too prudent to stray. To see what Voltaire superimposed on Newton is always instructive. In Newton there is nothing about free will in man, and no occasion to evade its reconciliation with divine omniscience. That is pure Voltaire. The remark in *Micromégas* (1752) about writing history was based on wide experience: "I am going to recount ingenuously just how it happened, without adding anything of my own, which is no small effort for an historian"[14]—an effort always beyond his own strength.

Two further theological chapters deal with natural religion and the problem of soul and body. In the former, Voltaire defended Newton as a supporter of natural religion. Such, indeed, Newton was, although it would be simplistic to imagine him motivated by theology to study physics. Voltaire, for his part, worked through a discussion of community in ants, bees, and savages toward the proposition that Newton had believed in natural religion for the sake of the welfare of society, a consideration that was surely as foreign to Newton as it was central to Voltaire. Turning to soul or mind and body, Voltaire said that Newton had thought consciousness to be beyond the scope of experimental philosophy, soul being no property of matter, which is all that experiment can touch.

Throughout the *Élémens de la philosophie de Neuton,* Descartes appears to be the obstacle in physics and Leibniz in metaphysics, and for the same reason. They both presumed to project their dogmas upon nature or God in the guise of necessities, whereas Newton had generalized his laws from phenomena, confirmed them by experiments, and restricted them within the scope of a description of nature, albeit a mathematical one. The question is one of dogma against fact, and therefore of freedom, for the enemy of freedom is dogma and never fact. Never mind that Voltaire was no theologian and no philosopher. He was acute enough at seeing the point at issue between philosophers. It would be difficult to better in one sentence his summary of the issues in the Leibniz-Clarke correspondence:

> "But there is no such thing as better or worse in things that are indifferent," say the Newtonians; "But there are no things that are indifferent," reply the Leibnizians.[15]

Never mind that what he was fathering on Newton was opinion, and not philosophy or science. It may be that one of the differences between philosophy and opinion is that philosophy is a product of coherent reflection on the deepest actualities, whereas opinion is programmatic and not thus constrained somehow to be faithful to itself. Never mind, therefore, that Voltaire's derivation of free will in man from free choice in a putative Newtonian deity is not only unconvincing but in itself uninteresting in a way that even the most unacceptable reasonings in a Descartes or a Leibniz never are. What is interesting is Voltaire's strategic insight into enlisting science in his armoury of weapons for the battle against dogma.

III

For that, we shall need to turn to his purposes in the 15 years, from 1734 to 1749, when he resided with Madame du Châtelet at Cirey, providing himself in early middle age with the self-education through which the versifier and wit of his youth was transformed into the *philosophe* of a maturity enduring through old age, unique perhaps among those who set up as the conscience of their time in that he was incapable of boring it. It is convenient to associate the phases of Voltaire's intellectual development with his residences. The life of the young Voltaire about Paris prior to 1725 ended in the critical and enforced English sojourn from 1726 to 1729. The early mark he made was that of a poet, tragic in the threadbare theme of *Œdipe* (1718), redeemed from banality though not mortality by the style, and epic in the *Henriade* (1723), a literary foretaste of his historical bent. The first excursion into science occurred in the *Lettres philosophiques*, published in 1734 but drafted in England. A famous passage lightly compares the Cartesian plenum of Paris to the Newtonian attraction of London. Among the many dignities of English life that he contrasted with the indignities of French society was the esteem accorded to men of science and letters beyond the Channel. It is clear, however, that it was only upon settling down to work at Cirey after 1734 that Voltaire actually learned anything about the science he had come to admire in English thought and culture.

Departing from Cirey in 1749 after Madame du Châtelet's death (in the sad absurdity of giving birth to the unwanted child of Saint-Lambert), Voltaire tried the court of Frederick II at Potsdam for a time, in search of tolerance and civility. Disenchanted there, he was further disappointed in his expectation of surrounding his life with those qualities in the city of Geneva, where he maintained a house from 1755 to 1759. He settled finally in France again, at Ferney, close enough to the Swiss border in case of persecution, and did not again set foot in Paris until his triumphant return to die in 1778. Can it have occurred to him to compare that apotheosis to Newton's, whose funeral just over half a century before had so moved him in Westminster Abbey? That is to indulge in a fancy that would have been not untypical of him. To return to the ground he celebrated, that of fact—*droit au fait*—it was from these successive vantage-points that Voltaire conducted the campaign that gives him a place in history comparable with that of one who admired him and continued it in later times, Bertrand Russell. It was the campaign of a civilized intelligence at war with the follies, the stupidities, and the brutalities that make life more miserable for men at large than one might think, on looking out at it from the world of books, it really needs to be.

Let us recall for a moment the most famous occasions on which he raised his voice and made infamous episodes that would otherwise have been soon forgotten, or perhaps not even noticed beyond the several localities. On the evening of 13 October 1761 Marc-Antoine Calas, the eldest son of a Protestant clothier, Jean Calas, in the largely Roman Catholic city of Toulouse, was found dead, evidently by hanging, on the ground floor of the shop above which the family had just finished dining. It appears overwhelmingly probable that the case was a suicide, and that in order to avert odium the family gave out that they had found Marc-Antoine dead on the floor rather than there in the embrasure of the windows where he had in fact hanged himself. The tragedy touched off the witch-hunt instincts in that not notably devout city, where many were inclined to believe that the Protestants in their midst were capable of any evil. It was soon asserted that the young man had been murdered by his family in a Protestant plot to thwart his intended conversion to the Roman Church. Arrest and accusation followed within 24 hours, and the case was evoked to the Parlement itself, the sovereign court of Toulouse. Under its procedures Jean Calas was tortured, but he never weakened in repudiating all allegations. He was, nevertheless, condemned and sentenced to death by breaking on the wheel.[16]

In 1765—to recall a further episode in which Voltaire constituted himself publicist for the vindication or defence—a 19-year-old lad called the Chevalier de La Barre was convicted of desecrating a rural crucifix near the town of Abbeville in Picardy, and of blaspheming in word and song. La Barre does appear to have been an idle, ignorant, and foul-mouthed lout, easily led into adolescent bravado by the companion who was accused with him but who, being more enterprising, escaped. They belonged to the perennial type of young men from whom mature and civilized people would do better not to take offence because that is what they seek to give by their appearance and behaviour. The reaction of the court at Abbeville exceeded what might be allowed to outraged respectability, however, and was itself not to be distinguished from savagery. It condemned La Barre and sentenced him to do public penance, upon which his tongue was to be cut out, his right hand severed from his body, and he himself then burned at the stake. An appeal against the verdict was carried from this provincial tribunal to the Parlement of Paris, where—to the horror of enlightened opinion—it was upheld. In the most civilized country in the world and in the most civilized century of its history, that sentence too was carried out while the crowd applauded the executioner.[17]

These travesties of justice, which in the full power of his pen Voltaire made notorious, had certain features in common with the episode in Voltaire's own life that, although its fortunate consequence had been to require his removal for a time to England, opened his eyes to worldly realities. In 1726, at the height of the *mondain* success of a daring new poet, he exchanged insults with a young nobleman with whom he thought himself to be on terms of camaraderie and badinage, the Chevalier Gui Auguste de Rohan-Chabot. A few days later Voltaire was called to the door from a dinner party in the Hôtel de Sully in the rue Saint-Antoine. There, instead of finding a message, he was set upon by Rohan's lackeys, hauled into the street, and thrashed with sticks under the eyes of their master, who never descended from his carriage at the kerb. Voltaire's host of the evening, the Duc de Sully, gave no help to the scribbler, whom he thought amusing enough company at table but no gentleman eligible to enter upon some affair of honour involving a kinsman.

Voltaire determined to have satisfaction. He even engaged a fencing-master to instruct him in an art of which he knew nothing. Instead of facing his enemy, however, Voltaire found himself incarcerated in the Bastille under *lettre de cachet* issued by the Regent at the request of Cardinal de Rohan, lest he further trouble the tranquillity of the nobleman who had had him beaten.[18] At bottom a realist, Voltaire appears to have volunteered to absent himself in England in return for his release. So it happened that he became aware simultaneously of Newton, science, and the prospect that things might be different, however generously he may have then idealized the country that, for its part, proved capable of executing Admiral Byng—"pour encourager les autres," Candide is told—and kept Tyburn in full swing.

Amid such incidents, grave or trivial, must a man somehow steer his life. What they exhibit in common are those strands of prejudice, privilege, intolerance, delusion, superstition, and dogma that history had woven together into a tissue of falsehood masquerading in the guise sometimes of religion, sometimes of law, and sometimes of morals; and in the name of their abstract principles the authorities never hesitated to crib, manipulate, oppress, and often murder real persons, and most of all through dogma. That is what Voltaire meant by infamy.

In science there was none of that, none at all. All Newton purported to do was to say how the world *is* made, in contrast to Descartes and Leibniz, who in their different ways proceeded from dogmas to the pretence of science, and presumed to lay down how the world must be made if it was to satisfy their metaphysics. Not that Voltaire looked to science for comforts: in the closing line of *Micromégas*, the giant visitor from Sirius, having heard all the philosophies and smiled approvingly at Locke's—of which he finds mildly that "it's not the least wise"—promises the earthlings a book that will teach them admirable things and show them the "good in things." He leaves it before his departure for his galactic home, and it is taken round to the Academy of Sciences in Paris. But when the elderly permanent secretary—that is, Fontenelle—opens it, the pages are blank.

It was neither science nor his own relatively harmless encounter with the Rohan that turned Voltaire into a crusader. He was not a self-centred man, and it took the world at large to do that. Nor was it his years of study at Cirey. What does appear to have been the product of those years and of his new knowledge of science was the invention of the one fully original contribution he made to literature itself, the genre of the *conte philosophique*, first exemplified in *Micromégas* and fully realized in *Candide* (1759). True, there is scarcely a dull page among the tens of thousands that he wrote in his life. Yet, somehow, only in *Microméyus* and *Candide* did he achieve altogether individual works of art transcending time and idiom. *Micromégas* draws on the tradition of scientific fancies about inhabited worlds, in the moon or beyond, that derived through Bernard de Fontenelle's *Entretiens sur la pluralité des mondes* (1686) and Johannes Kepler's *Somnium* from Plutarch's *Face in the Moon*. Voltaire adopted this physical or cosmological conceit to achieve displacements and contrasts of scale of the sort that Jonathan Swift had employed to bring out moral relief in *Gulliver's Travels* (1726). Voltaire gets the planetary dimensions and stellar proportions approximately right. Micromégas, the space voyager from Sirius to our blob of mud, is eight leagues (roughly 24 miles) tall. The dwarf from Saturn whom he picks up *en route* is only 2000 toises (a bit over two miles), and the latter's new wife, a pretty little brunette of only 668 toises, complains that after only 200 years in her arms, he wants to go vagabonding through the solar system.

Voltaire admired Swift and certainly had Gulliver in mind, but the tone is altogether different, entertained at what mites these mortals are rather than outraged at what knaves. It is a light-hearted tale, more spoof than allegory, conveying merely a hint of the sadness that underlies much 18th-century gaiety, and no bitterness of soul. It is a most amusing tale, despite its being philosophic fiction. The philosophic aspect here pertains rather to man's place in the scheme of things than to his moral destiny, and Professor Wade, in preparing the modern critical edition, has argued convincingly that, although *Micromégas* was published in Berlin in 1752, the date of composition was 1739, the year after the first edition of the *Élémens de la philosophie de Neuton,* when Voltaire's mind was as full of scientific information as it would ever be.[19]

Only *Candide* retains the standing of a masterpiece, however, immortalizing Voltaire's rising indignation with the stupidities and deceptions that rule the world itself. There is nothing of science in it. But we need to know Voltaire's standpoint with respect to science in order to take the full thrust of his assault on Leibniz in the caricature of Dr. Pangloss. No doubt it was quite unfair, and even perhaps a touch philistine, of Voltaire to let it be supposed that so high a philosophy could decently be coupled with sanctimonious and mealy-mouthed rationalizations of the existing state of things personal or political. There Dr. Pangloss always is, however, with the word for everything out of his specialty of metaphysico-theologico-cosmologicology, and it is always the same word, whatever the grotesque or bloody misadventure that Candide has barely survived: his own illegitimacy, his expulsion into the world, the floggings called military training, his beloved Cunégonde raped and left for dead, whole villages mindlessly massacred by passing armies, the city of Lisbon demolished by an indifferent shrug of nature's shoulder. Always and inanely, the word is that things are as they are because they could not be otherwise, and "all is for the best in the best of all possible worlds." There Pangloss prates, the personification of dogma justifying and denaturing any absurdity, any horror, in the name of principles that will leave untroubled the enjoyment of their beneficiaries.

Voltaire's technical mastery of the tale is a perfectly concealed triumph. The contrast being between fact and illusion, he nowhere casts Candide upon the reader's sympathies. The technique is one recently discovered again in certain theatrical and cinematic experiments. The story is told less in narrative than in a succession of exposures. One vignette of horror is followed by another of chicanery, without any but the most deliberately banal and ludicrously inadequate expressions of emotion. The secret is in the pace; things happen so fast that the reader has no time to be affected. Scenes flash by like stills in an accelerated slide show, and the reader is never asked to identify himself with the spectacle. Instead of engaging his emotions, it leaves an intellectual recognition. Voltaire had the skill to create and the wisdom to put more trust precisely in the intellectual recognition that the world is a hard, indifferent scene, external to us, and that all the affective abstractions—about the nobility of nature, or its oneness with human nature, or the loving care of providence, or the statesmanship of the King of the Bougres (Frederick the Great), a protector of mankind only in the sense of preferring his own soldiers to women—all that, and all the other things we are told by the authorities, and taught by the tutors they engage, to persuade us that misery and cruelty and oppression are something else—all of it is as specious as Dr. Pangloss, forever parroting, "Whatever is, is right." Yes, says Candide at the end—settling down to make what he can out of a few remaining realities, and replying a little wearily to Dr. Pangloss's ultimate proof that all events were causally related in the best of all possible worlds in order to have brought the company where they were—"that is well said, but we must cultivate our garden."[20]

Freedom is acceptance of fact and emancipation from dogma. Candide has to learn that by experience. Mankind has science, however, which is the same thing, and once emancipated is on its own, denied the comfort of illusions even about its own resources, among which remain conscience, work, and jest.

IV

Johann Wolfgang Goethe (1749–1832) preferred acting out the illusion that science, like art, can be made from some direct correspondence between nature and human perceptions of beauty and fitness.[21] His conversations with Eckermann in the last years of his life record one of the most astonishing self-deceptions in the history of literature: "As for what I have done as a poet," he said, "I take no pride in it whatever. . . . But that in my century I am the only person who knows the truth in the difficult science of colours—of that, I say, I am not a little proud, and here I have a consciousness of superiority to many."[22] Can he really have come to believe that? It may be so. His writings on scientific subjects occupy 13 volumes in the Weimar edition of his works,[23] and in them his reaction to the scientific materialism of the Enlightenment exhibits a consistent emotional pattern first evident during his student days. He then once hurled aside Baron d'Holbach's materialist *Systéme de la nature* (1770), later recalling of the book that he and his fellow-students, all warm with life and feeling, could hardly endure its presence. Falling ill soon after, he turned for read-

ing in his convalescence to older writings of another sort, among them
books of alchemy and the works of Paracelsus. That the conception of
Goethe's Faust—"Thee, boundless nature, how make Thee my own?"[24]—
was modelled upon its author's apprehension of Paracelsus, rather than
upon the semi-legendary namesake himself, seems well established in
modern scholarship.[25] On returning to university life, the young Goethe
mingled mainly with medical students, drawn to their company by their
esprit de corps and by his own preoccupation with courses in chemistry
and botany under Spielmann and anatomy under Lobstein. The recollection finds its place in the opening pages of *Faust* (Pt I, 1808; Pt II, 1832).
In the youthful *Sturm und Drang* years, Goethe was making his literary
reputation with *Die Leiden des jungen Werthers* (1774; The Sorrows of
Young Werther) and *Götz von Berlichingen* (1771–3), and lived mainly at
Wetzlar. There in 1774 he became friendly with Johann Lavater, whose
studies in the supposed science of physiognomy convinced him that fa-
cial expression is an index to character and a function at the same time of
bone structure. This cranial and psychic work sustained his interest in
anatomy. He collaborated in it, furnishing descriptions of the heads of
noted men and also of certain animals.

Ten years later, by which time Goethe was already 35 years old and installed at Weimar, he announced in letters to Gottfried Herder and to Frau
von Stein what he then and ever after believed to be an actual scientific
discovery, the first that he had made. He had determined to his own satisfaction the existence in the upper jaw of the human skull of the premax-
illary or intermaxillary bone. (In many vertebrates that structure is distinct from the maxillae on either side, the former carrying the four incisor
teeth, and the latter the canines and molars.) All excitement, he composed
an anatomical memoir and circulated copies to authorities in the science,
notably to Loder at Jena, Blumenbach at Göttingen, Camper in Holland,
and Merck at Darmstadt, who sent it on to Sömmering in Mainz. It refuted, so Goethe thought, the teaching of many anatomists, that among
the factors differentiating men from apes is the absence of an intermaxillary bone in the former and its relatively defined development in the
higher anthropoids.[26] Thus had he come upon an osteological link in
the chain binding man to nature. Peter Camper, in company with other
anatomists, had adduced the absence of a human intermaxillary bone in
countering certain writers who tended to cause offence by pointing out
the close similarities between men and apes. On receiving Goethe's paper
in manuscript, Camper re-examined his findings carefully but was unable
to accept Goethe's, except in the incidental matter of the presence of the
bone in question in the walrus. Neither did Blumenbach or Sömmering

accept Goethe's findings, contrary to legend, which—embroidering on Goethe's ensuing sense of grievance—has it that they reversed their initial agreement under pressure in order to make common cause with the professional community of specialists erecting a wall of solidarity against the ingenious outsider.

The facts are otherwise. Except in the matter of the walrus, Goethe did not adduce any evidence not already familiar to his correspondents in the anatomical literature. He was mistaken in supposing the intermaxillary structure to be so well developed in man as to constitute a distinct element of skeletal architecture. The sutures that trace its junction with the maxillae are reduced in man to the status of vestigial markings. It is true that the literature was itself mistaken in attributing the bone to the great apes, for neither is it to be distinguished in the adult anthropoid.[27] Not that Goethe would have thought his finding refuted by such a lack of literal evidence: in accordance with the vertebral archetype, the bone must be present in both men and apes, evident to the mind if not clearly to the senses. Otherwise, certain forms would violate the principle of unity in the plan on which all the vertebrates are variations. But it will follow the sequence of his own thinking to turn rather to his botanical studies for the development of his conception of archetypal biology.

Botany was Goethe's favourite among the sciences, as indeed it has been for many other literary persons, among them Rousseau, whose example in the study of plants he followed. He was able to do so in the lovely ducal gardens at Weimar and at Jena, as recreation from the administrative duties of minister and privy counsellor in the small grand duchy. His experience was unlike that of Rousseau in another respect. Instead of being gratified at the improved accuracy with which a command of the Linnaean system of classification enabled the amateur to identify and collate the plants, Goethe found himself increasingly irritated by its artificiality and unable (no doubt because unwilling) to remember the arid Latin names it assigned to the forms he loved. By what right had Linnaeus singled out the number and arrangement of stamens and pistils to be the discriminants of one species from another? By what right, indeed, did he assert even the identity and boundaries of species, varying as they do with such abandon from one soil and location to another? For the differences and divisions between plants never interested Goethe; it was their similarities and variations upon a common form that entranced him.

What that form, that *Urpflanze* (primal plant), might be was revealed to Goethe only after he had left Weimar for Italy in 1786. Of all his writings his *Italienische Reise* (1816–29; Italian Journey)[28] is perhaps the most accessible to members of the English-reading public, many of whom (like

other northerners) have found their own Arcadia alive in a first exposure
to the Italian sun and landscape, Italian painting, and the Italian people.
Art, architecture, churches, the language, the theatre, olive trees, the Colos-
seum by moonlight, Sicily—Goethe was that most satisfactory of tourists
who notices everything. The hints of a botanical revelation in the offing
are interspersed amid the thousand adventures and episodic discoveries
of alerted travel. "It is really and truly a misfortune to be haunted and
tempted by so many spirits," he recorded of an early morning walk in the
public gardens of Palermo. He had meant to meditate on poetic dreams,
but finding himself surrounded by plants growing freely in the open air
and not in pots or under glass, an old fancy came back: "Among this multitude might I not discover the Primal Plant? There must certainly be one.
Otherwise, how could I recognize that this or that form was a plant, if all
were not built upon the same basic model?"[29] On the road from the temple
of Segesta back to Alcamo he noticed the young fennel: how, although
there is a difference between upper and lower leaves, "the organism is one
and the same, but it always evolves from simplicity to multiplicity."[30] A
letter to Herder from Naples reflects on Homer and the new immediacy
he feels for the *Odyssey* in a classical seascape, and confides that having
solved the main question, where the germ is hidden, he is very near to the
secret of plants.

> The Primal Plant is going to be the strangest creature in the world,
> which Nature herself shall envy me. With this model and the key to
> it, it will be possible to go on forever inventing plants and know that
> their existence is logical; that is to say, if they do not actually exist,
> they could, for they are not the shadowy phantoms of a vain imagi-
> nation, but possess an inner necessity or truth. The same law will be
> applicable to all other living organisms.[31]

Actually, so he recorded a little later, while putting the finishing
touches to *Egmont* (1787), the secret had come to him "like a flash" during
that walk at Palermo: "... in the organ of the plant which we are accustomed to call the *leaf* lies the true Proteus who can hide or reveal himself
in all vegetal forms. From first to last the plant is nothing but leaf, which
is so inseparable from the future germ that one cannot think of one without the other."[32] Such was the pregnancy of this idea in Goethe's mind that
he recognized it himself for an obsessive passion, one that would occupy
him for the rest of his life. For botanical though it was in origin, the principle was applicable to everything—a Columbus's egg by which he also
now saw how to interpret works of art of the Renaissance and even of an-

tiquity. "These masterpieces of man were brought forth in obedience to the same laws as the masterpieces of Nature. Before them, all that is arbitrary and imaginary collapses: *There* is Necessity, *There* is God."[33] Such was the genesis of Goethe's *Metamorphosis of Plants,* an essay of some 86 pages published in 1790 after his return to Weimar.[34]

Goethe confined the discussion to flowering plants, the phanerogams. The argument consists in exhibiting the occurrence of forms that appear to be transitional between that of the leaf and what have been mistaken for organs of a different nature—cotyledons, petals, stamens, pistils, and even stems—but are in truth metamorphic variations upon the archetypal leaf. Thus, in certain roses the outer petals of the flower take the form of a leaf of the calyx or sepal, and often one of them foliates into an ordinary leaf. Thus, too, gardeners have reversed the normal progressive process of metamorphosis in creating double flowers in plants such as dahlias or chrysanthemums wherein the additional petals crowded into the centre are anatomically metamorphosed stamens. In Goethe's appreciation, however, the plant is to be studied biographically, so to say, and not in some static anatomy. Its divisions—leaf, flower, and fruit—pertain to its tri-partite phases of existence. The goal of the initial phase of growth and leaf formation is the contraction of the plant's inwardness into bud and calyx. Thence bursts outward the glory and the climax of the flowering, within which, concentrated in pistils and stamens, the consequence of the reproductive act outlasts the full-blown petals to produce the final swelling and ripening of the fruit, its flesh dropping away in turn to leave the seed. Thence in the next generation the shoots will sprout, opening into the paired cotyledons that begin the next cycle in the slow, perpetual motion of life, expansion ever followed by contraction, withdrawal by return.

From botany and the *Urpflanze,* Goethe's interests moved back to osteology and comparative anatomy. Walking upon the dunes of the Venice Lido one day in 1790 he came upon the skull of a sheep, and it appeared to him that the bones composing it—the palatine bone, the upper maxillae or jawbone, and his favorite intermaxillae—all resembled exploded vertebrae. In another flash of intuition, comparable to that which had revealed the primal form of the plant in the leaf, he conceived the idea that in vertebrates the skull consists in a forward extension of the spine, and that its component structures are nothing but metamorphic forms of vertebrae enclosing the brain and higher functions of sense as the lower segments of the backbone enclose the spinal cord. The osteology of the skull and spinal column fascinated him in the years that followed. He drew up elaborate tabulations of comparative vertebrology in which the bones from the front of the skull to the tip of the tail, and even out to the extremities of the limbs, are ranged vertically and the species of animals horizontally. His purpose was to trace the modifications of each structure in all species and of all structures in each species. He never did consider that he had brought the vertebral theory of the skull to the same perfection as that of his theory of leaf metamorphosis in plants. Variations of it had a very considerable vogue in the early nineteenth century, however, among what may be called the archetypal school of comparative anatomy. To what extent these views derived from Goethe's speculations or *vice versa* is a question unresolved by scholarship and one that produced charges and counter-charges of plagiarism among several of its German adepts in the tradition known as *Naturphilosophie*. The most widely read was Lorenz Oken, and the most famous in England was Darwin's arch-opponent, Richard Owen.[35]

Despite what is often said, it should not be thought that this approach to plasticity in species had anything methodological or theoretical in common with the Darwinian theory of evolution by natural selection. No two approaches to knowledge of nature could be more dissimilar in spirit or in method, the one symbolic and metaphorical, the other literal and circumstantial. Even the concurrence on mutability of species is illusory if taken for agreement. In Goethe's view a simple form was one that was primal and deepest in principle, though not necessarily earliest in time. Indeed, it need never have existed at all in fact: it need only be required in idea. Later in life Goethe tended to draw back from the vertebral theory of the skull, but never from that of archetypal comparative anatomy. Its purpose is the delineation of the type, whether of particular organs and structures or of entire organisms, in relation to which existing examples are to be judged as variations determined by the interplay between the internal drive of the organism and the external moulding of climate and habitat. The last essay of his life was the draft of a piece championing the part of Etienne Geoffroy Saint-Hilaire in his famous encounter with Baron Cuvier before the Académie des Sciences in Paris. For although Goethe's views had nothing in common with the Darwinism that lay ahead, neither did they accord with the theological explanation of adaptation in organisms in accordance with the notion of special creation and design. Ultimately, his view of nature derived through Spinoza and Leibniz from a mingling of Plato with Stoicism. It was altogether incompatible with the natural theology that had superimposed the doctrine of divine creation on Aristotelian natural history.

It is in Goethe's optical work, however, that the student must reckon with the full measure of the artist's repudiation of both the mathematical and the experimental aspects of that mode of mediating between thought and nature that constitutes the modern scientific enterprise; and here, too, Italy was the catalyst in the luminescence of its landscapes, in the mastery of its greatest painters. To penetrate their secrets (especially in the work of Raphael) had been a leading purpose of Goethe's journey. He consorted with their successors, among them Angelica Kauffmann, convinced that those secrets lay in colour. He tried his hand, only to find that he had no talent for drawing. Dimly he recalled lectures at the university on the Newtonian theory of colour. Italy was no place to search things out in libraries, however, and in this one respect he returned to Germany with his curiosity unsatisfied.

At home in Weimar he looked up a manual that confirmed his recollection of having been told that colours are contained in light. Finding this statement no help, he borrowed a set of prisms from a physicist in Jena, one Büttner, and then, in the press of other interests and duties, forgot all about the matter for months until Büttner, exasperated by Goethe's repeated failures to return his apparatus, sent a servant with instructions not to come back empty-handed. Thinking to have just one look, Goethe glanced through a prism at the white wall opposite—which to his astonishment remained white! Only along the window-bars did a fringe of colour appear. At once Goethe recognized that a boundary or edge is required to bring out colours, and immediately said aloud, by a kind of instinct, that the Newtonian doctrine was false. He prevailed on Büttner to let him keep the prisms for some experiments, which he now threw himself into. At the same time, he went back to the source of error itself, to Newton's *Opticks*, and worked through Book I (which deals with the prismatic production of the colours of the spectrum), trying the experiments and rearranging them in natural order. Not that it was necessary to retain them all, for Newton had deliberately created a labyrinth of complications in order to confuse those who came after, with his *experimentum crucis*, his separations and recombinations of colours, his artificial isolation of a single beam, his parade and charade of geometry, his pettifogging measurements of angles and widths. It does not appear that Goethe then noticed that in Book II and the fragmentary Book III of the *Opticks* Newton had been examining the phenomena that would now be described as interference effects and diffraction with different questions in mind from those posed in Book I. Even had he done so, however, it would not have modified the fundamental nature of the difference between them. It comes down to this: in Book I of the *Opticks*, Newton described the production of colours by passage of light through a prism, whereas Goethe was interested in the perception of colours by the human eye.

The latter has become a rewarding aspect of optics, intermediary between physiology and physics, ever since the work of Helmholtz in the middle of the nineteenth century. It must be acknowledged of Goethe's excursion into these matters that the distinction he makes in *Zur Farbenlehre* (1810) between what he called physiological colours, physical colours, and chemical colours was capable of development. The first set of problems concerns sight, the second the appearance of colours in light itself by refraction, diffraction, and polarization, and the third the production of the surface colours of opaque objects. The purpose of the work is set out in the final chapter, entitled "The Physical-Moral Effect of Colours," which is intended to achieve the unification of physics, aesthetics, and ethics. Although it cannot be said that the content of Goethe's discussion in these chapters contributed to science in any professional sense, the merit of an insistence on the functioning of the eye in respect to colour does incontestably belong to his treatment of the first of these subjects.

Optical illusions—optical truths he thought them—were well observed and well reported. It interested him that a white disc on a black background looks larger than a black disc of identical diameter on a white one. He studied very carefully the after-images that persist in the eye following brilliant illumination. When we look at a dark object against a bright background and close our eyes, we "after-see" darkness and brightness reversed. The way the images disappear depends upon the circumstances. If we glance at the sun, the residual spot is first yellow, then purple, then blue if allowed to fade in the dark. Against a white background, however, the sequence is blue, green, and yellow, going off into grey in both cases. In general, the eye functions in a polarity between light and darkness, white and black, the primary colours (in his theory blue, yellow, and red) and the complementary colours (orange, violet, and green), which he preferred to call demanded colours. Their relations can be exhibited in a circle[36]:

Red

Orange   Violet

Yellow   Blue

Green

Presented with a colour, the eye demands its opposite in the after-image: orange for blue, violet for yellow, and green for red. The notion of demanded colour is the heart of Goethe's theory of optics. Just as the functioning of the eye holds the key to physiological optics, so light itself is an active agent in the world. "Colours are the deeds of light," he says at the outset, "deeds and sufferings."[37]

Goethe accompanied this first, "didactic" part of *Zur Farbenlehre* with a second, "polemical" part, which those who profess admiration of his physical insights have been less inclined to celebrate. In it Goethe indulged the animus he had developed, not only against Newtonian physics but against Newton himself, in the 20 years that had intervened between his glancing through that prism and the publication of his treatise. Experiment by experiment, even sentence by sentence, Goethe follows Newton through Book I of the *Opticks,* exhibiting how in his view Newton had either misconceived or misrepresented his results, often by what Goethe takes to be verbal sleight of hand. Physicists who over the preceding century had been taken in by this imposture must be distinguished from rational people.[38] The very title is a misnomer when transposed into Latin or German. In English the word "optics" connotes only the geometrical description of propagation and transmission, in the sense of Newton's *Lectiones opticae,* and therefore promises what the book cannot fulfil. To carry it over into Latin or German is to claim that these abstractions are capable of embracing the whole subject of colour, light, and vision.[39] It was because physicists had been thus tricked and bewitched by mathematics that his own first paper on the subject had been ill received in 1791. For in it, and indeed in all his work, he had attempted to demonstrate that only by ridding itself of mathematics could physics become true to nature.

Thomas Mann observed somewhere that there was so monumental an egoism in Goethe's personality that it beggars the adverse judgment of character one might make of an ordinary man. He was his own first idolater, and the reputation for universality that he started about himself has reached proportions equalled only by the Leonardo tradition. To the reader who has escaped the magic it might seem that in the succession of his writings about scientific subjects we see the enthusiastic openness to nature of a young writer gifted for every sort of observation giving way to the obsession of an old sage with a point of view. Longevity helped— or hurt, depending on one's own point of view. It is surely significant that at the time when Goethe first advanced his favourite ideas, about skulls or plants or colours, they held no scientific interest for those best placed to judge them; the scientists themselves in these several specialties were men who had not yet had time to fall under the spell of Goethe's reputation. Goethe, of course, put their resistance down to the inveterate obduracy and exclusiveness of pedants defensive about their pet preserves, a judgment always easy to make of professional scholars and scientists and one in which contemporaries and posterity are often ready to join. Even scientists in later times are prone to impute to their predecessors corporate resistance to new and meritorious ideas. The reasons for their willingness to share in such imputations go deep into the psychology of science, and have to do with cushioning themselves against possible disappointment.

A further complication arises from the manner in which Goethe presented his scientific writings. The first statement was generally in the form of what he liked to call an *aperçu* conveyed in a privately circulated memoir, in letters, or in a brief essay. Only after absorbing the initial rejection would he draw together into a finished work the reflections that in the meanwhile he had been turning over in his mind: for 20 years in the case of *Zur Farbenlehre,* for 30 years in the case of his writings on general morphology, which he published as a collection in 1820. In the meanwhile a new generation had grown up, whose members were by then ready to join in the putative adoption by science of one who combined the qualities of literary genius and unappreciated pioneer. The claim for Goethe as a scientist is always that he was a seer, too far ahead of his time to be appreciated by it, one whose views would come into their own "only in our own day," as has been said at many intervals since. Nor is it surprising that scientists themselves should have concurred in this romantic legend of the way that science grows, once it no longer threatened them with confusion in their actual problems, but merely lent an air of cultural respectability to the scientific enterprise. It is natural that such a tendency should have been particularly marked in German science, and that it should have taken on new strength since World War II. Those who lived through the dark epoch in Germany's recent history have been the more eager to associate her scientific past with her cultural past. The tradition is a testimonial to the anxiety of German scientists not to be separated from culture by virtue of being scientists and not to be separated from civilization by virtue of being German.

There is often an advantage in looking at cultural history dialectically, and probably the most general interest presented by Goethe's lifelong incursion into the realm of science is that it permits us to observe in the most illustrious instance how science since Newton's synthesis has produced an antithesis that, though sometimes claiming to be science—Goethe's advice to a young writer was that he must soak himself in science, though what he meant was nature—was in fact its opposite. The antithesis was Romanticism, and that its provocation did in fact lie in science in a way that was anterior to any other of its protean manifestations in literature, art, philosophy, or politics will appear when it is appreciated that only in respect of science were its attitudes consistent. In point of literary form, as opposed to content, Goethe may be classified as belonging to German Classicism once *Sturm und Drang* were behind him along with his youth. But this is not so in his philosophy of nature, wherein the elements of Romanticism may be identified categorically enough. They comprised, first, repudiation of the mechanistic, the atomistic, and the mathematical; second, adherence instead to the "organismic" and to the sort of explanation that descends to particulars from the hypothesis of unity and continuity in nature taken as a whole; and third, attainment of knowledge through the participation of consciousness in some sense of a world alive rather than through quantification, classification, and abstraction from the personal.

The term "idealism" is sometimes applied to such a personification of knowledge and of nature. Goethe himself employed it in conjunction with his archetypes and primal forms. So also did the school of *Naturphilosophie*, with which his views had much in common. Archetypal botany and anatomy in the life sciences and a resort to polarities in physics were the keynotes of *Naturphilosophie*, together with the hypothesis of the interconvertibility of forces deriving from an underlying identity. It is true that the latter emphasis does not come out strongly in Goethe's writings, and that the proponents of *Naturphilosophie*—Oken, Treviranus, Nees von Esenbeck, Novalis, and Ritter (whose collective influence has probably been much exaggerated by historians eager themselves to romanticize science)—looked to Friedrich Schelling as the founder of their school.[40] The relation of Goethe's thought to theirs was one of resonance rather than of participation. He was too great a man for schools, was half a generation or more older, and differed from them in literary taste.

An ambiguity in the word "idealism" sometimes produces a more misleading confusion between Kant's philosophy of science and the naturalistic Romanticism exhibited alike by Goethe and by the school of *Naturphilosophie*. The distinction is one about which it is important to be clear, because Kantianism and Romanticism were idealistic in very different ways. Kant was a philosopher: his realm was knowledge. Goethe and the school of *Naturphilosophie* were writers: their realm was sensibility. Kant's thought was rigorous, whereas that of the Romantic writers was sentimental (when it was not grandiose and speculative). The difference comes out most clearly in the central Kantian assertion that there is only as much of science in a subject as there is of mathematics. For Kant the mind's capacity to constitute mathematics is what makes science possible at all. The condition of personal freedom in the moral world is that we operate as persons within a strict system of physical causality rendering us responsible for our choices, and not that we should presume—as did the Romantics—to constitute science in ways conformable to the thrust of wish and will. For Kant, the idea is in the mind, and science is a function of the structure of reason. For Goethe, the idea is in nature, and we make it what we will. Its attributes are those of personality, and not the categories of geometry and mechanics. The mediator is the artist, not the mathematician. It is no wonder that Goethe could never truly comprehend Kant, much less accept him.

For Goethe was a writer first and last—an artist as Voltaire and all those other writers of the French Enlightenment, even Rousseau, had never been. Perhaps there has been no other writer in any language whose ear for its every idiom was so true and who could speak in such a range of voices. There was no limit to the virtuosity with which he adapted the German language to his purposes, and nothing of which the language is capable that he could not do with it. No wonder he should have been incapable of stopping at the limits of literature or of language itself, and should have been bound to treat science and even nature as instruments of his creative powers.

There lies the difference from Voltaire. Everything Goethe touched, he brought to life: Faust, Marguerite, Mephistophiles, Götz, Egmont, Wilhelm Meister, Werther—they people the world of all who have read him. Not so the *philosophes,* whose Enlightenment was a movement in ideas and criticism but not in creation. Voltaire—fluent, accurate, witty, humane critic that he was—never brought a real man alive with his pen. His characters are historic personages or silhouettes serving purposes. It may be that those who regard the liberal legacy of the Enlightenment as the closest civic approximation that Western history has yet achieved to true civilization, those who think individual liberty the paramount question, those who consider the exaggeration of personality a threat in practice to persons, will put their confidence in the sceptical view of the human condition to be read out of Voltaire's lifelong defence of mankind-as-it-is from deceptions of the past or delusions of transfiguration in the future. Certainly Goethe took no comparable interest in the impact of politics, the equity of judicial proceedings, or the liberties of ordinary people. It may be that from the limited perspective of such ordinary people, Goethe's treatment of individual persons will appear—like Faust's of Marguerite or Werther's of Albert—a distasteful instance of the ruthlessness of genius or the self-acceptance of personality unbounded by kindness, guilt, or fairness. Nietzsche's contempt for compassion lay not far in the future along the same line, and if persons have no rights when confronted with the requirements of personality, why should the findings and limits of science be exempt?

How well the literary imagination consorts with science depends upon the kind of literary imagination with which one has to deal, and (among other attributes) upon whether it works with outer fact or with its metamorphosis into some inner recognition.

Notes and References

1. Voltaire, *Élémens de la philosophie de Neuton mis à la portée de tout le monde* (Amsterdam 1738). Voltaire published a revision in 1741 and a second edition in 1745; the latter is the version included in most editions of his works, and it forms the basis of the discussion that follows. The standard (though never completed) work on Newtonianism in France prior to Voltaire is Pierre Brunet, *L'introduction des théories de Newton en France au XVIIIᵉ siècle avant 1738* (Paris 1931). Ira O. Wade, *The Intellectual Development of Voltaire* (Princeton 1969), draws upon the author's earlier studies, in which the importance of scientific themes to Voltaire is carefully explored. An unpublished dissertation in the Princeton University Library by one of Professor Wade's students, Robert Walters, "Voltaire and the Newtonian Universe," is a study of the *Élémens.* See also Martin S. Staum, "Newton and Voltaire: Constructive Sceptics," *Studies on Voltaire and the 18th Century,* LXII (Geneva 1968), pp. 29–56; Henry Guerlac, "Where the Statue Stood: Divergent Loyalties to Newton in the 18th Century," *Aspects of the 18th Century,* ed. by Earl Wasserman (Baltimore 1965), pp. 317–34; and Henry Guerlac, "Three 18th-Century Social Philosophers: Scientific Influences on Their Thought," *Daedalus* LXXXVII (Richmond, Va. 1958), pp. 12–8.

2. For an introduction to the literature of Newtonian explication in the eighteenth century, see I. Bernard Cohen, *Franklin and Newton* (Philadelphia 1956), especially Pt III.

3. A useful modern edition is *The Leibniz-Clarke Correspondence,* ed. by H. G. Alexander (Manchester 1956).

4. René Descartes, *La dioptrique,* Discours 1, in *Œuvres,* ed. by Charles Adam and Paul Tannery, Vol. VI (Paris 1902), pp. 83–6.

5. Printed in Isaac Newton, *Correspondence,* ed. by H. W. Turnbull, Vol. I (Cambridge 1959), No. 40, pp. 92–107.

6. *ibid.,* No. 146, pp. 362–92.

7. Isaac Newton, *Mathematical Principles of Natural Philosophy,* trans. by Andrew Motte (1729), ed. by Florian Cajori, 2 vols. (Berkeley and Los Angeles 1962), Book I, Section 14. I, pp. 226–33.

8. *ibid.,* Book I, Preface, pp. xvii–xviii.

9. *ibid.,* Book III, Foreword, II, p. 397.

10. Madame Du Châtelet, *Principes mathématiques de la philosophie naturelle* (Paris 1759).

11. In 1736 the Académie des Sciences in Paris chose "The Nature of Fire and its Propagation" as the subject of an essay competition for a prize to be awarded in 1738.

Voltaire's entry is entitled "Essai sur la nature du feu et sur sa propagation," and Madame du Châtelet's "Dissertation sur la nature du feu." The former is published in the Kehl edition of Voltaire in the second volume of *Physique* (pp. 83–164), together with an abstract, "Mémoire sur un ouvrage de physique," that Voltaire had made of Madame du Châtelet's essay for publication in the *Mercure de France* (June 1739).

12. See René Taton, "Madame du Châtelet," *Dictionary of Scientific Biography* (New York 1971), Vol. III.

13. On Voltaire's deism, see René Pomeau, *La réligion de Voltaire* (Paris 1956).

14. Ira O. Wade (ed.), *Voltaire's Micromégas* (Princeton 1950), Ch. 4, p. 133.

15. Voltaire, *op. cit.,* Pt I, Ch. 3.

16. The definitive account is David Bien, *The Calas Affair* (Princeton 1960).

17. For an account of the La Barre affair, see Peter Gay, *Voltaire's Politics* (Princeton 1959), pp. 278–81, and Marc Chassaigne, *Le procès du chevalier de La Barre* (Paris 1920).

18. Theodore Besterman, *Voltaire* (New York 1969), pp. 106–9; and, more fully, Lucien Foulet (ed.), *Correspondance de Voltaire, 1726–1729* (Paris 1923), pp. 211–32.

19. See Wade (ed.), *op. cit.,* Professor Wade's introduction to the text is a classic of Voltaire scholarship.

20. See the admirable discussion of *Candide* in Peter Gay, *The Enlightenment: an Interpretation,* Vol. I (New York 1966), pp. 197–203.

21. The literature on Goethe's relation to science is immense and, for the most part, adulatory. The most frequently cited survey is Rudolf Magnus, *Goethe as a Scientist,* trans. by Heinz Norden (New York 1949). A more comprehensible (though apostolic) account is Ernst Lehrs, *Man or Matter* (London 1951), subtitled "An Introduction to a Spiritual Understanding of Nature on the Basis of Goethe's Method of Training, Observation, and Thought." For a study sceptical of the positive value of Goethe's scientific writings, see J. H. F. Kohlbrugge, "Historischekritische Studien über Goethe als Naturforscher," *Zoologische Annalen* V (Würzburg 1913), pp. 83–228. Excellent critical studies with extensive bibliographies are George A. Wells, "Goethe and the Intermaxillary Bone," *The British Journal for the History of Science* III (Cambridge 1967), pp. 348–61, "Goethe and Evolution," *Journal of the History of Ideas* XXVIII (New York 1967), and more generally "Goethe," *Dictionary of Scientific Biography,* Vol. V (New York 1972), pp. 442–6; see also Stanley L. Jaki, "Goethe and the Physicists," *American Journal of Physics* XXXVII (New York 1969), pp. 195–203, for a review of the idolizing of Goethe by German physicists.

22. J. P. Eckermann, *Gespräche mit Goethe in den letzten Jahren seines Lebens,* ed. by F. A. Brockhaus (Leipzig 1925), p. 261.

23. *Goethes Werke herausgegeben im Auftrage der Grossherzogin Sophie von Sachsen,* II. Abteilung, "Naturwissenschaftliche Schriften," Vols. I–XIII (Weimar 1890–1904).

24. Johann Wolfgang Goethe, *Faust,* trans. by Bayard Taylor (Boston 1870), Pt I, "Night."

25. Harold Jantz, *Goethe's Faust as a Renaissance Man* (Princeton 1951).

26. The text is in *Goethes Werke . . . ,* II. Abteilung, Vol. VIII, pp. 91–139, followed by appendices and supporting observations added at later dates.

27. For a careful, accurate, and dispassionate account of this episode, see Wells, "Goethe and the Intermaxillary Bone."

28. Johann Wolfgang Goethe, *Italian Journey,* trans. by W. H. Auden and Elizabeth Mayer (New York 1968).

29. *ibid.,* p. 251.

30. *ibid.,* p. 256.

31. *ibid.,* pp. 305–6.

32. *ibid.,* p. 363.

33. *ibid.,* p. 383.

34. *Goethes Werke . . . ,* II. Abteilung, Vol. VI, pp. 23–94.

35. *ibid.,* Vol. VIII, contains Goethe's writings on comparative anatomy.

36. *Zur Farbenlehre* fills the first five volumes of the "Naturwissenschaftliche Schriften" in the *Goethes Werke . . .* Vol. I contains the "Didactic Part"; for the colour circle, see p. 364. Vol. II consists of the "Polemical Part," and the three other volumes contain a lengthy history of optics from antiquity until the eighteenth century.

37. *Goethes Werke . . . ,* II. Abteilung, Vol I, p. ix.

38. *ibid.,* Vol II, p. 255.

39. *ibid.,* pp. 5–6.

40. Gerhard Hennemann, *Naturphilosophie im neunzehnten Jarhhundert* (Freiburg 1959).

9

The Neesima Lectures I:
The Coming of Age of American Science, 1910–1970

I HAD THE HONOR of delivering this lecture and the one that follows on 18 and 19 September 1981 in the Divinity Hall Chapel of the Doshisha University in Kyoto.[1] They constitute the third in the series of lectures commemorating the founder of the Doshisha University, Joseph Hardy Neesima (1843–1890), Amherst College, Class of 1870. His was one of the most extraordinary lives, Japanese or American, of the nineteenth century.[2]

Ten years old when Admiral Peary sailed into the Bay of Yedo (now Tokyo), Neesima Sheemeta was the eldest son of a samurai family. His ancestors were retainers of the Daimio (feudal lord), of Joshu, a province in the interior. In a memoir of his youthful years Sheemeta tells of the excited rumors that circulated after the American landing. By age sixteen he had learned Chinese. A Chinese-language Atlas of America then fell into his hands. He recorded the impression it made in a diary he kept before perfecting his English: "I was wondered so much as my brain would melted out of my head, because I liked it very much; picking one president, building free schools, poor-houses, houses of correction and machine working and so forth, and I thought that a government of every country must be as President of United States, and mourned myself that a governor of Japan, why you keep down us, as a dog, or a pig?"

Young Neesima wanted to learn "American," but could find no teacher. Because of the centuries-old Dutch trading concession in Nagasaki, Dutch was the only western language tolerably widespread in Japan. So he learned Dutch. He liked to study and picked up algebra, geometry, and elementary navigation from Dutch textbooks, educating himself while in the service of the Daimio, who was a kind man. In the course of his reading, he came upon a Chinese abridgment of the Bible. Having there learned that a "Heavenly Father" had created the world, he began to have a dim sense of God and of the notion of some Truth. What that might be he felt he must learn. He was certain it must be current in America, and he determined to find his way there. The chance came in 1864 when he was 20. A friend knew the master of a junk engaged in coastal commerce and headed for Hakodate, a port where the Shogun allowed international trade. He bade his family farewell and talked himself aboard.

In Hakodate he enlisted the help of a Russian missionary priest and a friend with a little English. Between them they arranged for Neesima to be taken aboard an American clipper bound for Shanghai, the *Berlin,* Captain William T. Savory of Salem, Massachusetts. Emigration was forbidden. His friend got a rowboat, hid him under a tarpaulin, and rowed him to the anchorage at midnight. Had the customs officers found him, he would have been executed. He had agreed to work his passage as a cabin boy. The captain and the one passenger taught him a few words of English. Before returning to Japan from Shanghai, Savory transferred him to another clipper, this one headed for Boston, the *Wild Rover,* Captain Horace S. Taylor of Chatham. Neesima again offered to work his passage without pay and presented Captain Taylor with the longer of his two samurai swords. Going ashore when the ship put in at Hong Kong, he found that his remaining Japanese coins would not pass. He then asked Taylor to buy his short sword for eight dollars. In Manila he found a New Testament in a Chinese bookstore. It was a devoutly Christian young Japanese, called Joe by Captain and crew, who four months later disembarked in Boston, thrilled by the sight of the golden dome on the State House.

The Captain hurried off to his family in Chatham, leaving Joe with the crew, a samurai forced to labor in emptying bilge and cleaning out the ship, and at a loss what to do next. On returning from Cape Cod some weeks later, Taylor told the shipowner, Alpheus Hardy, of the Japanese lad who had shipped on the *Wild Rover* in search of Truth. Hardy and his wife resolved at once to take him under their wing and provide for his support. Theirs was a deeply religious faith of an evangelical bent. They saw Neesima as he saw himself, sent providentially in order to prepare for advancing the cause of Christianity in Japan. He saw them as surrogate parents. Hence the name Joseph Hardy Neesima. Once his English permitted, the Hardys sent him first to Andover, then to Amherst, and finally to Andover Theological Seminary. His ambition now was to found such a Christian college as Amherst in Japan. The Meiji Restoration in 1867 had eased matters. He spoke of his desire to members of the American Board of Missions. They were not encouraging, but invited him to attend the annual meeting of 1874 in Rutland, Vermont. He forgot his prepared speech, but spoke so movingly from the heart that the members present pledged some $5,000 on the spot to finance the college of his dreams. Neesima returned home that same year, ten years after he had stolen out of Hakodate in the dark of the night hidden in a rowboat. Such was the origin of the

Doshisha, now the foremost private university in Japan with some 19,000 students, Japanese and international. In the midst of its campus in the heart of Kyoto is a federalist-style brick welcoming center that would be at home on the Amherst campus.

Its like, indeed, is there.

This lecture and its sequel on French science, 1780–1820, deal with the motif of professionalization in the context of the relation of science to society and culture. I chose to exemplify it in the two areas I was then working on. The latter had been and remained my principal field of research. I had decided a few years earlier to branch out, or better to return home, and develop an undergraduate course on history of American science. This lecture is an abstract of its main themes.

Notes

1. I was and am much indebted for the opportunity to my colleague, Professor Eikoh Shimao of the Doshisha University, who translated one of my books, *The Edge of Objectivity* (1960), to two of my former graduate students, Professor Sasaki Chikara of the University of Tokyo and Professor Yoshida Tadashi of the Tohoku University, and to my former undergraduate student, the late Jackson N. Huddleston, Jr., Princeton 1960, then the manager of American Express in Japan.

2. Neesima wrote a memoir of his youth in Japan, *My Younger Days* (Kyoto, 5th edition, 1959: The Doshisha Alumni Association). A slightly younger contemporary, the Rev. Jerome D. Davis, Professor of Theology in the Doshisha University in its early days, wrote a very personal memoir, *Joseph Hardy Neesima, a Sketch of His Life* (Kyoto, 5th printing, 1936: Doshisha University).

NINE

The Neesima Lectures I:
The Coming of Age of American Science,
1910–1970[*]

—✺—

T he subject of my two lectures is the relation of science to society and culture in the two historical sectors that I have studied most closely—the one as an historian specializing in French science in its greatest days; the other as an American scholar interested in the formation of a vigorous scientific enterprise in his own country in the century since 1865, the end of the Civil War. I am lamentably ignorant of the history of Japan, and must leave comparisons, if any are pertinent, to the hosts who have done me the great honor of inviting me, and to this audience that is so kind as to attend. But I may begin by observing that the creation of professional science in the United States largely coincides historically with the adoption of western modes of science and technology in Japan. Perhaps both may be studied as problems of acculturation. I shall take the American case first, therefore. In a personal, if not a scholarly, sense, it is more familiar to me than is the French, and for the reason just suggested, it may also be closer to your own experience. That plan also allows me to introduce these remarks with the reflections of a very great scholar, one whose understanding of America grew out of the detachment of the foreigner sharpened by the perception of the friend, and whose relation to America was the reciprocal of mine to France. I can pretend to no advantage in the comparison.

Alexis de Tocqueville completed his famous book, *Democracy in America*, in 1840. Chapter IX, Part II, has the title: "The Example of the Ameri-

---

[*] This lecture and the following one were published jointly under the title, *The Professionalization of Science: France, 1770–1830, Compared to the United States, 1910–1970* (Kyoto, 1983: Doshisha University Press).

cans Does Not Prove That a Democratic People Can Have No Aptitude and
No Taste for Science, Literature, or Art." The first sentence fails to reassure: "... In few of the civilized nations ... have the higher sciences made
less progress than in the United States, and in few have great artists, distinguished poets, or celebrated writers been more rare." Still more damaging is it that Tocqueville attributes this low state of affairs not, as many
would have it, to our being democratic, but rather to our being American.

Preoccupied with exploiting the continent, we were incapable of meditation. Forever in pursuit of gain, we had no taste for seeking truth. Indus-
trious in applications, we could draw upon the great store of pure science
continually created and renewed in the heart of the European culture on
whose periphery we pressed restlessly west. For in Tocqueville's account
of knowledge, the theoretical and abstract parts of science are qualified as
high and noble. Pure science, like philosophy which is its wellspring, pre-
supposes the provision of leisure for disinterested reflection on the part
of gifted persons, and pertains, therefore, to aristocracy.

Tocqueville has been taken at his word, and a body of literature has
grown up explaining or deploring American indifference to basic science
in the last century. Moreover, the few scholars who have disputed the finding have labored under an embarrassment. For even when the historian
travels down the decades, ten, thirty, fifty years on, and arrives at the opening of the twentieth century, he will encounter few Americans in the
world-wide company of leading men of science. He will, in fact, meet only
two who clearly qualify: Benjamin Franklin and Willard Gibbs. And even
now, when you mention the latter name to students, or to their educated
elders, very few will ever have heard of the founder of statistical mechanics and the formulator of the Phase Rule of chemical thermodynamics.

And yet, one somehow feels that this cannot be the whole story of 19th-
century American science, this tale of near illiteracy. Other names come
to mind: Robert Fulton and the steamboat; Eli Whitney and the cotton
gin; Samuel Morse and the telegraph; Commodore Perry and his flotilla:
their machines and a thousand others are insufficiently defined as applications surely, and too vigorous to be classified as parasites drawing their
vitality from a body of science with its real life elsewhere. Having begun
somewhat personally, perhaps I may also be permitted a recollection of
how American science felt to one of its students in the 1930's, a century
after Tocqueville wrote.

Memory is treacherous, but of one statement I am relatively confident.
There was then none of the ambivalence about science that is so marked
a feature of attitudes today. Respect, but neither awe nor dread, was the
prevailing sentiment. Apart from the very few with some inborn voca-

tion, science was something you studied if you needed to and if you could. By "need" I mean, not poverty, but the combination of temperament and family circumstance that made it important to know where you were going and to maximize the probability that your studies would equip you to enter a well-defined career. Looking out upon the world of the depression, we had the sense that the more promising areas would be open to those qualified in chemistry, in electricity, in aerodynamics, in geology, and in their engineering complements, and also in medicine. The old industries, ordinary business, banking and finance, and law looked a lot less promising, and those uncertain prospects were all that lay before people whose studies were literary, historical, economic, and political. Humane and social studies attracted the strongest students—I speak of men now— mainly in the older, private universities serving wealthier families whose children could afford to consult mere inclination.

As for our professors, they were certainly more than teachers in our eyes. We sensed which of them were really doing research in their laboratories, and which were going through the motions. The ethos was such that we well understood that the business of science is the advancement of knowledge, and not only its transmission and application. Most of us had never heard of Tocqueville. None of us had a more formal sense of the state of American science than would a school of fish of the condition of the stream in which they live and breathe. We would have been astonished to be told that we belonged to a culture that held fundamental science in low esteem—astonished and unconvinced. And now that I am aware that science itself can be the object of studies of which such statements are the findings, I remain unconvinced. I do not believe that our attitudes and expectations were the creation of our generation of the Depression, nor even of that of our fathers. Our positive attitude must have had a history, and the problem is to identify its elements and to situate them in the American experience at large.

How, then, were these expectations possible? When and how did science become a profession in America? Where and how were its people educated? What did they actually do and where did they work? How good was their work? When did it begin to contribute to worldwide science? How and why did American science take on leadership? What was its distinctive quality? And what have been the costs, or some of them? And what were the benefits?

First of all education. In this respect, it is certainly true that American science and learning drew upon Europe, both intellectually and institutionally. Neither the training of scholars nor the practice of research formed any part of the function of American universities until a century ago. Before the 1870s any American who wished research training had to complete his education in Europe, and the great majority of the several hundred persons who did so studied in Germany. Historically, American graduate education stems from the grafting of the German research institute, with its instruction in seminar and in laboratory, onto the stem of the American undergraduate college. The first such institution, The Johns Hopkins University, opened in 1876, and its provision for doctoral training was quickly emulated in older institutions—Harvard, Yale, Princeton—and in the leading state universities.

With it, the spirit distinguishing the German university penetrated into American academic life: the commitment to research no less than teaching, the practice of teaching by research, and—perhaps most strongly—the competitiveness that entered into the complex of scientific motivations. Organizationally, the two systems were well suited to emulation. Both in Germany and in the United States, the university system was federal rather than centralized. In both countries, there were many institutions of comparable distinction rather than—as in France and England—only one or two of great prestige dominating and draining vitality from newer, provincial bodies.

Once the process of competition was in full swing, and I should put the date around 1910, it took two forms, institutional and personal. Universities, and other research institutions, competed for talent, thus bringing scientific careers into greater demand. Personally and individually, scientists both collaborated and competed with each other. In the academic sector, the rewards were reputation; in the growing industrial sector—and I shall say more about that—they were also economic. At the outset, universities were the more prestigious forum, however, and still may be. In yet another respect American experience replicated the German model. Scientists made their careers in the universities. But they made their reputations in the several scientific disciplines—mathematics, astronomy, zoology, etc. Science thus came to be organized horizontally in universities and vertically in disciplines, the latter partaking more than the former of an international character.

I shall have more to say of the disciplines in my second lecture, for the modern form is first recognizable in France in the first decade of the nineteenth century. Historians have begun to refer to their appearance as a second scientific revolution. In the United States it first manifested itself in the form in which knowledge came to us, that is to say in the treatises dealing with bodies of subject-matter. The word "discipline" has come to connote both the content and practitioners of a science, however, and in America the institutional development was clearly instrumental in making the transition between our ability to apprehend and our capacity to advance the various sciences. The need was felt. The American Mathematical Society, the American Physical Society, etc., took form in the 1890s, in a matter of ten or fifteen years after the initiation of graduate schools, and as the complement to the same movement toward incorporation of research into the fabric of intellectual life. That development opened possibilities of another sort. For it has been through the medium of professional societies that government has had access to scientific knowledge and reciprocally that scientists have exerted influence upon government.

Industry forms the second matrix for scientific careers, and here the mode of development was much more largely indigenous. Thomas Edison opened his laboratory for invention in 1876, almost simultaneously with the founding of Johns Hopkins. The two sets of people were very different, and would have mutually despised each other if they had ever mingled. Nevertheless, in their organized commitment to invention and discovery we can see them in retrospect as complementary enterprises. They are to be differentiated from each other by social, cultural and economic rather than by intrinsic factors. In substance Edison's analysis of the electric current was as much physics as anything done by the professors who were beginning to call themselves physicists, and certainly, the difference between invention and discovery is more conventional than real. Edison represented the rugged individualist pioneer of industrial research and development, however. The corporate phase followed. The Bell Telephone Laboratories go back to the 1890s; General Electric came on slightly later. Dupont, International Business Machines, Radio Corporation of America, the major oil companies and the rubber companies, all established important laboratories after the First World War. Consulting companies arose of which the whole business was research—Arthur D. Little was the prototype. In general, provision for research had become a characteristic feature of the science-based industries, the new industries, by the time that the depression administered such shocking reverses to the older, heavier industries.

Here, as in the university system, the evolution paralleled that in Germany. There the chemical laboratories in the pharmaceutical and aniline dye industries contributed enormously to the development of medicines and of organic chemistry, which was virtually a German science by the time of the First World War. Even more dynamic, however, have been the technologies of the electric power, the communications, and the information industries. And there American industrial research has clearly been the leader—getting its start, no doubt, in the imperative necessity of establishing far-reaching communication across the entire country.

In addition to industry and universities, the third principal sector that has made provision for science in all countries in modern times is government. In the case of American science, however, nothing systematic may be said about that until the onset and then the aftermath of the Second World War. From the time of the Civil War, branches of government did create agencies with scientific competence for ad hoc purposes: agricultural research stations, weights and measures, public health, mapping and navigation, etc. The work was largely regulatory rather than innovative, but the personnel formed part of the scientific population. As for the First World War, the engagement of scientists was a relatively minor aspect, despite the notoriety of poison gas and the premium on submarine detection. Moreover, the mechanism for enlisting scientists in the service of the war effort was dismantled like the rest of the military structure after 1918.

Neither before nor after the First World War was there anything that could be called a science policy on the part of any American administration. The developments I have indicated in universities and in industry occurred without guidance or participation by the governments, federal or state. Under Franklin D. Roosevelt, the New Deal drew social scientists into federal service and consultation, but took virtually no interest in natural science. Nothing in the nature either of advice or statistics guided people of my generation in our choice of studies. And yet the statistics assembled since the study of science in society became a specialty show that we judged well in our sense of what was going on. This is not the place to recite numbers in detail. Let us simply take the commitment of Bell Telephone to Bell Laboratories. In 1912, its budget was $1,000,000 and its payroll 50 scientists. In 1919, the figures were $6,000,000 and 200 people; in 1925, $12,000,000 and 350 people; in 1934, the depth of the depression, the budget increased 25% to $16,000,000 and 500 scientists were employed. Similar, though perhaps less spectacular, patterns in the other industries I have mentioned made of research, if not the unique, then the surest area of growth in the American economy in the 1930s, when contraction was the general rule. Figures for the number of doctorates in physics and chemistry, and for advanced degrees in engineering, show the same exponential increase in a kind of tacit reaction to reality.

Assessment of quality is always more tendentious than of quantity, and sometimes social historians avoid it altogether. At what points, and in what sciences, did American research begin to contribute to the fundamentals? And here it is clear that in the early stages physics was not the leading edge of American entry into basic science. Those points of entry, at which American innovations contributed fundamentally to modifying the form and content of a discipline, may be identified with some confidence. They occurred in genetics, in physical chemistry, and in astrophysics, and all at about the same time, in the years just prior to the First World War. That is why I have put the opening date at 1910 in the title of this lecture. It is perhaps noteworthy that all three subjects have in common that they were hybrid sciences, arising not so much on the boundaries of the parent disciplines as in the junction between them. May we conclude that it was easier to cross disciplinary lines in the early stages of professional science, before the boundaries became too rigid? And that the relative fluidity and flexibility of American institutions gave us a temporary advantage there at a time when American science was scarcely able to compete in the established fields?

It may be so. At any rate, the modern science of genetics emerged from the synthesis of evolutionary biology with Mendelian heredity in the laboratory at Columbia University where T. H. Morgan and his associates studied the breeding of drosophila, or fruit flies. That program, one might say that campaign, of experiments occupied a team of colleagues and associates at a high pitch of intensity for a span of five years, and at a lower but steady rate of productivity until 1925. Out of that work came the gene theory of heredity in the form of experimental verification of the Darwinian theory of evolution. Further down the line are the entire discipline of population biology, the mathematization of the theory of natural selection, and—internal to the gene—the origin of molecular biology. These latter developments occurred internationally, of course, but the point of departure was Morgan's laboratory, and in my opinion it was the most impressive of the new departures marking the entry of American research into the world-wide forum of basic science.

The second instance, the formation of a new school of physical chemistry, is less dramatic and much less famous, but not less interesting. Like genetical research at Columbia, it is associated in the first instance with a single institution, Massachusetts Institute of Technology, and with a considerable group of scientists. Its leader was the chemist, W. K. Lewis. The thrust here was the conversion of chemical engineering into an abstract rather than an applied discipline: topics like heat exchange, high temperature chemistry, reactions at high pressure, and the process of gas absorption, were treated as problems of physics and chemistry rather than as exercises in the control of industrial machinery. The motivation was partly a taste for elegance, and partly the sense that industry had learned the value of research all too well, and that its commissions were turning the institute into its hand-maiden. The significance, apart from the entry into systems research in general, is that the summons to generality should have been heard, not in the conventional reaches of basic science, but in a realm of industrial science—heard and answered.

As for my third topic; it is significant that the professional journal of American astronomers has been named astrophysical from the date of its foundation in 1899. The term signified commitment to what was called the New Astronomy, physical as distinct from positional. It studied to know, not where the stars are, but what they are physically, what their processes are, and how they have evolved over the life of the universe. The most fundamental relationship in classification of the stars is that between energy output and surface temperature, and the correlation of those parameters was in part worked out in 1911 by Henry Norris Russell at Princeton University. Its importance in subsequent work is comparable, perhaps, to the technique in genetics for mapping the location of genes upon the chromosomes developed almost simultaneously at Columbia. But gathering and verifying the information depended on another, perhaps more characteristically American contribution. I have in mind the construction of ever larger telescopes, financed by a kind of scientific extortion practised upon newly rich industrialists: The 40-inch Yerkes Telescope in 1899; the 60-inch Mount Wilson reflector in 1908; the 100-inch reflector in 1917; the 200-inch Mount Palomar in 1947. Except for the last, each was started before its predecessor was even completed, let alone in service, and each penetrated orders of magnitude further into space. The 100-inch had 2 1/2 times the light gathering power of the 60-inch—and seven times that of any other instrument in the world.

The last development, along with the relative importance of science done in industrial laboratories, makes evident that already by the 1920s it is possible to identify the emergence of a distinctively American style in science. It was nothing philosophical; not the counterpart of Cartesianism in French science, Baconianism in British science, or Kantian idealism in German science, and not at all, therefore, what Tocqueville had been looking for. No, what was distinctive in American science was the scale: the ability to bring large masses of men, money, and equipment together over great distances to achieve a result. Indeed, throughout all of American history, one of the leitmotifs has ever been the enlargement of scale. Nowhere has its operation been more apparent than in science, where finally the dimensions have become coextensive with the dimensions of industry and technology.

So it clearly was in the science of physics, to return to the discipline that I believe to be the central nervous system in the whole body of science in modern times. It was clearly not by way of theoretical contributions that American physicists began affecting the development of physics. Neither in relativity nor in quantum mechanics were the fundamental formulations originated by Americans prior to our entry into war. In the 1920s and 1930s our contributions were of two sorts, institutional and instrumental. Institutionally, the scientific establishment, and particularly the universities, proved capable of welcoming a large proportion of the élite of European physicists, most of them Jewish, and driven from Germany and Central Europe by the Nazi regime in Germany. That migration occurred, of course, not only in physics, but across the entire scholarly and scientific spectrum. Indeed, the episode was perhaps unique in history: the migration at a single moment of an entire intelligentsia from one continent and culture to another. The effect—again I speak for a generation that had the benefit of their presence on hundreds of college campuses—was enormously stimulating, I may say elevating, to the tone of intellectual life.

Tone and substance are not the same, however, and testimony is virtually unanimous, even among the immigrants, that excellent working physics was already a going concern in the United States, and that only the mathematical edge, the conceptual imaginativeness, was still wanting by comparison to what had been largely destroyed in Europe. And if we look to the decisive contribution from America in the 1930s, it was not some body of new or refined theory produced by the immigrants. It was the invention and development of the cyclotron at the hands of E. O. Lawrence and in the laboratories of California. The story repeats the motifs of the rapid succession of the giant telescopes of astrophysics: the first in 1931 was a table-top machine with a 4-inch diameter; the second was an 11-inch machine that got over a million electron-volts; the third was a 27-inch machine weighing 85 tons, capable of 6,000,000 electron-volts. Its diameter was later enlarged to 37-inches. The fourth was a 60-inch machine that cost the Rockefeller Foundation more than $300,000 and weighed 200 tons. It was the first scientific instrument on the scale of an industrial dynamo. All that was accomplished by 1939 in the span of 8 years, by never waiting for the last to be completed before planning its successor. The next machine projected in 1940 was to weigh 4,000 tons. By the time of the war, 14 cyclotrons were in operation in America, and only one, a small but clever one, in Europe, at the Collège de France.

Neither Lawrence nor anyone then involved in atomic physics foresaw that their instruments would one day have to do with warfare. In five years time, they and all informed persons would certainly have accepted the proposition that the relation of science to military force posed the most urgent moral and practical problem faced by our entire civilization. I know no more striking example of the sense in which historical vision is a one-way street looking back. In retrospect the record always appears to be a closely knit fabric, but those whose fingers work the loom can never see what the pattern is going to be. It is true that there were scientists in the United States who, soon after the outbreak of war in Europe, thought to equip the government with the means to draw science into a program for national defense. They were mainly science administrators, however, rather than persons engaged in research—Vannevar Bush, President of the Carnegie Corporation; James B. Conant, the President of Harvard University; Karl T. Compton, the President of M.I.T. Guiding themselves on British experience they expected that the major contribution would come from physics; more specifically, the physicists and military experts they consulted welcomed the British mission headed by Sir Henry Tizard in August, 1940, which imparted to American authorities the design of the cavity magnetron, the device that applied short-wave radar detection to gaining victory in the aerial Battle of Britain.

The further development of radar formed the first great commitment of American physics to the war effort. An entire organization called the Lincoln Laboratory, administered first at M.I.T., was created to that end. Its director was the Nobel Prizewinner, I. I. Rabi. I recur to the question of scale: at the end of the war, its scientific personnel numbered over two thousand physicists and engineers, without counting the still larger number employed on subsidiary missions contracted out to other laboratories and to industry.

Only later did atomic energy become of comparable and then, of incomparably greater moment. As everyone knows, the possibility was first pressed upon President Roosevelt over the signature of Einstein by three Hungarian refugee physicists in the summer of 1939. They were intimately acquainted with the capacities of German physics and fearful lest Hitler secure the weapon first, if indeed such a weapon proved possible at all. It required two years of rather desultory investigation before the prospect was taken seriously, and then in consequence of parallel but more urgent research in Britain, where too the refugees from Germany had alerted the military and scientific authorities.

I do not know who has the moral stature to impose judgment, either about what ensued, or about the antecedent events. Certainly not I, an American scholar invited to visit in Japan. Still, one has to try to satisfy one's own mind. As a scholar, I have read fully in the record. As an historian, I am persuaded that science is not to be abstracted from history, and that so far in human history neither is war. I cannot see that the scientists who employed their knowledge in war are to be held more accountable to some higher humanity than are men in other walks of life, whether soldiers or politicians or businessmen. For what it is worth, my tentative conclusion about the small group of persons responsible for American actions, both politically and scientifically, is that at one juncture they were at fault beyond the bounds of normal fallibility. I feel they had no responsible choice but to develop atomic energy. But I also think they could and should have found a way to demonstrate it before employing it. There are difficulties with that conclusion, and I would not wish to urge it upon the conscience of others.

Allow me to conclude in another vein. Throughout history, of course, warfare precipitates developments of many sorts, and generally tends towards the progressive aggrandizement of the state over aspects of life other than political. It was no accident that America arrived at great-power status in science not quite a generation after doing so in politics. In America, it was never possible, as it had been after the First World War, to return to the practice of science as simply individual research, nor even as serving mainly industrial development in the sector of private enterprise. Few people even wanted that, either from the side of science or of government, and it appears to me that a major consequence of its enormous involvement in the war was to create very suddenly in America the structure of relations between science and government that had evolved more gradually between the older scientific and political establishments of Europe. I shall be arguing in the next lecture that this system was first fully recognizable in French polity at the end of the Old Regime, if I may borrow the wording that serves as the title of a book I have recently completed on the subject.

Let me simply say here what I think the relations have been. Science and politics are clearly enterprises of very different sorts, and to try to make politics scientific, or science political, simply produces confusion. At the same time there are interdependences and interactions. What is it that statesmen have generally wanted of science? They have not wanted admonitions or collaboration, much less interference, in the business of government, which is the exercise of power over persons, nor in the political maneuverings to secure and retain control over governments. From science all the politicians and statesmen want are instrumentalities, powers but not power: weapons, techniques, information, communications, and so on.

As for scientists, what have they wanted of governments? They have certainly not wished to be politicized. They have wanted support, in the obvious form of funds, but also in the shape of institutionalization and in the provision of authority for the legitimation of their professional status. Indeed, it has ever been a feature of the scientific enterprise that it has

needed to draw upon authority—externally in the manner I have just indicated, and also internally or intellectually, in the interest of maintaining
standards of rigor and discipline and in creating incentives. The relation
of science to the state, then, has generally been one of partnership, rather
than one of partisanship, and has normally exhibited a certain independence from the strife of faction and party that constitutes the political process itself.

I do not think that this mutual dependence of science and government,
of knowledge and power, is inconsistent with a liberal form of political authority, though neither does it entail such a form. The legitimacy of the authority is the limiting factor, and that (I think) has been specially evident
in the association of science with politics in America since they were thus
suddenly brought into interdependence by the prolongation of wartime
relations into peacetime. I do not mean that science has been under military domination—on the contrary, the assertion of civilian control over
atomic energy in the immediate aftermath of the war was one of the triumphs of the liberal tradition. But the flourishing of science in the period
of economic growth in the 1950s, the national consensus in favor of further stimulating our scientific effort in the decade after sputnik in 1957,
the space program in the 1960s—all that bespeaks a sense of purpose, almost a general will. And it is significant that the movement of dissent, specially among the younger generation in the Vietnam years, should have involved revulsion both from science and from politics. That would carry
me beyond my closing date, however, and I am already too close to the
present to feel comfortable as an historian. I must impose no further on
your kindness, and—not to leave the sense of an adversary relation with
Tocqueville, whom I so admire—I shall stop with a reflection at which he
arrives at the end of his own discussion, where he admits that aristocracy
may not after all be the precondition of science:

> "It is, therefore, not true to assert, that men living in democratic
> times are naturally indifferent to science, literature, and the arts;
> only it must be acknowledged that they cultivate them after their
> own fashion, and bring to the task their own peculiar qualifications
> and deficiencies."

TEN

The Neesima Lectures II:
The Flourishing of French Science,
1770–1830

—⚋—

In my lecture on American science, I ventured to characterize a dynamic structure of relations between science and the state, between knowledge and power, as one of the features defining modern political systems. I also suggested that it first took on identifiable form during the period of French scientific eminence at the end of the eighteenth century. The appropriate place to begin developing that proposition is with the ministry of Turgot, the statesman and encyclopedist who drew upon science and systematic knowledge in formulating policies intended to rehabilitate the French monarchy on the accession of Louis XVI in 1774. During the next half-century, say between the last years of d'Alembert and the death of Laplace in 1827, the French scientific establishment predominated in the world to a degree that no other national complex has since done or had ever done.

Its eminence persisted through the lifetime of two generations rather well marked off one from the other. The earlier was that of Lavoisier, Laplace, and Monge in their most creative years, of Lagrange in his maturity, of Coulomb, Buffon and the young Lamarck, to name a few of the more famous—in a word, of the final generation of the old Royal Academy of Science, founded in the seventeenth century along with the Royal Society of London. Their successors were the first generation of the Institut de France and the École Polytechnique, still the two senior technical bodies in France; I mention here a few names that appear in all our current textbooks—Ampère, Dulong, Petit, Fresnel, Fourier, Poisson, Cuvier, Bichat, Sadi Carnot, Cauchy.

There was a contrast in spirit between those two generations. The outlook of the former was encyclopedist and pertained to the eighteenth century and to the Enlightenment. The outlook of the latter was positivist and pertained to engineering and to the nineteenth century. The succession in generations corresponds to large-scale political phases. The earlier extended from 1774, the beginning of attempts at reform, through the opening of the Revolution in 1789, down to the overthrow of the monarchy late in 1792. The Reign of Terror that ensued in 1793 and 1794 was something of a hiatus in science, though certainly not in politics. The second phase, like the careers of our second group, extended from the reorganization following the overthrow of Robespierre in mid-1794 through the Napoleonic period and the Restoration down to the July Revolution of 1830.

In the work of this half-century of science, I believe that French cultural leadership in Europe reached its zenith. I mean that statement very broadly. The critical movement of classicism that distinguished the French intellectual spirit first made itself fully felt in the reign of Louis XIV in the seventeenth century, and in the realm of letters, architecture and manners. Thereafter it took the form of the system of rational ideas about nature, humanity, and society called the Enlightenment. Passing over from thought to action, that movement issued in the incorporation of science into polity amid the circumstances that are the subject of this lecture. Throughout, the sectors in which science came to be of moment to the state were broadly those in which its relation with society have transpired generally in modern history. They were three, as we observed in the case of the maturing of American science over a century later: first, administration and public works, civil as well as military; second, education, as to both training and recruitment of élites; third, technology, in respect to industry and agriculture, to engineering and invention.

I shall consider first the sense in which science figured in the measures introduced into administration by Turgot and by his entourage. Like their predecessors, the writers and *philosophes* of the type of Voltaire, Diderot and Rousseau, the enlightened reformers of the last decades of the Old Regime were critical of authority. Unlike literary people, they were critical of it rather in its exercise than in its existence, however, and proposed indeed to avail themselves of the instruments of authority to bring about rational changes. Expert knowledge was their hallmark rather than propaganda and philosophy, and the enemy was rather routine and ignorance than despotism in the abstract.

First, public affairs: perhaps the most striking, though not the most effective, illustration was the application of mathematics to social phenomena and governmental processes themselves. I have discussed this aspect of the subject with students specializing in the history of science at

Tokyo University, and here I will simply quote Turgot's conviction. It was "that the moral and political sciences are capable of the same certainty as those that comprise the physical sciences, and even as those branches . . . which, like astronomy, seem to approach mathematical certainty." Condorcet was Turgot's principal man of confidence in the Academy of Science. In service to his patron's belief, Condorcet began the application of the calculus of probability to analysis of the selection and voting procedures of electoral bodies and of judicial panels. Even more important for the development of the theory of probability itself, Laplace started the statistical analysis of population figures which had been collected systematically in France throughout the eighteenth century, and thus initiated a mathematically based science of demography. Thereupon the taking of a census as a normal act of government followed early in the Napoleonic period.

Of all the reforms stemming from the Revolution, the one that most regularly affects the daily doings of scientists and other men was certainly the creation of the metric system of weights and measures. It is a common mistake, however, to suppose that it began in the early phase of a revolutionary passion for basing social arrangements on uniformities in nature. Like most things in the Revolution, its origins go back to reforms attempted by the Old Regime, and specifically to Turgot's appointment of Condorcet to be Inspector of the Mint. His instructions were to develop a proposal for standardizing and unifying commercial and scientific units of measurement, preferably on a decimal basis. Again like many such improvements, only the impetus of the Revolution invested government with the authority to carry the change into effect. Another misinterpretation, of a more technical nature, concerns the definition of the meter as the ten-millionth part of the quadrant of the meridian. That decision is also often attributed to ideology, or else to the desire of astronomers to secure government funding for an ambitious geodesic survey. The alternative would have been adoption of the more easily determined length of a pendulum with a frequency of sixty oscillations per minute at the 45th parallel. In fact, the preference for the meridional survey had a perfectly good mathematical justification; Laplace was the one responsible for it. Only so could celestial and terrestrial measurements be numerically interchangeable, so that unit angle on the celestial sphere subtended unit length on the earth. Unfortunately, angular measurement is the one sector where decimal subdivision failed to catch on.

Armaments also pertain to public works, and of all Turgot's administrative reforms the most immediately successful was his reorganization of the gunpowder industry. Everywhere in Europe fabrication of powder had been a sovereign function, like coinage or justice, since the introduction of explosives from Asia in the fifteenth century. In the late seventeenth century, the scale became so large that the French government began leasing out the facilities to private enterprise. Speculators took over in the eighteenth century. An ancient trade-guild had a monopoly of exercising the royal right of entry to scavenge in private property for saltpetre, the principal ingredient. The saltpetremen preferred living from bribes by owners who paid them not to search, and the financiers, who were supposed to monopolize their product to produce gunpowder below cost, concentrated on speculating with the advances from the crown instead of on manufacturing munitions. The situation was a scandal, therefore, typical of the entire regime of privilege and exemption, and blamed by the military for their loss of the Seven Years War against England from 1756 to 1763.

In correcting it, Turgot turned first to the Academy of Science, and charged it to institute an urgent and exceptional study for chemical production of saltpetre. He dismissed the powder farmers and substituted a commission responsible directly to the Ministry of Finance. More important, he put Lavoisier in charge of the commission. Thus did the science of chemistry and its leading light enter into the sort of relation to the military which is familiar to our times, and to America, from the example of the science of physics and its leaders in the 1940s. One consequence is seldom noticed by historians of science. It is that the chemical revolution, out of which emerged the modern science, occurred in the Arsenal of Paris. Lavoisier's laboratory there was the nerve center, the social center, and the experimental center of French science in the years down to 1792 and his fall from political favor in the political revolution. Lavoisier succeeded brilliantly in restoring the munitions industry to self-sufficiency, with capacity adequate to furnish arms also to the American armies in our revolutionary struggle. It is true that no new theory was involved, and that the triumph was one of responsible and expert administration rather than of science—unless, as I believe, that itself can be called scientific. Moreover, Lavoisier was arrogant and high-handed and made enemies of the suppliers, the working men, and the population in the quarter around the Arsenal. His unpopularity contributed to his arrest and execution by the guillotine at the climax of the Terror in 1794, at the very time when the revolutionary armies were being supplied with ammunition by methods he had devised.

Only as a consequence of institutions created in the French Revolution could education become one of the principal sectors wherein the place of science was important to the state. France had admirable secondary schools in the eighteenth century, and many of them taught science well. They were all religious foundations, however, and the one institution providing higher technical education, was the small military engineering school at Mézières. For scientific education in general, all we have in the early phase of the Revolution is a series of projects. They are interesting, however, in showing what the leaders of scientific opinion—Turgot, Lavoisier, Dupont de Nemours—had in mind.

They considered, first, that education at all levels must be taken out of the hands of the church and assumed by agencies of the state. They considered, secondly, that citizenship would pre-suppose an educated population, at least up to the level of universal literacy. They considered, finally, that instruction in science and its applications must replace Latin and the humanities as the staple subject-matter in the formation of responsible and productive citizens. Sensibility must be oriented to change and to shaping the future, and no longer to tradition and conserving the past with its agelong accumulation of abuses. The Academy of Science was itself to be reorganized and enlarged and turned from the fount of honor and supreme tribunal of research into the ultimate authority in education, prescribing the curriculum, writing the textbooks, examining the qualifications of teachers, inspecting the operations of the classroom. That never came about, but that notion of a proper role for science in society tells much about the mentality of the leaders of the reforming generation among French intellectuals.

In the area of technology, to turn to our third sector, science had not yet reached the stage where basic theory was proving immediately applicable to processes of production, as it began to be after the middle of the nineteenth century. Science was related to industry rather in a descriptive and regulatory capacity. Throughout the eighteenth century, the French government called on scientists to undertake studies of industrial processes, not so much with a view to finding new ones, as to determining what were the best ones. Officials had in view bringing French industry abreast of more progressive and productive technologies that had evolved in the free enterprise system of the British economy, and also in Germany and Scandinavia, mainly in metallurgy. The result was a kind of natural history or encyclopedia of industry, of which the prime examples are the great multi-volume *Description of the Arts and Trades* published by the Academy of Science from 1750 through the 1780s, and the technical parts of the Diderot *Encyclopedia* itself. Under government encouragement, scientists produced a literature of manuals bringing into the open processes that had always been handed down as trade secrets—and identifying the best of them.

More specifically, the government also made of scientists the arbiters of its policy of encouraging and subsidizing inventions. Patents were unknown to French law before the revolution, but the government did make a practice of awarding prizes, subsidies, and sometimes monopolies, for a term of years to inventive entrepreneurs. When an artisan with an invention or a manufacturer with a new process approached the government with the request for a grant, the proposal would be referred to a committee appointed by the Academy of Science. It was their duty to pass upon its originality and merit. Many members of the Academy themselves had direct technological responsibilities in government regulatory agencies and state enterprises. In effect, the Academy of Science was thus the highest technological authority in the state, and the exercise of its responsibility inevitably created resentments. The great majority of so-called inventions were chimerical and valueless. In keeping with the haughtiness and disdainfulness that often characterized class relations in the Old Regime, scientists developed the reputation among artisans for arrogance, pride, unfairness, and even exploitation of the ideas of practical men. Those resentments were politically the most powerful aspect of the hostility to the Academy of Science that manifested itself as the revolution moved to the left, and the champions of the sansculottes, or working classes, gained power.

At the same time, a current of intellectual and emotional hostility to exact and abstract science emerged among certain writers, politicians, and demagogues in the early 1790s. That attitude, as a kind of unformulated, deeply felt anti-thesis to scientific rationalism, goes far back into the cultural history of the eighteenth century. Rousseau was its greatest exemplar. The attitude exhibits a psychology of alienation from the structure of modern society that is always latent, in my judgment. It emerges, or becomes influential, in times when authority is called into question, and is then often associated with radical political movements, whether of left or right. Other examples are the world-wide anti-scientism of the 1970s and the anti-rationalism that was so prominent a feature of German culture in the Weimar Period before the rise of Hitler. It is a further feature of these mentalities that they couple rejection of authoritative, abstract, mathematical or mechanistic science based in physics with enthusiasm for what they imagine to be an alternative science based in biology and in romantic sympathies between man and nature.

The contest between these conflicting attitudes, and the empathy between the latter and political radicalism, makes sense out of what the historian of science would otherwise find to be a very puzzling paradox in the French Revolution. For, in the summer of 1793, as power came into the hands of the Jacobin left and the dictatorial Committee of Public Safety, the enemies that the Academy of Science had made in its domination of technical affairs carried the day. On 8 August 1793 the Academy of Science was declared to be a body whose existence was incompatible with a republic, and it was abolished, along with other privileged academies. Those who had exercised leadership since the time of Turgot—Lavoisier, Cassini, Condorcet, Bailly, Laplace, Vicq d'Azyr—went into prison, and some to the guillotine, or into hiding. But at virtually the same time, the Convention, which had thus dismissed the guardians of rigor in exact science, lavished favor on the institution responsible for the biological sciences. The old Royal Botanical Garden—the famous *Jardin des Plantes*—was democratized and confided to the responsibility of its own staff, and twelve professorships of natural history, botany, anatomy, geology, paleontology, and so on were created. Twelve chairs: there was no scientific institution in the world in the 1790s with so munificent a provision of professorships. In the contrasting fate of institutions, the reality of radical hostility to exact science and enthusiasm for life science is perfectly evident.

Equally interesting is the conduct of the scientific community when under attack. In no way did it react like a political faction. There was no opposition from science to the regime that had proscribed its corporate existence. Indeed, throughout the entire period that concerns us, from the Turgot Ministry to the July Revolution of 1830, the political behavior of scientists makes striking contrast with that of other groups among the intelligentsia, such as artists, writers, philosophers, and social scientists. The scientists pressed into the service of each regime, without regard to political or constitutional distinctions, and from each they drew a measure of advantage for science. The French Revolution was the crucible in which modern politics was formed, and not only for France. For that reason the behavior of the scientific community, the most developed in the world, is the paradigm case of what I have called the relation of partnership rather than partisanship that obtains between science and government, whatever the momentary strife of party or of principle.

France was at war from the spring of 1792. The suppression of the Academy of Science coincided with extreme military urgency in the summer of 1793. Far from rallying to their fallen leaders, remaining members of the Academy rallied to the war effort in the autumn, winter and spring of 1793–1794—the year II in the revolutionary calendar. They took the lead in devising and overseeing methods for emergency production of saltpetre, gunpowder, firearms, cannon, and clothing for the armies. It would be an exaggeration to attribute French victories to the con-

tributions of science, but the scientific commission on armaments that
served under the governing Committee of Public Safety certainly marks
the entry of science into the military service of the modern nation in arms.
It will convey the spirit of the effort if I quote from the account Cuvier left
in the official éloge of Berthollet, the leading chemist in France after the
arrest of Lavoisier:

> Everyone recalls that prodigious and sudden effort which as-
> tounded all Europe, and aroused admiration even among the enemy
> it thwarted. Monsieur Berthollet and M. Monge were the moving
> spirits. It was according to their instructions that this immense
> movement was directed. The chemists who were commissioned to
> conduct tests for so many new procedures worked only by their in-
> structions; and it is said that, if they had wished to follow up all the
> secrets they came upon, weapons more powerful than any we pos-
> sess would have emerged from their laboratories.
>
> It would be wrong to suppose that the use of such inventions is
> at last accounts as harmful to humanity as their effects are alarm-
> ing. Exactly the contrary is the case. It is not only that science, in fur-
> nishing civilized peoples with these means of defense, has been the
> sturdiest shield of civilization itself;
>
> Nor is it merely that science has been able to count on the sup-
> port of government only since it became one of the essential ele-
> ments of the art of war.
>
> But, paradoxical though the assertion may appear it would be
> easy to prove that the means of destruction furnished by science, in
> rendering combat more decisive, have made wars less frequent and
> less murderous.
>
> As for M. Berthollet, what he primarily saw in these extraordi-
> nary developments of human industry, motivated by the greatest of
> interests were simply chemical experiments on a large scale.

Not for nothing was science thus involved in the events that changed
the world. Traversing them consummated its own transformation into the
professional enterprise that it has been ever since, and that it fully became
in the careers of the second, the Napoleonic generation of French scientific leadership. Political reaction following the liquidation of the Jacobin
regime in July 1794—the 9th of Thermidor in the revolutionary calendar—never lessened the enthusiasm with which the scientific community served the state. Napoleon was following Jacobin precedent when, the
youthful general on the rise, he named a small staff of scientists to ac-

company him in the invasion of Italy in 1796. Monge, Berthollet, and Geoffroy Saint-Hilaire were the principle parties, their mission being to appropriate whatever of scientific and artistic value might serve the republic. Such was the satisfaction that Napoleon took in their company in the respites of the campaign, that he conceived the far more elaborate notion of attaching an entire scientific and archaeological expedition to his task force when he set forth on the conquest of Egypt in 1798. The camp-following of sixty men of science and learning that he there organized into the Institute of Cairo founded the science of Egyptology. Monge and Berthollet were the leaders once again, but they were getting to be elders. The younger men, notably Fourier, Geoffroy, and Savigny, had the exciting experience of a summons to participate in a great affair at the impressionable beginning of their careers. The entire adventure presaged the resonance that scientific eminence would lend to the Napoleonic regime in later years.

In speaking of professionalization, we are concerned with the second scientific revolution, as historians call the transformation that came over science in the first decade of the nineteenth century. It had to do both with the content and the organization of the various disciplines. It will exhibit what happened if we compare the classification of knowledge as it appeared in the *Encyclopédie* of Diderot and d'Alembert in the 1750s to that of Auguste Comte, the founder of the positive philosophy, who was educated at École Polytechnique from 1817. D'Alembert divides physical science into mathematical science and general plus experimental physics. The former consists of rational mechanics with a little astronomy and optics. Problems of sound, electricity, magnetism, heat, chemistry—all that belonged only to experimental physics whereas the life sciences and earth sciences pertained to natural history except for the medical parts of anatomy and physiology.

Auguste Comte, on the other hand, elaborating his philosophy out of the very practice of the first generation of the nineteenth century, conceived the order of the sciences to be that which every schoolboy now supposes to be simply natural: mathematics, astronomy, physics, chemistry, biology, and sociology. It is important to note that the d'Alembert classification was reflected in the organization of the old Academy of Science, and the positivist in that of the body that replaced it in the reorganization of science after the Terror, the Institute of France.

What had brought the change? To a very considerable degree, of course, factors internal to the sciences, in the obvious sense that experimental knowledge and mathematical analysis had developed to the point that the mathematical physics of Poisson, of Fresnel, of Fourier, of Ampère was in fact possible. The same might be said for biology, considering the analysis of zoological information by the techniques of comparative anatomy and of the creation of experimental physiology in the new practice of clinical medicine in teaching hospitals. It is virtually a law of the evolution of science that its direction has been toward more sophisticated and extensive modes of quantification, and at any stage the qualitative is outranked or superseded by the quantitative.

In a sense, the statements I have just made beg the question, however, since what we wish to know is what had changed the circumstances of men who can now properly be called professional scientists to bring about those developments just there, just then, and in that way. It is as anachronistic to speak of professional science in the eighteenth century as it is of mathematical physics, and perhaps it would be well to be definite about what we mean by professional. A profession, I take it, is distinguished by three attributes. First, it is an association more definite than an occupation, in that its practice presupposes mastery of a body of knowledge, and thereby qualifies for the prestige attaching to the cognitive. In the second place, however, a profession does share the economic character of an occupation. It is legitimately followed for gain and is not a status held of right, even though terms like fees or honoraria are preferred to words like wages or profits. Finally, and most distinctively, a profession is self-governing, in that it exercises jurisdiction over the education, qualifications, and conduct of its members, usually by tacit or actual delegation from the state, supposedly in accordance with the public interest.

If this definition of professional is acceptable, it follows that the word when applied to vocations prior to the French Revolution must still be reserved for divinity, law and medicine. Science did not yet qualify. It was coming close, closer than the humanities or social sciences, and much closer than pedagogy, engineering, or the military. But the French community of science, though more advanced than any other under the Old Regime, could fully satisfy only the first of these criteria, the possession of natural knowledge. As for the second, livelihood, scientific breadwinning was an unsystematic affair. For the most part, members of the old Academy did hold positions of some sort involving technical knowledge and supporting a modest style. But if they were not sinecures, and some were, they generally entailed purposes other than research and engaged the scientist's ability only obliquely to his investigations and well behind their frontier—school-teaching, examining military cadets, pharmacy, consulting on porcelains or dyes, and so on.

Turning to the third sector, self-governance, the Academy did exercise a large and adequate measure of control over its own membership, though botanists might yet have a voice in the selection of chemists. But it was in the field of education that the situation remained almost wholly unprofessional before the Revolution and before the foundation of the École Polytechnique and the École Normale in 1794 and 1795. That occurred in the immediate aftermath of the Terror and amid the continuing pressures of the war. The École Normale was conceived in too generous a fit of enthusiasm at first. Intended to train teachers for a nation full of schools, it tried gathering some 1400 pupils into the auditoriums of the *Jardin des Plantes,* and they ranged from virtual illiteracy to the virtuosity of the young Fourier.

The École Polytechnique, on the other hand, was a success from the very outset. Conceived by Monge, Prieur and Carnot, Polytechnique was intended to open to general competition the opportunity for the kind of education formerly given to the military engineers at Mézières, though on a larger scale and carried to a far more advanced level. The school long bore the marks of its origin. Its regime was paramilitary. The numbers were significant but manageable—392 students at the outset.

They were chosen by competitive examination. The course required three years of intensive application. For the first time in the history of science, students were being put through a systematic scientific and mathematical curriculum under the foremost scientific minds: Lagrange on the theory of analytical functions; Laplace on the theory of probability; Monge himself on descriptive geometry; Fourcroy, Berthollet and after him Gay-Lussac on modern chemistry. The students were able and eager. They lived their days at Polytechnique exhilarated by the sense of being conducted to the very forefront of scientific conquest, and being told that the future of mankind, of the Republic, and not least of themselves depended on how they performed in so exposed a situation. The mood was messianic, and this was the spirit turned into philosophy by Comte, for whom science would know only in order to predict and predict only in order to control. Indeed, positivism in scientific thinking was a proper creation of the Revolution, not of this or that faction or party, but of its consciousness and action, whether directed left or right. It was the philosophy of that thrust which Revolution and Empire made in common, the philosophy of that science which would fulfill itself in engineering—civil engineering, social engineering, perhaps the engineering of humanity itself.

Yet when they graduated, Biot, Malus, Fourier, and the rest, they did physics. (So, too, did those formed in the courses of the regenerated *Jardin des Plantes* and the schools of health do biology. The story there is congruent, if less dramatic.) To see how it was that these, the modern scientific disciplines, emerged out of this crucible, we must look to the combination of the new pedagogical and educational modes with the older tradition of encyclopaedic rationalism revived and redirected in the Institute of France—a living encyclopaedia, its founders called it. A revival of the academies in republican form, the Institute carried over into the new order the responsibility of an enlightened state for science, arts and letters. The difference in the dispositions reflects what had happened in the cultural scale of values. There was as yet no replacement for the literary rule of the *Académie Française.* The Institute consisted of three "classes," science coming first in precedence and numbers with sixty resident members, moral and political science second with thirty-six, and fine arts and literature third with forty-eight. Each class was subdivided into sections according to a modern definition of the disciplines: mathematics, mechanics, astronomy, etc. The first class resumed the functions of the Academy, serving science as the goal of young ambition and the guardian of standards, and the state as high court of technical resort. Laplace and Cuvier became the lawgivers in their respective spheres, and for Laplace the creation of a mathematical physics was no mere function of the logic of development of his subject: it was a matter of conscious policy, to be elicited from the young men, to be favored in the publication of this memoir and not that, in the award of prizes or places to this person and not the other.

These, the sociological and institutional factors in the professionalization of science, may be specified quite explicitly, I think, and they lead me to feel some confidence in a final remark that I'll make bold to venture. It has to do with the intangibles, the qualities of pride and collective self-confidence that distinguish the professional man from the retainer, who lives from patronage and serves a master. The founders of Polytechnique and its first generation of students saw themselves in a heroic guise, continuing at a high level of mathematical sophistication the enlistment of science in the service of the republic that had begun in the Levée-en-masse and the revolutionary production of weapons, material and even strategy in the extremity of the year II. That sense of having played a worthy part in the great events of the time—it made men of them, and not merely professional men.

ELEVEN

Science in the Eye of the Beholder, 1789–1820*

—⚋—

O ver the last forty-odd years I have been occupied with preparing two large books and several spin-offs on French science during the long half-century of its preëminence, from the 1770s into the 1820s.[1] So far as science internally is concerned, the argument of the final book is that the scientific movement of the first two decades of the nineteenth century was the seedbed of mathematical physics, a rigorous and deterministic biology, and physical chemistry. With respect to physics and biology, we have to do with the formation of new disciplines. Not so chemistry, an established discipline wherein the focus and emphasis shifted from the affinities and properties of reagents, taken as givens, to the physical factors affecting the course of reactions.

The term physics was then displacing natural philosophy and entailed experimental, not mathematical procedures. Phenomena newly to be subjected to mathematical analysis, largely at the hands of Pierre-Simon Laplace his disciples, their opponents, Joseph Fourier, and Sadi Carnot, were those pertaining to optics, electricity and magnetism, acoustics, heat, work, and energy. In optics Etienne Malus formulated expressions governing double refraction. Augustin Fresnel worked out the equations entailing the wave theory of light. Simeon-Denis Poisson mathematicized the forces of electrostatic and magnetic attraction and repulsion. André

* This essay is a slightly modified version of one that I am contributing to a Festschrift for Paolo Rossi to be published by Leo Olschki in Florence and edited by John Heilbron. I am including it in this collection with Professor Heilbron's kind permission. I am grateful also to Paul Forman, whose well-taken criticisms are responsible for the modifications. They concern mainly the discussion of whig history and the distinctions adduced in the concluding paragraphs.

Ampère accomplished an experimental and mathematical construction of electrodynamics. Poisson and Sophie Germain developed competing formulations for the behavior of sound waves. Fourier invented the analysis that goes by his name in order to give a mathematical account of the propagation of heat. Sadi Carnot's heat cycle was the starting point of thermodynamics. Gustave-Gaspard Coriolis, finally, defined the quantity work and equated it to what would be called kinetic energy.

Although Jean-Baptiste de Lamarck and Gottfreid Reingold Treviranus had independently coined the word *biology* in 1803, the new term came into currency only after Auguste Comte employed it in *Cour de philosophie positive* (1830–1842) to cover the new disciplines of comparative anatomy and experimental physiology. Until then zoological anatomy had been a branch of natural history wherein it served to classify species in accordance with their external characteristics. Employing dissection and not mere observation, Georges Cuvier and Etienne Geoffroy Saint-Hilaire, scalpel in hand, inaugurated the transformation of anatomy and of zoological taxonomy by analyzing how the internal organization of animals related species to each other and fitted them to occupy the environmental niches that they do. Such was the program of research that occupied much of the staff of the Muséum National d'Histoire Naturelle in the first two decades of the nineteenth century. In the same period Xavier Bichat, Julien Legallois, and foremost among all François Magendie inaugurated experimental physiology with the practice of vivisection, interfering with or excising particular organs or systems of organs in the living animal in order to determine their contribution to the functioning of the whole organism. Until then physiology, introspective and literary in vein, had been largely a philosophical part of medicine.

The essay that follows is an inquiry into how the body of science wherein those changes occurred appeared in the eyes of contemporary practitioners. The motivation is simple curiosity in the first place. In the second place, my sense is that the currently enjoined prohibition of whig historiography is probably overdone.[2] It may well be impractical if not impossible to write historically of science and other activities solely in the light of contemporary knowledge, awareness, and standards. I shall attempt here to determine what one could and what one could not say about early 19th-century French science if it were possible to induce in oneself an experimental amnesia blanking out awareness of what came after.

Two sorts of evidence are relevant. The first exhibits what two exceptionally well-placed observers did in fact report of their own times. The second consists of a sampling of contemporary journals.

THE DELAMBRE AND CUVIER REPORTS

J.-B. Biot's interesting *Essai sur l'histoire générale des sciences pendant la Révolution francaise* (1803) is both too early and too focused on the political, military, and institutional context to be pertinent. Let us begin instead with the reports on the progress of science that the Permanent Secretaries of the First Class of the Institut de France, Delambre for the exact sciences and Cuvier for the natural sciences, were required to submit to the imperial government in 1808.[3] Delambre, it should be emphasized, had been educated as a classicist. His six-volume *Histoire de l'astronomie* (1817–27) remains the fullest, most reliable, and most informative treatment of the science up to his time that has ever been written. The introductory overview of his report singles out the main developments: in geometry, Mascheroni's demonstrations depending on use of the compass alone; in trigonometry, measurement of the meridian and Prony's decimal log and trig tables; in theory of numbers, Euler's and Lagrange's proofs of several of Fermat's theorems; in mechanics, the derivation of its laws from the principle of virtual velocities and creation at the hands of Lagrange and Laplace of celestial mechanics; in astronomy, accurate tables of the motions of sun, moon, and planets, progress in bringing comets into computational camp, and discovery of asteroids. All this is straightforward enough, and we need not follow Delambre into the development he gives these topics in the body of the report. There Delambre also has a lengthy section on voyages of discovery and geographical science. For it is clear that by "mathematical" we are to understand exact and not abstract or theoretical science.

The opening paragraph of the chapter on "Physique mathématique" confirms the point. As the sciences progress, Delambre observes, the space separating them diminishes, and it becomes more difficult to discern the lines of demarcation. The development he gives the remark, however, is almost the contrary of what we might expect. If on the one hand a science makes conquests, it may also lose parts of its domain, which pass into that of its neighbor. Thus light, gravity, and laws of motion and impact are today almost entirely in the realm of geometry. There has even been an attempt to subject phenomena of magnetism and electricity to the calculus. It would appear, however, that galvanism, born in current times, may in part compensate physics for its losses. Volta's pile, however, one of the most ingenious of his inventions, has passed entirely into the hands of chemists, for whom it is the instrument of the most difficult and least expected discoveries. The new direction taken by researchers leads them to abandon fields that are almost exhausted in order to cultivate those that promise a more abundant harvest. Scientists have thus in recent times neglected investigations of the sort that constitute physics properly speaking. But even if the science no longer has the brilliance it once did, we can still cite work that is worthy of attention.

The exact physics to which Delambre points is, in the first instance, Coulomb's experimental demonstration of the inverse square forces of electrical and magnetic attraction and repulsion. Its most ingenious application was to Cavendish's measurements of the density of the earth. Instrumental determinations of this sort occupy the remainder of the chapter: Coulomb's perfected compass, measurements of magnetic inclination and declination, Biot's estimates of the location of the magnetic poles and equator, his and Gay-Lussac's demonstration in their balloon ascent of the invariance of magnetic force with altitude, Biot's and Arago's determination of the refractive indices of various gases, Borda's and Lavosier's measurement of the relative expansibility of metals with temperature, and Lefèvre-Gineau's and Fabbroni's measurement of the mass of the standard kilogram. Such was the stuff of physics. The superb development of rational mechanics at the hands of Euler, d'Alembert, Lagrange, and Laplace had found no application to real procedures of arts, trades, and construction. Only in hydraulics did a rapprochement come close. Prony's *Architecture hydraulique* (2 vols., 1790–96) gives the first elementary demonstrations of the most general equations of the equilibrium and motion of fluids as well as a treatment of motors and machines more complete than those of any predecessors "tant pour la partie rationelle que pour la partie expérimentale."

Like Delambre, Cuvier also had had a classical education, in his case in the Höhen Karlschule in Stuttgart, and was versed in historical thinking as well as in the *Naturphilosophie* he later rejected. In retrospect it appears that chemistry was in the process of forming the subdisciplines of physical chemistry under the impetus of Berthollet's *Essai de chimie statique* and of industrial chemistry in response to examples set by both Berthollet and Chaptal and further encouraged by the latter's widely circulated *Chimie appliquée aux arts* (4 vols., 1807). The title of Cuvier's report, *Chimie et sciences de la nature,* indicates that in 1808 chemistry was not generally considered to be an exact science, so that chemistry in company with natural history fell to his lot instead of to Delambre's. Cuvier's report, indeed, is evenly divided between those two main branches followed by a cursory section on applications. Placed between "sciences mathématiques" and "sciences morales," the natural sciences "commencent où les phénomènes ne sont plus susceptibles d'être mesurés avec précision, ni les résultats d'être calculés avec exactitude."[4] In a full account of the reforms worked by Lavoisier and his entourage, Cuvier emphasizes combustion theory and nomenclature and corrects Lavoisier on oxygen as the principle of acidity. He nowhere mentions Lavoisier's practice of gravimetric precision nor his aspiration that chemistry become a mathematical science, unless that is what he had in mind in one obscure paragraph about the need for chemists to qualify themselves as "physiciens et géomètres."

Cuvier considered, as indeed Berthollet had in the lectures he gave in 1795 before the short-lived École Normale de l'an III, that the future of chemistry would lie in the perfecting of affinity theory. Though coached by Berthollet, Cuvier appears not to have understood that his colleague had since modified the notion of affinity in such a way that it no longer looked to the construction of tables of "elective affinity," or relative tendency of sets of reagents to combine. Instead Berthollet had come to suppose that chemical combination consists in an equilibrium between forces of cohesion inhering in the physical properties of particular reagents and forces of attraction acting at short distance between the particles of one reagent and another. The latter model was Newtonian, of course, rather than classificatory. Even if Cuvier had understood the point, however, there was no way to quantify the model, and he goes on to consider the imponderable fluids, heat, light, and electricity, as chemical rather than physical agents. Only magnetism has no evident effect on chemical processes. The body of the essay consists of a well-informed inventory of the properties of the classes of chemicals, with a brief summary of what had been learned with great difficulty by research into "des produits des corps organisés." Still in a nascent stage, organic chemistry was either vegetable or animal chemistry.

Anatomy and traditional physiology occupy forty of the 120 pages Cuvier devotes to his proper domain, natural history. The subdivisions are still distributed among the three kingdoms. Anatomy itself he does not see as pertaining to zoology. He discusses it rather under the two traditional rubrics of "Anatomie générale" and "Anatomie particulière des divers organes." Physiology also has two aspects, "Physiologie générale ou théorie des forces vitales" and "Physiologie particulière des diverses fonctions." Bichat is mentioned for his general anatomy while his experimental physiology and that of others, though it was well under way, is passed over. True, Magendie made his appearance only in 1809, and despite the richness of his research in the ensuing dozen years, recognition of the new discipline of experimental physiology in the form of foundation of a journal came only in 1821.

Cuvier's own masterpiece, *Leçons d'anatomie comparée,* had appeared
in five volumes from 1799 to 1805. Of his later writings only *Recherches
sur les ossemens fossiles des quadrupèdes* (1812) was comparably rigorous. He comes to comparative anatomy in the concluding pages of his report, and there explains that his method represents the extension to zoology of the system of natural classification that A.-L. de Jussieu and his
three uncles had developed in botany. The context is a peroration in which
he calls for an entirely new *Systema naturae.* The encouragement of his
Imperial Majesty will motivate French naturalists to redouble their zeal
and, leading with the frontispiece NAPOLEON, create a system that will
surpass the achievements of the great Aristotle in the same measure that
their patrons do the deeds of the Macedonian conqueror. He failed to ask
himself, perhaps, whether Napoleon would have relished the sentence
that points out how Aristotle's creations survived while Alexander's were
soon destroyed. Despite this pirouette, what remains in the modern
reader's mind is the impression of the amount that Cuvier and those who
coached him did in fact, whatever their theories, know about the properties and actions of chemicals, minerals, plants, and animals.

THE JOURNAL LITERATURE

A survey of contemporary journals confirms that impression in overwhelming measure. It happens that the library where I have the good fortune to work has on its shelves runs of the *Annales de chimie* (which began publication in 1789), the *Journal des Mines* (1794), the *Journal de
l'École Polytechnique* (1794), and the *Bulletin de la Société d'Encouragement pour l'Industrie Nationale* (1803). The *Annales du Muséum National
d'Histoire Naturelle* (1802) is nearby at the Academy of Natural Sciences
in Philadelphia. These are among the earliest fully professional scientific
and technological journals. I say survey. It would clearly be impossible to
read them straight through. The first series of the *Annales de Chimie,* completed in 1815, consists of 96 volumes of an average length of 335 pages.
The contents include short notes, correspondence, reports on foreign research, papers read before the Société Philomathique, reports to the Institute, research articles with diagrams of apparatus, and short treatises.
The first series of the *Journal des Mines,* a monthly published by the Conseil des Mines, consists of 38 volumes averaging 475 pages each. In 1804
it added the subtitle *Recueil de Mémoires sur l'exploitation des mines, et
sur les sciences et les arts qui s'y rapportent,* which carried over into the
new series when the title changed to *Annales des Mines* in 1816. Although

the dimensions of the other journals are more modest, none is slim. Either the historian will scan them to get a sense of the general drift of the subject matter or will use the indexes, which are excellent, to find articles that he or she already knows to have been important.

ANNALES DE CHIMIE

Scanning produces some nice surprises. One knew, for example, that Monge did chemistry on occasion. Pieces he contributed to *Annales de Chimie,* however, would seem to lie outside what we consider the boundaries of chemistry. One is a monograph on the causes of meteorological phenomena such as atmospheric pressure, rain, heat, wind, clouds, and thunder (5, 1790). These are to be analyzed in terms of molecular interactions in the atmosphere and not on a simplistic mechanical model of masses of matter in motion. True, he does say that this is something natural philosophers ("physiciens") need to learn from chemists. A later paper by Monge makes no such gesture towards chemistry. It concerns the effect on vision of viewing variously colored surfaces through tinted glass or prisms of different colors (26, 1798). From this it appears that a subjective element enters into perception. A lengthy extract translated from Goethe's *Farbenlehre* (89, 1812) makes a similar point. An earlier volume (26, 1798) contains the earliest report in France of Chladni's *Entdeckungen über die Theorie des Klanges* (1787).

Far from revealing a correlation between the progress of chemistry and specialization, as might have been expected, these and other entries like them, such as two memoirs on polarization of light and one on double refraction by Biot (94, 1815), remind us that boundaries of the sciences were porous. In 1816, indeed, the journal became *Annales de Chimie et de Physique* under the joint editorship of Gay-Lussac and Arago. Before and for some years after that, geometric optics was the only mathematical aspect in the physics reported. Except for crystallography, the same was true of Haüy's *Traité élémentaire de physique,* the publication of his course at the ephemeral École Normale of 1795. One recalls, too, the apparent paradox that in the early curriculum of the École Polytechnique, the nursery of mathematicization, chemistry had a major role while physics was a trivial course concerning the properties of bodies, and further that both gave ground, with physics vanishing (even as Delambre noted), before the imperialistic advance of mathematics. All this may serve as a reminder that chemistry was not yet generally taken for an exact science. The unmodified word physics still connoted what was natural rather than exact whereas physique mathématique referred to what was experimentally exact. Optics, for its part, had long been geometric without being called a physical science.

A survey of the identity of contributors to *Annales de chimie* over time gives a rough sense of the growth of the population, and perhaps of democratization, in the field of chemistry. Publication ceased in 1793 during the Terror and was resumed only in 1797. In the earlier five-year period (1789–1793) most of the articles were by members of the Board of Editors, Guyton de Morveau, Lavoisier, Monge, Berthollet, Fourcroy, Dietrich, Hassenfratz, and Adet. About a dozen other contributors, mostly foreign or well-known people such as Coulomb, Séguin, and Chaptal, furnished one or two articles apiece. From 1811 to 1815 the volumes are a little larger, but the board of editors consisted of 16 people, while the number of additional contributors was approximately ninety, mostly little known and some among them Italian, German, and British.[5]

As to subject matter, an able research assistant, Connie Malpas, has made a quantitative analysis of the primary emphasis in the substantive notes, reports, and articles in the first six volumes (1789–1791) of the *Annales de chimie*. I have recently followed suit for the last six (1814–1815).

	Vols. 1–6	Vols. 91–96
Reactions	31	14
Properties of reagents	10	62
Theory	9	2
Applications	102	36
Organic chemistry	4	9
Physics	2	10
Priority claim	0	1
Total	**158**	**134**

(For the sake of clarity the terms physics and organic chemistry designate the subject matter in the modern sense.)

The fewer, longer, and more thorough articles by a much larger number of contributors at the end of our period bespeak, not only a growth in the population of the discipline, but a deeper, more precise, and more analytical mode of research and reasoning. Use of the new nomenclature increased gradually as the older, more amateurish generation gave way to new, often highly trained people. It was virtually complete by 1800. The paucity of theoretical considerations may be thought to reflect what Comte would later see as transition from a metaphysical to a positive stage in development of a science. The proportion of investigations concerned with reagents increased at the expense of description of reactions and, still more strikingly, of interest in applications. Those factors reflect discovery of a host of new elements and compounds and reinforce one's sense, which is otherwise largely stylistic, of a deepening science. What is most surprising, perhaps, is the nature of the applications both in the earlier and later period. With the exception of dyeing, none of them concern what we would call chemical industry. Instead they pertain to the traditional fields of arts and crafts, metallurgy, pharmacology, medicine and physiology, viniculture, and mineralogy. With the exception of a series of essays on organic chemistry by Berzelius in 1814–15, there is little or nothing programmatic in all this large body of scientific writing to give the historian any clue to the direction that this rapidly expanding science, or the physics accompanying it, would take in the future.

MÉMOIRES DE LA SOCIÉTÉ D'ARCUEIL

Modern historians, including this one, are wont to see the *Mémoires de la Société d'Arcueil* (3 vols., 1807, 1809, 1817) as a harbinger of things to come. All that was apparent at the time was that the members of the Society were handpicked by Laplace and Berthollet and known to be promising. The third, and largest, volume appeared when the society was in decline, years after most of the papers had been composed and delivered. Of the 93 papers published, 52 concern chemistry, 36 physics, and five botany or geography. Among the chemical papers, a slightly larger proportion deal with organic and physical chemistry than was the case in *Annales de Chimie,* but the difference is hardly striking. Gay-Lussac published the Law of Combining Volumes of Gases (though he does not yet call it that), in Vol 2. "Mémoire sur la combinaisons des substances gazeuzes, les uns avec les autres." The problem is anticipated in an investigation of 1802 published in *Annales de chimie,* "Sur la dilatation des gaz et des vapeurs," (43, pp. 137–175). No references to his findings appear in further issues of that journal prior to 1815. Among the papers on physical phenomena in *Mémoires de la Société d'Arcueil* are Malus on double refraction and Biot on the passage of polarized light through thin plates. Neither of these reports is mathematical in any but a geometric sense. Only two among the physics papers contain analytical expressions. Biot incorporates elementary algebraic equations in "Recherches sur les lois de la dilatation des liquides à toutes les températures," (3, pp. 191–197). Laplace employs the calculus in "Sur le mouvement de la lumière dans les milieux diaphanes" (2, pp. 111–142).

JOURNAL DES MINES

The *Journal des Mines* is quite another matter. Published by the Agence des Mines, it was an official rather than a scientific journal and regularly printed governmental directives and regulations as well as individual memoirs. In the early years the Agence distributed copies free to proprietors of mines, mineralogists, and foreign experts. The editor, Charles-Etienne Coquebert de Montbret, divided his energies between the consular service and an amateur's involvement with mineralogy and natural history. The early issues are a miscellany comprising installments of a continuing inventory of the mineral resources of France, region by region and department by department; detailed descriptions of particular deposits of coal, peat, iron, and other minerals; accounts of individual mines and mining machinery; methods of extraction; and safety precautions. Though nothing systematic, the vein is technological rather than scientific, informational and educational rather than experimental. A small and gradually increasing proportion of the memoirs did deal with chemical and metallurgical analysis by such well-known persons as Clouet, Vauquelin, and Guyton de Morveau until, after 54 monthly issues, the journal foundered in February 1799.

It came to life in March 1801 at the instance of Jean-Antoine Chaptal, Minister of the Interior during Bonaparte's Consulate (1800–1804). The Ministry provided subsidy by subscribing to and distributing a large number of copies. A chemical manufacturer, Chaptal had directed the revolutionary production of saltpeter and gunpowder in 1793–94. During his time in office, 1800–1804, he organized the Napoleonic civil administration giving special attention to education, science, and industrial development. It would appear that resuscitating and supporting the *Journal des Mines* was related to a policy of promoting the still-fledgling École des Mines and new Corps des Mines to the importance they have since attained, which is equivalent to that of the École and Corps des Ponts et Chaussées for civil engineering.

While continuing to be an official publication, the revived journal was a very different periodical. It still carried governmental regulations and extended the mineralogical inventory of the country. Two academicians, Haüy and Vauquelin, and four mining engineers, among them Collet-Descotils, initially constituted the editorial board. No longer addressed merely to miners and metallurgists, the journal welcomed memoirs on the theoretical aspects of mining, namely geology, mineralogy, and docimasy or mineral chemistry. Contributions from "mathematical sciences," would also be welcome, provided they be applicable to subterranean geometry and the design of machinery. Memoirs along these lines in the ensuing numbers were both briefer and more focused than the often rambling reports of the earlier years.

As time went on, however, the *Journal des Mines,* without losing sight of its original purpose, often included articles of general scientific interest. Werner's theory of diluvial origin of rock formation and classification of strata by mineralogical criteria found many followers, though I have come on no allusions either to the biblical flood or the nascent geological conflict with James Hutton's vulcanist disciples. Cuvier and Brongniart published their first draft of the 1822 treatise on the Paris basin, "Essai géographie minéralogique des environs de Paris," in No. 138 (23, pp. 421–458, 1808). Though detailing the differing fossil content of the various formations, there is no hint that fossil content could serve as indices in determining the sequence of strata in the geological column, nor is there any mention of catastrophism in earth history. An announcement of the fourth edition of Laplace's *Système du monde* in vol. 34 (1813) details the changes since the earlier editions. Vol. 38 (1815) reprints the essay "Sur l'application des probabilités à la philosophie naturelle" that Laplace read before the Institute and published in *Connaissance des Temps.* The same issue has a paper by Biot on rotation of the plane of polarized light in certain homogeneous fluids.

BULLETIN DE LA SOCIÉTÉ D'ENCOURAGEMENT

Even like the revived *Journal des Mines,* the *Bulletin de la Société d'Encouragement pour l'Industrie Nationale* and its parent society were launched at the instance and with the support of Chaptal in the Ministry of the Interior. In company with the technological papers in the *Annales de chimie* and even more in the *Journal des mines,* the contents of the *Bulletin* consist of reports on technical innovation rather than the description and classification of the best existing techniques that had been characteristic of the literature on arts and trades in the eighteenth century. Ten to twelve issues each year were bound into volumes, circulated to subscribers, and offered for sale to the public. A single number would contain accounts of fifty to sixty inventions, procedures, devices, requests for information, and announcements of prizes and awards. To list the contents of any one issue would take more space than is appropriate in the present essay. To know which items ever entered into production and made a difference would require difficult research that has never been done. The necessity of catching up with England is the only consistent theme. The very account given of the foundation of the society is misleading. Its organizers

conceal governmental patronage and make it appear the fruit of spontaneous combustion among a public-spirited set of financiers, businessmen,
inventors, engineers, and academicians.[6]

JOURNAL DE L'ÉCOLE POLYTECHNIQUE

The *Journal de l'École Polytechnique* is not a journal at all. The title of the
first two issues, 1794–95, carries the original name of the school, *Journal de l'École Centrale des Travaux Publiques ou Bulletin du Travail fait à
l'École.* The subtitle continues through Cahier 6 (1799) and then disappears. Those early cahiers are largely pedagogical. Cahier 5 (1798) does
contain a serious mathematical memoir by Fourier, his earliest published
work, "Sur la statique, contenant le principe de vitesses virtuelles." The
demonstration is a novel one. What it does, we can now say, is to trans-
pose virtual velocity into virtual work. It belongs to a genre that was soon
to be called theory of machines. The approach had started with Lazare
Carnot's little noticed *Essai sur les machines en général* (1783), which
Fourier was one of the very few to mention.

A period of editorial confusion set in following Cahier 6, and the next
six issues are a bibliographer's nightmare. Neither then, nor earlier nor
later, is there any consistent relation between the sequences of *cahiers* and
tomes in which they are bound. Sometimes there is one issue to a volume,
sometimes there are two. Cahier 11 (1802) picks up where Cahier 6 (1799)
had left off and opens with a report on the curriculum, regulations, and
administration of the École during the intervening three years. The avant-
propos to Cahier 11 (1802) further says that Cahiers 7–8, which had appeared, and 9–10, which would soon appear, are entirely consecrated to
Prony's *Mécanique philosophique.* In fact 7 and 8 had not appeared. When
in 1812 they finally did, they contained, not Prony, but the courses that
Laplace and Lagrange had given at the École Normale of year III (1795).
As for Cahier 9, it consists of Lagrange's *Théorie des fonctions analytiques.*
The title page bears the date of year V (1797). It was bound with Cahier 10
in 1810. The latter is a miscellany of memoirs on geometry (Brianchon),
topology (Poinsot), use of differential equations in hydrography (Prony),
the hydrography of Italy (Prony), geography (Humboldt), and Andrieux's
Cours de Grammaire Générale et Belles Lettres at the École.

Evidently publication had lapsed entirely between 1799 and 1802 and
did again between 1802 and 1806, when Cahier 13 appeared. That issue
consists entirely of mathematical memoirs by Monge, Poisson, Ampère,
Poinsot, and Biot. The *Journal de l'École Polytechnique* thereupon became
a collection of research memoirs, mainly though not exclusively mathe-

matical, published at irregular intervals. Reporting on the affairs of the school was left to the informal *Correspondance,* started by Hachette, which also included papers by its current students.

In ensuing numbers of the journal one or two among the ten or twelve memoirs in each issue contain papers that to later eyes would appear to be mathematical physics. The remainder consist of mathematics of one sort or another and are categorized as Analyse or Mécanique by running heads, which in every case indicate what the editors took the subject matter to be. Malus had two papers in Cahier 14 (1808). "Mémoire sur l'optique" is a forerunner of *Traité d'optique* (1813), a work of differential geometry. "Mémoire sur la dioptrique" follows. It is an experimental piece of geometric optics that turned out to be a prelude to the debate that culminated in the eventual triumph of Fresnel's wave theory of light in 1822. Both are classified as Géométrie analytique. Poisson's "Théorie du son," which appeared in the same issue, is labeled Analyse. It was not an investigation, he says, but a demonstration of several general theorems that may interest both physicists and mathematicians. A Poisson piece in Cahier 17 (1815), on the radiation of heat on the Laplacian model of forces operating on point masses at finite distances is labeled Mathématique. So too is his response to Fourier on heat in a lengthy later paper, "Mémoire sur la distribution de la chaleur dans les corps solides," (Cahier 19 1823).

In Cahier 16 (1813) Petit's "Théorie mathématique de l'action capillaire" is Mécanique. The above Poisson memoir on distribution of heat in solid bodies is Mathématique. Cauchy on a particular case of fluid motion is Hydrodynamique. Only three memoirs labeled "Physique" had appeared by 1831: Malus, "Pouvoir réfringent des corps opaques" (Cahier 15, 1809); a translation of Cavendish's experiments to determine the density of the earth (Cahier 17, 1815); and Poisson, "Mémoire sur les équations générales de l'équilibre et des mouvements des corps solides élastiques et des fluides" (Cahier 20, 1831).

ANNALES DU MUSÉUM NATIONAL D'HISTOIRE NATURELLE

To turn now to the life sciences, research in progress appeared in memoirs contributed to the house organ of the Muséum. The series consists of 20 quarto volumes and runs from 1802 to 1813, when the title changed to *Mémoires.* Twelve professorial chairs had been established in 1793, when the ci-devant Jardin Royal des Plantes was transformed into the Muséum. The designation of those chairs may be taken as an indication of the categories of subject matter constituting the field of natural history. In 1802 the staff consisted of (1) Haüy in Minéralogie, (2) Faujas de St.-Fond in

Géologie ou Histoire Naturelle du Globe; (3) Fourcroy in Chimie Générale, (4) Vauquelin in Chimie des Arts; (5) Desfontaines in Botanique au Muséum; (6) A.-L. de Jussieu in Botanique à la Campagne; (7) Thouin in Culture et Naturalisation des Végétaux, (8) Geoffroy Saint-Hilaire in Zoologie, Mammifères et Oiseaux, (9) Lacepède in Zoologie, Reptiles et Poissons (10) Lamarck in Zoologie, Insectes, Coquilles, Madrépores, etc., (11) Portal in Anatomie des Hommes, (12) Mertrud and Cuvier in Anatomie des Animaux. Mertrud, an elderly holdover from the old regime, died after two years. Fourcroy died in 1809 and was replaced by Laugier. No other change in staff or designations occurred during the period of the *Annales*.

As in the case of the *Annales de Chimie,* most of the papers in the early volumes were written by members of the staff, who constituted an informal editorial board. In the *Annales du Muséum* memoirs by the professors appear first in each issue. The following tabulation of subject matter in the first and last five volumes exhibits the distribution of memoirs in the fields of natural history. The several categories are taken from the designation of the respective authors' chairs. Contributions by other contributors are listed in the same order of subject matter. As will appear, the number of memoirs by the staff dwindled in the last years, at which time Lamarck and Cuvier were compiling their major works on taxonomy. Neither Mertrud nor Portal contributed anything.

MEMOIRS BY STAFF

		Vols. 1–5 (1802–1804)	Total	Vols. 16–20 (1810–1813)	Total
Mineralogy	(Haüy)	11	11	4	4
Geology	(Faujas)	15	15	4	4
Chemistry	(Fourcroy)	12			
	(Fourcroy-Vauquelin)	3			
	(Vauquelin)	1		18	
	(Laugier)		16	1	19
Botany	(Desfontaines)	12			
	(Jussieu)	23	35	7	7
Horticulture	(Thouin)	12	12	4	4
Zoology	(Lacepède)	4			
	(Geoffroy)	12		8	
	(Lamarck)	26	42	7	15
Anatomy	(Cuvier)	29	29	7	7
All staff			160		60

Memoirs by Other Naturalists

	Vols. 1–5 (1802–1804)	Total	Vols. 16–20 (1810–1813)	Total
Mineralogy	4		2	
Geology	4		3	
Chemistry	0		11	
Botany	7		10	
Horticulture	0		1	
Zoology	7		17	
Anatomy	0		3	
Entomology	6		2	
Ornithology	2		0	
Paleontology	0		1	
Pharmacology	0		2	
All others		30		52

The number of extramural naturalists was thirteen in the earlier period and thirty in the later. Certain well-known names are among them. Volumes 1 and 2 contain memoirs by the Swiss botanist Alphonse de Candolle and the entomologist Pierre-André Latreille, who contributed the third volume to Cuvier's *Règne animal.* Early papers of the prolific and long-lived chemist Michel-Eugène Chevreul are responsible for the large presence of that science in the later period. The correspondence includes letters from Humboldt about his explorations in the Andes, from Charles Willson Peale about the opening of his museum, and two from President Thomas Jefferson, one on a deep-cutting plow with a blade designed to minimize resistance of the soil, the other on the plans for the Lewis and Clark expedition in the wake of the Louisiana Purchase.

As for the several disciplines, given Haüy's specialization, crystallography took up a disproportionate share of the space on mineralogy, which to be sure was fully covered in *Journal des Mines.* These were early days for the term *geology.* Faujas's papers and others like them were nothing historical and were mostly concerned with vulcanology and topography. A much larger proportion of the papers on chemistry dealt with the composition of animal and vegetable substances than was the case in *Annales de chimie.* Jussieu's botany was a continuation of his life work of fitting plants into a natural system. With respect to zoology, Geoffroy and Lamarck, but not Lacepède (a holdover from Buffon's day), did taxonomy in accordance with the methodology—subordination of characters and correlation of parts—that Cuvier had adapted from Jussieu in his *Leçons d'anatomie comparée*. The few writings on anatomy other than Cuvier's were not on the whole comparative.

It is more difficult to follow current perceptions of experimental physiology since no journal existed prior to Magendie's foundation of *Journal de physiologie expérimentale* in 1821. Indicative of the reception of his program, however, is the identity of the commissioners named by the Institut de France to referee his successive memoirs, beginning in 1809. In the early years they were all medical members. By 1820 they were mostly chemists and physicists.

Biology in Hindsight

No one reading through or in these volumes would have had any reason to suppose that the ancient study of natural history was on the verge of fissuring into the new earth sciences, the largely unchanged practice of botany, and a comparative anatomy that in partnership with a newly vivisectionist physiology would multiply into the set of sciences soon to be called *biology*. When Lamarck and Treviranus severally coined that word in 1803, their purpose was to legislate a denominative distinction between laws of animate and inanimate nature. I do not know about Treviranus's milieu in Germany, but the word appears little if at all in the literature in France prior to Auguste Comte's positivist classification of the sciences in *Cours de philosophie positive* (1830–1842). The sciences of life, Comte declared, are capable of forming general laws in the measure that they address themselves to analyzing the relation between the diversity of organic forms and the physical milieu in which each form exists. The nature of the milieu being known from the other sciences, it follows that "Le double problème biologique peut être posé, suivant l'énoncé le plus mathématique possible, en ces termes généraux: *étant donné l'organe ou la modification organique, trouver la fonction ou l'acte, et réciproquement.*"[7] The body of practice from which he had abstracted this definition was the zoological taxonomy developed by means of comparative anatomy in the previous thirty years at the hands, above all, of Cuvier for vertebrates and Lamarck for invertebrates. Both of them had died before Comte published that lecture. So far as I have found, the first scientist to use the word in Comte's sense was François Magendie. In the fourth edition of *Précis élémentaire de physiologie* (1836) he writes of "Physiology, or Biology, that vast natural science which studies life wherever it exists and investigates its general characters."

Only since Comte is it apparent that an incipient convergence of the new techniques of comparative anatomy and the new practice of vivisectionist physiology were early instances of the gathering of a family of sciences under the rubric biology. At the time comparative anatomy was the instrument for establishing a natural rather than artificial system of zoological classification while experimental physiology had yet to achieve independence from medicine. As late as 1865, after all, Claude Bernard entitled his covering masterpiece *Introduction à l'étude de la médecine expérimentale.* What the two disciplines had in common in the second and third decades of the century were an emphasis on rigor and mutual awareness of and respect for each other's work on the parts of Cuvier and Magendie. Cuvier considered that the principle of correlation of parts was as determinate as any mathematical proposition. Magendie considered that the goal of experimental physiology was to reduce the explanation of vital phenomena to laws of chemistry and physics.

Cuvier and Magendie stated those regulative principles firmly and clearly in their current publications and exhibited their cogency in practice. What has preoccupied historians of zoology, however, is the three-way clash of theories as between Lamarck's transformism, Cuvier's fixity of species, and Geoffroy Saint-Hilaire's unity of type. That is much less evident in the contemporary literature. Lamarck was mainly admired at the time for *Histoire naturelle des animaux sans vertèbres* (7 vols., 1815–1822). He does not write of biology but regards his transformism as pertaining to zoological philosophy, the title of his earlier book on that subject. *Philosophie zoologique* (1809), however, was less noticed than was *Histoire naturelle.* As for Cuvier's static "embranchements"—verterbrates, mollusks, articulata, radiata—they nowhere figure in *Lecons d'anatomie comparée.* He first proposed them in "Sur un nouveau rapprochement à établir entre les classes qui composent le règne animal,"[8] a memoir published in 1812, the same year when *Recherches sur les ossemens fossiles des quadrupèdes* appeared. The latter was prefaced by the famous "Discours sur les révolutions de la surface du globe" which was provoked, there is every reason to suppose, by what Cuvier regarded as Lamarck's pantheism. The branchings reappear in the prefatory remarks to *Le Règne animal, distribué d'après son organisation, pour servir de base à l'histoire naturelle et d'introduction à l'anatomie comparée* (1817).

Geoffroy Saint-Hilaire, for his part, a somewhat lesser part, did primarily zoology in the memoirs he published in the *Annales* prior to 1818, when the first volume of *Philosophie anatomique* appeared. That is a work rather of morphology, a word not yet coined, in which the theme is unity of form. The notion first appears in 1807 in three memoirs on the anatomy of fish in the ninth volume of the *Annales*. The no-doubt latent and growing hostility between Cuvier and his two colleagues did not burst into the open, however, until Cuvier's denigratory éloge of Lamarck in 1830, the year in which he also initiated the acrimonious debate with Geoffroy over unity of type. The dispute took place on the floor of the Academy of Science and was trumpeted to the world by Goethe. It is, finally, indicative of what seemed important scientifically that the eleven disciples who published the second edition of *Le Règne animal* (22 vols., 1836–49) paid little attention to the branchings. In like manner, of the two principal editors of the second edition of Lamarck's *Histoire naturelle des animaux sans vertèbres* (11 vols., 1835–43), C.-P. Deshayes and Henri Milne-Edwards, the former had never believed in transformism, and the latter renounced it, writing "As to the physico-physiological theory on which the hypothetical views of our author depend, it appears useless to us to discuss it." What mattered was the unequaled accuracy of those two taxonomic masterpieces.

MATHEMATICAL PHYSICS IN HINDSIGHT

The relation of work that reads to us like mathematical physics to contemporary perceptions is somewhat different from that in the life sciences. Cuvier, Magendie, and their colleagues knew themselves to be doing the disciplines they and others defined as comparative anatomy and experimental physiology respectively, even if they did not see their work as building blocks of biology before Magendie took up Comte's usage. Not so among those who later appear to have been doing mathematical physics. Comte did not regard the application of mathematics to physical phenomena as positive physics, which in his eyes was exclusively experimental, observational, and metrical. Laplace, the principal instigator of a program of mathematicization, may have been more prescient. He and Lagrange had been there and done that with astronomical phenomena, and celestial mechanics remained astronomy. His own analyses of capillary action, speed of sound, and atmospheric refraction pointed out the path taken by physics. The four famous prizes he had the Institute set were for mathematical analysis of double refraction (1808), heat diffusion (1810), theory of elastic surfaces (1811), and diffraction of light (1817). Whether Malus, Fourier, Sophie Germain, Augustin Fresnel, and unsuccessful entrants, or others such as Poisson, Biot, Arago, and Ampère, regarded themselves as doing physics is unclear. Certainly they knew themselves to be analyzing physical phenomena mathematically while performing or drawing on experiments and data.

As we have seen above, however, down to 1831 only two of the papers that read to us as mathematical physics were labeled physics. The rest were analysis, analytic geometry, mathematics, or rational mechanics. Of those authors and others like them, none considered themselves to be natural philosophers. They had been trained as engineers. What they did was usually called analysis, not mathematics. Throughout the eighteenth century and still, people who did what we see as mathematics were "Géomètres," whether or not they did geometry, and most did not.

Nor had the elements of 19th-century classical physics yet come together. Except for an abortive memoir by Biot, the French gave no mathematical development to the properties of Volta's battery or of the electric current until Ampère, who had never done anything but mathematics and philosophy, reported his extraordinary series of experiments between 1820 and 1825. Even then his électrodynamique had nothing to do with force in general or with energy but only with forces of attraction and repulsion between conducting circuits. Such was the schizophrenia of research into heat that Fourier and Sadi Carnot paid no attention to one another. In the preface to *Théorie analytique de la chaleur* (1822) Fourier asserts that, whatever the extent of mechanical theories, they apply in no way to the effects of heat. These are a special order of phenomena that can never be explained by principles of motion and equilibrium. There is no question but that his treatise is intrinsically mathematical physics, but its importance in the development of physics was negligible while its analysis became a major mathematical specialty in the nineteenth and twentieth centuries.

The roots of thermodynamics are in Sadi Carnot's *Puissance motrice du feu* (1824), but not until later was it recognized as a constituent of the physics of work and energy. Only following Coriolis's formulation of kinetic energy in *Calcul de l'effet des machines* (1829) did the physics of work and energy emerge from what had been theory of machines since Sadi's father Lazare started it in 1783. Perhaps I may be permitted an anecdote here. In the early 1960s I had the privilege, as did my students, of participating in several seminars offered jointly with the late Salomon Bochner, one of the leading mathematicians of his generation.[9] He found Lazare Carnot to be a bore. When we reached *Réflexions sur la puissance motrice du feu*, however, he was astonished and fascinated. "It's verbal," he said, "but it is mathematics." That Sadi's treatise was mathematicizable was first apparent to Émile Clapeyron in "Mémoire sur la puissance de la chaleur," *Journal de l'École Polytechnique* (Cahier 22, 1834). That memoir itself went largely unnoticed until Kelvin and Clausius found in it the basis for the second law of thermodynamics. Not until their generation, perhaps, would such as they, Maxwell and the young Helmholtz, if asked what they did, have answered "Physics." Faraday, after all, was a natural philosopher.

Conclusion

If well founded, the foregoing considerations require some modification, not of the substance of the conclusions in my book, but of their statement. I there wrote that French science in the early nineteenth century was the nexus from which emerged the disciplines of comparative anatomy, experimental physiology, and mathematical physics. It would have been better to say the disciplines of comparative anatomy and experimental physiology and the practice of mathematical physics. I further stated the following: "The one thing that may be said of all the physics that mattered is that it was mathematical. The one thing that may be said of all comparative anatomy that mattered was that it was based on correlation of parts. The one thing that may be said of all experimental physiology that mattered is that it was based on vivisection." In each case I should have written "that mattered in the long run."

With respect to whig history, finally, I could not have reached those conclusions solely in the light of contemporary knowledge and practice. To be sure, certain other matters that were evident to contemporary observers would also be apparent to a historian, to wit the great and growing volume of knowledge acquired, the still greater growth in the population of people producing it, and the increasing factuality and heightening standards of proof in reports of research. The activity and production of science, furthermore, did resemble that which would transpire within the framework of the later disciplines once they were in place.

When that is said, however, and it needs to be said, it would have been impossible without benefit of hindsight to perceive which researches and which research programs would in the event lead to the evolution of science amid the changing circumstances of future times.

Notes and References

1. *Science and Polity in France, the End of the Old Regime* (1980, reissued 2004); *Science and Polity in France, the Revolutionary and Napoleonic Years* (2004).

2. The term, it will be recalled, originated with Herbert Butterfield's *The Whig In-*

terpretation of History (London, 1931). Butterfield there deplored treating British history in the Macaulay vein as if it were a story uniquely of progressive Whig politics prevailing over Tory regression, of liberty broadening down from precedent to precedent, all culminating in the liberal polity of the present. It is obvious how applicable such an approach would be to the historiography of science. A certain irony may lurk here since there is no better example of whiggery in scientific history than Butterfield's classic *The Origins of Modern Science 1300–1800* (London, 1950). Or did he perhaps consider that subject matter entails a difference between writing general history and writing history of science? Butterfield did not say. A book that critics would consider a more egregious instance of Whig history of science is Shmuel Sambursky, *The Physical World of the Greeks*, reviewed in essay 18 below.

3. *Rapports à l'Empéreur sur le progrès des sciences, des lettres et des arts depuis 1789* I, *Sciences mathematiques;* II, *Chimie et sciences de la nature.* The reports on French literature, on classics (history and ancient literature), and on fine arts were prepared by the Permanent Secretaries of the respective classes of the Institute, M.-J. Chénier, B.-J. Dacier, and J. Le Breton. In 1989 Editions Belin reprinted all five with prefaces by Denis Woronoff. Jean Dhombres contributed an introduction to Delambre's report, Michelle Goupil to Cuvier's account of chemistry, and Yves Laissus to Cuvier's account of natural history.

4. Cuvier, *Op. cit.* p. 38.

5. Maurice Crosland, *In the Shadow of Lavoisier: the Annales de Chimie and the Establishment of a New Science* (1994) gives the names of board members from 1789 until 1914.

6. For a discussion of mathematically trained people dealing with problems of technology and engineering, and an inventory of other journals, mostly in the 1820s and 1830s, see Ivor Grattan-Guinness, "The ingénieur-savant: a neglected figure in the history of French mathematics and science," *Science in Context* 14 (1993), pp. 405–433.

7. 40ième Leçon, 3, p. 269

8. *Annales du Muséum National d'Histoire Naturelles* 19, pp. 73–84.

9. See below, essay 16.

PART IV

Historians and Historians of Science

~~~ 12 ~~~

The Work of Élie Halévy: A Critical Appreciation

WHEN THE FOLLOWING ESSAY appeared I was just beginning to teach the undergraduate course on modern British history that was offered regularly until I became fully engaged in history of science some nine years later. After David Owen, my adviser at Harvard, suggested my writing out the reasons for admiring Halévy's approach, I set out to read everything I could find that he had written. A French colleague who was at the Institute for Advanced Study in 1950–51 told me that Halévy, who died in 1937, had left no survivors. Although my wife and I spent the academic year 1954–55 in Paris, where I was beginning research on science in the French Revolution, I never thought to inquire further.

What was my surprise—indeed I was thrilled—on receiving this letter six months later.

> La Maison Blanche
> Sucy-en-Brie, S&O
> 11 janvier 1956

Cher Mr Gillispie

C'est en octobre 1955 seulement que j'ai appris, par un ami qui fréquente les bibliothèques londoniennes et qui aimait beaucoup mon mari, Élie Halévy, que vous aviez écrit sur lui un très bon article dans le numéro de septembre 1950 du *Journal of Modern History* de Chicago. C'est hier seulement que j'ai enfin entré en possession de cet article, si complet, si consciencieux, si compréhensif. En le lisant, j'ai eu l'impression que vous aviez dû connaître mon mari—ce qui, je crois, n'est pas le cas?

Il faut, donc, que je vous dise merci, bien en rétard, mais n'oubliez pas que pour moi cette lecture est toute récente, et elle m'a profondément intéressée et émue. J'ai eu l'impression que vous connaissez admirablement votre sujet et que rien ne vous a échappé.

Je crois, toutefois, que sur un point vous vous êtes légèrement trompé. Mon mari éprouvait une vive affection pour le ménage Webb. Mais il est vrai que leur—comment dire?—fascisme? nous choquait parfois vivement. En voici un exemple. Quoique mon père fut français et que j'ai toujours été française, ma mère était italienne et je suis née à Florence. Je me rappelle une rencontre avec les Webb chez ma mère et l'enthousiasme fasciste manifesté par les Webb (par Béatrice surtout). Ils allaient de ville en ville, reçus et promenés par les autorités fascistes. C'était le moment où un grand nombre de nos amis étaient en grand danger d'arrestation et déportés.

Malgré cela, et d'autres "chocs" du même genre, nous avons toujours admiré ce couple étonnant—Béatrice surtout—dont le charme était grand.

Je me demande si, après tant d'années cette lettre vous trouvera. Mais j'ai senti le besoin de vous dire merci—d'avoir si bien compris, d'avoir si bien dit. N'oubliez pas que pour moi votre article date d'aujourd'hui.

Avec l'expression de ma vive sympathie,

Florence E. Halévy

I answered at once, of course, to say how thrilled I was and how I regretted having been misinformed prior to our time in Paris the previous year. Madame Halévy and I did meet once. The quadrennial International Congress in History of Science was held in Florence in late August 1956. I wrote to say that on my way I would be stopping in Paris. Again she was dismayed, for she was to be visiting in London and Oxford all that month. Naturally I stopped in London instead and went up to Oxford. There we had a three-hour lunch at the Mitre. She told of their frequent visits to England, and of their friendship with the Fabian circle, with Graham Wallas, who was a particular intimate, with Bertrand Russell, with the historian H. A. L. Fisher, and with the Webbs. As not everyone may recall, Sidney Webb came from a very modest background and Beatrice (née Potter) from a very wealthy if not quite aristocratic circle. Their joint labors on the history of local government, of the poor laws, and of other topics of the deep-down socio-political structure are monumental accounts of its detailed impact on the lives of the mass of the people throughout the nineteenth and into the twentieth centuries. Beatrice Webb fascinated Florence Halévy specially. A later letter tells of a ride together in the country, for Beatrice was a horsewoman among other things, in which she remarked that "the great thing about marriage is that it satisfies our need for companionship without waste of time."

Florence Halévy was a charming, urbane, warm, and cosmopolitan lady. She was fluent in German as well as English and Italian and published a translation of Goethe's *Wilhelm Meister*. We corresponded for another year and a half, a dozen letters on each side. She told of her many years happily collaborating with her husband in research, editing, and reading proof. During the time of our correspondence she was occupied with editing the letters Halévy had received throughout a close, lifelong friendship with Émile Chartier (the philosopher Alain).[1] She was looking forward to inviting my wife and me to the Halévy household, La Maison Blanche, in Sucy-en-Brie in order to show me his study and library. That was not to be. She died unexpectedly in November 1957.

Madame Halévy arranged for a translation of my article to be published in *La revue de métaphysique et de morale,* of which her husband had been co-founder in 1893.[2] I formed the impression that the editor agreed in order to please her, for I never heard a word from the journal. Neither she nor I was shown the translation, which is a bad one. No echo ever crossed the Atlantic.

The article did attract attention among students of English history, mainly because of the argument that the side effects of evangelical religion immunized England from revolution. A slightly abbreviated version leads off a collection of essays addressed to that question, among them a selection from Gertrude Himmelfarb's *Victorian Minds* (1968).[3] Among the fifteen other entries, some earlier than my article, some later, are analyses by J. L. and Barbara Hammond, Eric Hobsbawm, E. P. Thompson, and Bernard Semmel. In 1980 Myrna Chase, a student of Gertrude Himmelfarb's, published a study of Halévy's thought based on research in his papers.[4]

Interest in Halévy did revive in France late in the last century, not because of his treatment of English history, but (so I surmise with some confidence) because the scales finally fell from the eyes of a chronically fellow-traveling intelligentsia. Its representatives ultimately perceived the prescience of *L'Ere des tyrannies, études sure le socialisme et la guerre* published posthumously in 1938. At the time it was published, his associates tended to indulge as an aberration the judgment of their late colleague that there was nothing to choose between a communist and a fascist tyranny. Finally, a full and fascinating edition of Halévy's correspondence appeared in 1996 with an illuminating introduction by François Furet.[5]

At the time of writing the essay, I knew nothing of Halévy's biography. I now learn that his was an extraordinary family. He was the oldest son of Ludovic Halévy, a man of many talents best known as the librettist for Offenbach and for Bizet's *Carmen.* Georges Bizet was a cousin. Bizet's grandfather, and Halévy's great uncle, was Jacques Fromental Halévy, a composer of operas remembered for *La Juive.* The eminent chemist and statesman of science, Marcelin Berthelot, was also a cousin. Halévy's younger brother, Daniel, was a writer. The great-great grandfather, Élie

Lévy, took the name Halévy in 1807, after having immigrated from Bavaria some years earlier. Besides the above, those of the intervening generations were prominent, some in literature, some in liberal politics. It was a family of many religions. Ludovic converted to Catholicism. The custom was to bring up the children in the religion of their mother. Élie's mother, Louise Bréguet, came from a family of Swiss horologists. She was a Protestant, as were her sons. The multiplicity of religions and the maternal background of French Calvinism may, perhaps, have had some bearing both on the austerity of Halévy's judgment and his interest in the relation of religion to social and political behavior.

Notes

1. *Correspondance avec Élie et Florence Halévy* (Paris, 1957: Gallimard). Halévy's letters to Alain were burned during the German occupation of Paris.

2. No. 2, (avril-juin, 1957), pp. 157–186.

3. Gerard Wayne Olson, ed., *Religion and Revolution in Early Industrial England: The Halévy Thesis and its Critics* (New York, London, 1990: American University Press).

4. *Élie Halévy, an Intellectual Biography* (New York, 1980: Columbia University Press).

5. *Élie Halévy, Correspondance (1891–1937).* Madame Halévy's niece, Henriette Noufflard Guy-Loë, prepared the collection, which is annotated by herself, Monique Canto-Sperbere, and Vincent Duclert. So thorough are the annotations to each letter, and so informative the brief biographies of all the correspondents, that the collection amounts to a self-portrait of a considerable segment of the French intellilgentsia during the time of Halévy's career. My correspondence with Madame Halévy might serve as a footnote. Her letters with other of my papers are on deposit in the Library of the American Philosophical Society in Philadelphia.

TWELVE

The Work of Élie Halévy:
A Critical Appreciation

—⚮—

The publication late in 1948 of Élie Halévy's *Histoire du socialisme européen,*[1] the last of three posthumously published volumes by Halévy,[2] offers a good occasion for a critical appreciation of his contribution to historical literature. Halévy died in 1937, and *L'ère des tyrannies,* which appeared the following year, also deals with socialism. Halévy's evaluation of the socialist movement is thus, in a sense, his valedictory. The two books are interesting in themselves, but they take on greater significance if they are considered as the conclusion to a lifework which, until they appeared, had been devoted chiefly to the evolution of liberal opinion and liberal institutions in England.

More than is the case with the work of most professional historians, Halévy's writings can be regarded as a single body of thought in which is apparent a continuous and coherent development of mind, of method, and of attitude. In each of his major studies he reached a fundamental and general historical judgment which then—though this was probably not his intention in advance—naturally formed a starting point for the more comprehensive investigation into which it led him. His earlier work, therefore, apart from its relationship to his interpretation of the history of socialism (to which the concluding pages of this article are devoted), presents a number of problems which are also interesting for their own sake and which may be anticipated by a brief résumé of his more important publications.[3]

Halévy was a historian who had been trained as a philosopher at the École normale supérieure[4] and a Frenchman who made his reputation writing English history. His first book, *La théorie platonicienne des sciences,*[5] is a study in pure philosophy; but already Halévy was employing

* Reprinted from *The Journal of Modern History,* 22 (September 1950), pp. 232–249.

the distinctive analytical method that he later applied to historical research and exposition. Having become a historian, he chose to devote his attention to modern English history because he saw in England the best laboratory for the study of the historical problems that interested him most: the problem of liberty and, related to it, the problem of the relationship of ideas and beliefs to material circumstances in social motivations and hence in historical causation. Both these themes are central to his interpretation in the two great works which occupied most of his life, *The growth of philosophic radicalism*[6] and *A history of the English people in the 19th century,*[7] with which must be considered the two-volume *Epilogue,*[8] which treats the pre–World War I period. In the interval between these works, Halévy's interest was shifting from intellectual history to general history, a broadening of his approach which was as fruitful as was his earlier decision to write intellectual history rather than philosophy. *The growth of philosophic radicalism,* the definitive work on the subject, analyzes and describes the formation, development, and influence of Benthamite utilitarianism, the most powerful intellectual strain in the evolution of English liberalism. Halévy's strong sympathy for Bentham's individualist principles is apparent in his final judgment on the school. In the *History* Halévy undertook a comprehensive synthesis which, though he did not live to complete his plan, established him as the foremost authority on the period down to the 1840's. His central hypothesis is that the secret of British social and political stability lay in the moral influence of evangelical religion, without which British liberty would have been impossible. Apart from this basic hypothesis—and other subsidiary interpretations—the distinctiveness of his technique of narrative and the analytical quality of all his descriptive passages require notice, since the methodological approach itself contributes to his conclusions.[9] By the time he had completed his study of England, Halévy had arrived empirically at an implicit philosophical definition of liberty, together with an objective and general conception of the sort of political and social environment in which liberty is possible.

In order, then, to appreciate both the internal development of Halévy's thought and the perspective from which he finally surveyed the socialist movement, these are the aspects of his work which must be considered, and considered as growing out of one another: the influence of his philosophical training, his technique of abstract analysis, his view of historical causation, his tendency to utilitarian standards of judgment, his interest in English society and sympathy for its qualities of individualism and moral discipline, his method of historical narrative, and, finally, his conception of liberty as the highest practical political good.

Probably the most direct consequence of Halévy's formal philosophical training is to be found in the highly analytical quality which characterized all his work and in the particular technique of analysis which he perfected. To students of his historical writing, the method of analysis is the chief interest of *La théorie platonicienne des sciences,* though as his first book it is also important for the indications it gives of the bent of his mind before he became a historian. It is not a work of history—the treatment is, of necessity, relentlessly abstract, and I am not qualified to evaluate it as a piece of Platonic scholarship. It is, however, significant that Halévy should have chosen the Platonic dialogues as the subject of a dissertation. Except that he was never mystical, his own cast of mind was Platonic: austere, analytical, and logical, always seeking to penetrate to the central idea or conception, moral or intellectual, which gave form not only to bodies of doctrine but also to concrete political and social movements. Halévy always accorded primacy to systems of values, ideas, beliefs, and morals rather than to material interests in explaining human motives and social actions. This note runs through all his work. The issue does not arise as such in *La théorie platonicienne des sciences,* but he never had any sympathy for crude economic determinists, although later he gave great weight to the importance of objective social and economic circumstances in shaping ideas and opinions—which is to say that he himself was not a crude idealist. But his idealism was not of the romantic variety. He was, if anything, even less sympathetic to Hegelians than to Marxists.

Halévy's method of organization in *La théorie platonicienne des sciences* grew directly out of his reason for undertaking the book. His purpose was to systematize the apparent contradictions in Plato between what he called the "dialectique régressive" and the "dialectique progressive," and his statement of the problem may be taken as a descriptive forecast of his distinctive method of analytical exposition:[10]

> Ce serait un devoir pour quiconque se propose l'étude de cette philosophie ... d'interpréter la dialectique à la fois comme une méthode critique, ayant pour objet de détruire en nous l'illusion de la science, et comme une méthode positive, ayant pour objet de justifier toutes les sciences.... Que l'on se souvienne combien de contradictions se rencontrent dans les dialogues, contradictions relatives à des notions métaphysiques, logiques ou morales, et l'on conviendra que ce ne serait pas un médiocre avantage de ramener toutes ces contradictions à une seule contradiction fondamentale, de poser une problème unique là où les textes semblent poser cent problèmes.

In substance, he concludes that the function of the "dialectique régres-sive" was to clear away error in order that the "dialectique progressive" might erect the positive structure of truth.[11]

This pursuit of an interpretation by the device of resolving contradic-tions into systematic dichotomies is characteristic of Halévy's technique of analysis. It is itself almost a dialectical method, an analysis that un-folds a subject through layer after layer of conceptions until it comes to the core of basic attitudes and assumptions. Since very few, if any, systems of thought or doctrine are ever thoroughly self-consistent, Halévy's dis-section usually ends by laying bare two opposing aspects of a subject, around which he then organizes his own exposition and criticism, often uncovering subsidiary antinomies along the way. This was his method with utilitarianism, with classical political economy, with socialism, and to some extent, though less abstractly and hence less neatly, with his de-scriptive analysis of the social, economic, and political structure of early 19th-century England.

In *The growth of philosophic radicalism,* a portion of which formed his doctoral dissertation, Halévy took as the analytical framework for his his-tory the dichotomy between the principle of the artificial identification of interests, the basic assumption of utilitarian juridical and political theory, and the principle of the natural identity of interests, which underlay the economic opinions of the school. On the one hand, in the sphere of gov-ernmental theory, the utilitarians did not assume the existence of any spontaneous harmony between the self-interest of the individual and the welfare of the community. Quite the contrary, it was the whole business of the legislator to blend individual interests with those of society and to do so artificially, where necessary, by establishing positive rewards and punishments. On the other hand, in the sphere of economics, which was incorporated into utilitarianism after the juridical philosophy was already well developed, the Benthamites began with Adam Smith and with the principle of the natural identity of interests. Here they assumed that the hand of nature—the famous invisible hand—had so arranged matters as to make it impossible by definition for an individual to have any real in-terests which could be harmful to society. To Halévy the problem of the relationship of individual interests to the interests of society was a fun-damental one for any social philosophy. And in his view the two princi-ples of the natural identity and of the artificial identification of interests were ultimately irreconcilable so long as utilitarianism claimed (as it did) to be a consistent science of society which reduced *all* the phenomena of human behavior—political, economic, moral, or psychological—to a single set of universal laws. His discussion resolved all the utilitarian writings around one or the other of these basically contradictory principles. At the same time, he demonstrated how the deductions from both principles—in spite of their logical irreconcilability—were applicable in the circumstances of the time—in the one case, promoting political reform and, in the other, forwarding commercial liberalism.

This dichotomy is, however, only the skeleton of Halévy's account, essential to the form of the whole but not obtruded upon the reader's attention except at the critical junctures. The body of the material is a full and generally sympathetic, though critical, exposition of the writings of the utilitarian school over a period of fifty years. Halévy very skilfully allows the central analytical framework outlined in the preceding paragraph to emerge in the course of a treatment at once topical and chronological. By unfolding the evolution of utilitarian philosophy in time, he illustrates the interactions of ideas and the way in which new assumptions were insensibly introduced in order to open new areas of application to the doctrines and to suit them to changing circumstances and differing objectives. It was, perhaps, because Jeremy Bentham and his disciples concentrated on different aspects of their philosophy at different periods that they themselves were never aware of the contradictory dualism of assumptions which Halévy discovered at the root of their work. Halévy takes the three chief periods of Bentham's career as his chronological framework. In Part I, "The youth of Bentham" (1776–89), he discusses the background of the principles in English and in continental thought and their incorporation into a moral philosophy and a program of juridical reform, not yet associated with political radicalism. In Part II, "The evolution of the utilitarian doctrine from 1789 to 1815," the subject is Bentham's transformation into a democratic reformer and the incorporation of political economy into his creed, both developments taking place under the influence of James Mill. In Part III, "Philosophic radicalism," Bentham steps forth the acknowledged master of a full-fledged utilitarian school, almost a sect, which from 1815 to 1832 both elaborated and disseminated his principles until they had become the dominant intellectual influence shaping English social opinion toward the nineteenth century.

In this book Halévy was writing the history of a school of thinkers who consistently held that the dominant human motive both is and ought to be the rationally conceived self-interest of the individual, though they were inconsistent as to whether it required restraint. Inasmuch as in all his work, but particularly throughout his *History of the English people,* Halévy himself expressed the view that men are led by beliefs and opinions rather than by material interests, it is worth considering how this question stands in *The growth of philosophic radicalism*. Certainly, he would have maintained the idealist position in so far as he regarded the intellectual legacy of the philosophic radicals as the most important element in 19th-century liberalism. Utilitarianism touched and shaped reform in every department of British national life. Sometimes it retarded progress, but more often its influence was to systematize, rationalize, and modernize. The functions of the state were from some points of view, particularly the social and economic, narrowly restricted, but what the state did do it must do well and rationally. But in *The growth of philosophic radicalism* it is also apparent—and this is a point which Halévy makes explicit, though he does not draw the conclusion here suggested—that, whatever the *effects* of systems of ideas, their formation is to a great extent the product of environmental circumstances.

One may take as an example Halévy's treatment of the transformation of Bentham into a radical democrat after 1808. This is described as the crucial episode in the development of philosophic radicalism. Until then, utilitarian doctrines could, as in William Paley, equally well be enlisted in the interests of conservatism. In fact, Halévy goes to great lengths to show how this was so and how gradually all publicists, from Edmund Burke on the right to William Godwin on the left, were increasingly talking the language of utility.[12] This was partly because the French Revolution had discredited both progressive thought in general and the appeal to natural law in particular, but neither did the internal logic of utilitarianism necessarily move in the direction of radicalism. That Bentham became the high priest of the new radicalism was, Halévy thinks, partly an objective and partly a personal accident:[13]

One general fact was alone necessary for the principle of utility to become the mould in which all ideas of reform were to take shape—namely that, with the beginning of the war in Spain, the English people should once more become the champion of European liberty against the Napoleonic despotism. Henceforth, when liberal ideas once more gained some credit in England, was it not inevitable that they should be expressed in utilitarian language, since it was, to a greater or less degree, the language spoken by everybody? A particular circumstance, on the other hand, was necessary to make Bentham take over the leadership of the movement:— the meeting, in 1808, of James Mill and Bentham. James Mill, who had for a long time been an advanced Whig, converted Bentham to the cause of liberalism and then to political radicalism.

—⁓⁓—

In *The growth of philosophic radicalism,* then, although the discussion is very abstract, Halévy does not treat ideas independently of circumstances, nor does he regard the formation of opinions as necessarily anterior to material events. Since it was also under Mill's direction that classical political economy was merged with Benthamism, the accident described in the preceding quotation was also important in the identification of middle-class economic interests with political radicalism. Clearly, this circumstance—largely a chance one in Halévy's account—was an essential development for the later influence of utilitarian ideas. Ideas and beliefs may be, as Halévy always held, primary in historical explanation, but he himself seems to have had no unduly exalted conception of the possibility that ideas may be divorced from interests or that systems of thought are capable of being either objective or consistent enough to be true in any absolute sense. And, though he never put it this way, his work leaves the impression that the chief practical influence of ideas comes after they have hardened into dogmas and (though this makes the idealist interpretation of history a little difficult) that what often converts ideas into dogmas is their applicability to real interests in definite circumstances. The example of free trade suggests itself, a movement which originated as an intellectual construction, became a class interest, and finally expanded into a moral outlook transcending economics and shaping English life for a century.

In his concluding pages Halévy sums up the influence and the legacy of the utilitarian doctrine; and he outlines the channels through which it permeated the whole of English society in the period of reform: through the universities and the new facilities for popular adult education, through the press, through the radical members of parliament, through Bentham's own enormous correspondence, and through the activities of the administrative experts who had been his disciples and who were on hand to man the agencies required by the new reforms. The influence became so widespread that, like the assumptions underlying the doctrine, it moved in contradictory directions. To it the liberal tradition owed its emphasis upon the individual rather than upon the group as the basic social unit. At the same time and, though laissez faire was also utilitarian, to the extent that the individual could not naturally know or follow his interest, the state might be justified in intervening to enlighten or protect him. This was a derivation from the principle of the artificial identification of interests which requires the legislator to promote the greatest happiness of the greatest number. But, on the democratic side of the doctrine, the legislator must also represent the interest of the greatest number. Thus, both limitations on the state and the democratic reform of the state were utilitarian in the shape that they assumed in England. And though David Ricardo's economics has been modified out of all recognition, nonetheless economics as a science owes its origin to the classical school.

Serious criticisms may be leveled against many of the utilitarians' attitudes and contentions: their logical inconsistencies, the ambiguity of the idea of happiness, their want of human imagination, their philistinism, and the mechanistic quality of their social outlook. They may not have established the genuinely empirical social science that they proposed, but Halévy tends to qualify many of the usual criticisms. As he points out, the utilitarians were very far from preaching a hedonistic self-indulgence. For all that their morality purported to be a morality of pleasure, their emphasis was rather upon avoiding pain. Their idea of pleasure was a fairly prosy one, consisting chiefly of the sort of bourgeois security which may be purchased by abstinence and rigid self-discipline. But to Halévy— and, in view of his later treatment of liberalism, this is very important— it was exactly this aspect of the Benthamite doctrine that was its greatest strength. For in his view the success of liberal polity resulted more from the behavior patterns of self-restraint and self-discipline enjoined by the utilitarians than it did from a conscious attempt on the part of the majority of individuals to model themselves on the abstract economic man. It was in this sense (as will be more apparent when his interpretation of the very similar effects of evangelical religion is considered) that he seems to me to have held opinions and beliefs to be central to historical causation—not for their face value but for their results in behavior.

In *The growth of philosophic radicalism* Halévy has accomplished an all too rare feat, and one particularly rare in intellectual history—he has added to historical literature a secondary work so perceptive and so clear that it provides the reader more illumination on its subject than a study of the sources would do. A measure of Halévy's excellence may be found in a comparison of his book with Sir Leslie Stephen's *The English utilitarians,*[14] a three-volume study of the same subject which was written almost simultaneously and independently. *The English utilitarians* is a brilliant study in its own right, but Stephen never entirely managed to bring utilitarianism into focus as a philosophical synthesis. His approach was biographical, and the individual sections and particular judgments on this or that aspect of the figures whom he treats are often searching and illuminating. Stephen's book is the better on the utilitarians' lack of a historical sense. But he never penetrated beneath the surface to the contradiction between the artificial identification and the natural identity of interests— conceptions which are not mentioned in Stephen. Neither did he find any other central principle of organization. Consequently, he could not exhibit the process by which the utilitarians were able to apply their formulas to justify both positive action in their political program and, once artificial restrictions were abolished, abstention from action in their economic program.

It is, perhaps, unusual that, despite so thorough a dissection of the body of utilitarian thought, Halévy should nevertheless have admired his victim. Yet one feels him to have been sympathetic to the objectives of the Benthamites. Very occasionally he allowed himself to express his enthusiasm explicitly as, for example, in his comment on one of Thomas Babington Macaulay's onslaughts:[15]

Was the result of Macaulay's attacks to discredit the new school for ever? Did they not rather help to hallow its existence, and to fix once and for all—in a caricatured form no doubt, but no matter—the typical Utilitarian? There is no doubt that the young Benthamites had their faults and made themselves hated because of these faults. But were not their very faults respectable, if their exclusiveness and their pedantry can be explained by their fidelity to an idea which was the object of their considered allegiance? It may further be admitted that this idea was a narrow one, that it did not take into account all the facts in the moral and the social world which they thought to explain, and even that it systematically disregarded many aspects of human nature. If, in active life, it is the definition of courage to defend to the last extremity, in spite of all its risks, a position which was at first freely accepted, is it not likewise a sort of speculative equivalent of courage to dare to take an idea as the principle of all one's opinions and of all one's acts, and then to accept without flinching all the consequences which this original idea involves?

Halévy, in fact, was something of a utilitarian—or, at any rate, a pragmatist—himself. To him the important thing about ideas or dogmas seems to have been that they should lead to the right results. It was an intellectual duty to be unsparing in his criticisms of the confusions and mistakes of the utilitarians; but, though he criticized, he never blamed. They made no more mistakes than most men, and at least their work had fruitful and constructive consequences. Halévy was a convinced liberal in the classical sense, and the ideas which he admired were those that supported liberty. The fact that they were often logically contradictory was less important than that their conclusions were generally right if, indeed, a free society is the best society.

This may suggest that one of the reasons for Halévy's success in *The growth of philosophic radicalism* was an unusual combination of talents: he united great skill in abstract analysis with an ability to render judgments on a soundly utilitarian basis. Thus he was able to exhibit the anatomy of his subject with penetrating clarity while evaluating the doctrines that he described in the light of their effects rather than in the light of the confusions which he so patiently disentangled in their background. Other factors contributed to his success, of course. Halévy was, Graham Wallas remarked, the only person who had ever been through all the Bentham manuscripts.[16] And he was very well suited to his subject in his innate sympathy for the utilitarians' emphasis both upon individual discipline and self-reliance in society itself and upon diligent and rational inquiry in the science of society.

In the interval between the completion of *The growth of philosophic radicalism* and the publication in 1912 of the first volume of his *History of the English people,* Halévy's interest had shifted to general history. His study of the career of Thomas Hodgskin, published in 1903, leaves the impression that it was fortunate that he outgrew his exclusive interest in intellectual history.[17] *Thomas Hodgskin* is an unpretentious and in some respects an interesting book, particularly since Hodgskin has been an overly neglected figure. But this is one of the few places where Halévy fell victim to the besetting sin of intellectual history: treating ideas as if they descended like germ cells from host to host, affected by heredity but not by environment. The discussion is overintellectualized to the point of being almost bodiless. Halévy employed his usual dualistic technique of analysis. Hodgskin's work is regarded as having had two divergent tendencies, the anarchistic, culminating in Spencer, and the socialistic, in Marx. But here the interpretation seems wire-drawn and artificial, imposed upon the subject instead of growing out of it.

None of Halévy's later work is open to these objections. Throughout his *History of the English people* the central problem is to explain the combination of stability and liberty, of inequality and social solidarity, in English life. The approach now is that of complete historical description and narrative rather than of abstract philosophical analysis. The first volume, *England in 1815,* a comprehensive and concise panoramic study, is widely regarded as Halévy's masterpiece. The book is unique in both conception and execution. I do not know of any other historical work which arrests the stream of history at a particular moment in time, in order to portray the whole condition of a society at one critical juncture. Nor does any other work come to mind which, to put it a little flatly perhaps, includes so much information in so manageable a compass. Like the later volumes of the *History,* the first is at once a work of general synthesis and one of original research. Halévy uses the relevant secondary materials, but there is no subject which he does not treat primarily from the sources. Whatever scholars may eventually decide about his interpretation, the descriptive aspects of the volume are not likely to be superseded, and yet, though this may seem paradoxical, Halévy presented his interpretation not separately from his material but through it.

He begins with an analytical description of the legal, administrative, military, and political machinery of the government, both central and local. That, in spite of the tory reaction, England remained substantially a free country, Halévy attributed less to the strength of her constitution than to the inefficiency of her institutions. In 1815 a prophet who considered only political factors would have had to predict not the evolution of a strong representative democracy but the deepening of an aristocratic anarchy. A brutal and despised royal family; a chaotic, almost medieval administration studded with sinecures; a representative system which, if it was imperfectly responsive to public opinion and mass pressure, was nevertheless an antique absurdity—these were not materials which had within themselves the seeds of social stability. But, in spite of this unpromising picture, "the elements of disorder and anarchy inherent in the political tradition of the country, lost their character and submitted insensibly to the organization of a discipline freely accepted. . . . We must, therefore, seek elsewhere, in the character either of the economic organization or of the religious life of the nation, the secret of this progressive regulation of liberty."[18]

Proceeding by a process of elimination, Halévy next turned to a description of economic affairs, but there, too, he was unable to find a satisfactory explanation for the later solidity of English society. By 1815, enclosures had created a large and discontented agricultural proletariat—not that Halévy regarded the results of enclosures as being reducible to uniform sociological and demographic generalizations. Neither did his treatment of industrialization bear out the simple doctrine of increasing working-class misery, although that interpretation was then—around 1912, that is—very fashionable in the reaction against Victorian optimism. Despite these qualifications, however, a real class cleavage was developing in 1815, even if not in the simple Marxian sense. The economy, moreover, was innocent of any real control at all; and, if wealth increased, it did so interrupted by a series of appalling industrial crises and financial panics:[19]

If the materialistic interpretation of history is to be trusted, if economic facts explain the course taken by the human race in its progress, the England of the nineteenth century was surely, above all
other countries, destined to revolution, both political and religious.
But it was not to be so. In no other country of Europe have social
changes been accomplished with such a marked and gradual continuity. The source of such continuity and comparative stability is, as
we have seen, not to be found in the economic organization of the
country. We have seen, also, that it cannot be found in the political
institutions of England, which were essentially unstable and wanting in order. To find it we must pass on to another category of social
phenomena—to beliefs, emotions and opinions, as well as to the institutions and sects in which these beliefs, emotions and opinions
take a form suitable for scientific inquiry.

Here, in the practical effects of the prevailing religious movement,
Halévy found what he regarded as the roots of British social solidarity, and
this is the basic hypothesis of his *History:* that Englishmen were controlled not by material interests but by religious beliefs and moral sanc-
tions. In particular, the determinant factors were the evangelical current,
widening out of the earlier Methodist revival into the whole broad stream
of English life, and the pragmatic individualism which, itself strongly
reinforced by evangelical religion, characterized the British approach to
problems of every sort.[20] The influence of evangelicalism upon the whole
life of society was far more pervasive than would be apparent simply from
church history, important as that was. Its humanitarian effects were directly obvious in developments like the Sunday-school movement, the
abolition of the slave trade and of slavery, prison reform, and factory legislation—in all of which, Halévy noticed, evangelical philanthropy produced much the same urge for improvement and arrived at the same practical conclusions as did utilitarian radicalism. More indirectly, one may
trace to the Methodist revival the reform of manners and morals, the curious phenomenon of a ruling aristocracy itself adopting middle-class out-
looks, the genesis of the distinctively British attitude of moral superiority
to the continent, and even the techniques of mass agitation later adopted
by political pressure groups. Meanwhile—and most importantly—for
the individual, evangelical religion dictated philanthropy if he should be
successful and a resignation tempered by self-help if he should not, in
any case acquiescence in the dispensations of a providence which could
be expected to reward only individual effort. "Men of letters disliked the
Evangelicals for their narrow Puritanism, men of science for their intel-

lectual feebleness. Nevertheless during the nineteenth century Evangelical religion was the moral cement of English society. It was the influence of the Evangelicals which invested the British aristocracy with an almost Stoic dignity, restrained the plutocrats who had newly risen from the masses from vulgar ostentation and debauchery and placed over the proletariat a select body of workmen enamoured of virtue and capable of self-restraint."[21]

It is, therefore, to the place of duty, probity, and personal independence in British life rather than to some innate island genius for the practicalities of representative government that Halévy traced the origin of the behavior patterns which made Victorian Englishmen uniquely successful political animals. It was not only or even primarily through popular control of the political machinery of the state that England achieved self-government. The moral discipline, the self-restraint, and the individualist ethic arising out of the religious life of the nation—these rather than any external political authority, whether democratic or aristocratic, were what really governed the lives of Englishmen. It was in this deepest sense, "the moral and religious sense" (Halévy himself suggests the comparison to France), that the English were self-governing.[22]

Unfortunately, a summary can present only the bare outline of an interpretation; it cannot do justice to the wealth of information and detail worked into the presentation. One need not entirely accept Halévy's evaluation of political and economic institutions in order to profit from his description of them. Simply as an illustration of the range of treatment, it may be worth while to indicate some topics about which *England in 1815*—to take only the first volume of the *History*—offers an authoritative account and in many cases the best one available: the court and administrative institutions; the legal profession; local government; the army and navy; the structure and functioning of the unreformed parliament; the press; political parties; land tenure and agriculture in England and Ireland; the stage of development and methods of production in major industries; the immediate effect of economic warfare on British trade; the monetary and banking system; taxation and public finance; the church and the dissenting sects; drama and the fine arts; primary and secondary education; the condition and studies of the universities; the medical profession; and the cultivation of the natural sciences. As this catalogue suggests, one of the outstanding characteristics of all Halévy's work is the remarkable range of the subjects on which he could write with the authority of complete mastery. His success raises the question whether it is not better to be the historian of a limited area treated in all its aspects than the specialist pursuing his vocation of economic history or intellectual history or whatnot through a variety of times and places. Considering his beginning as an intellectual historian, it is perhaps particularly worth remarking that a further feature of Halévy's narrative is the close attention that he gives to governmental finance as one of the central aspects of practical, year-by-year politics.

The later narrative volumes of the *History,* although they presuppose and occasionally recur to the theme of the evangelical strain running through English history, do not insist upon the topic unduly. After *England in 1815,* Halévy's purpose was to follow the evolution of British civilization by an account of the stream of events through which it moved. His method of organization, both unique and successful, is particularly worthy of attention. Halévy did not facilitate his analysis by the usual device of slicing historical material up into topical sections—a method of the new history which, however convenient it often is, necessarily dissects, rather than portrays, the life of a society. In the hands of a master the old history, the chronological narrative, was capable of a better portrayal than the modern history often gives, but it was weak in analysis. The peculiar excellence of Halévy's method is that, having taken the whole life of a society as his subject, he is able both to portray and to analyze. He writes from the perspective, usually, of the central political scene, and, so far as practicable, he discusses developments in the order in which they presented themselves as major problems to the responsible figures and the contemporary public, whether the issue had to do with the church, Ireland, tariff policy, the budget, or whatever. But he casts a wider net than did the 19th-century historians, and, since he was aware of the existence of sociology and was, indeed, writing consciously within the framework of what amounted to a sociological hypothesis, Halévy's narrative has the quality of continuously analyzing the causation of the events which it simultaneously depicts. Where necessary, he weaves in the requisite background (much, it might be noted, as a responsible politician would have to bring himself up to date on an unfamiliar subject which suddenly becomes a pressing issue).

Halévy seldom exploited the vantage point of his own generation in order to be wise after the event—though one must except his treatment of foreign policy from this generalization. Instead, the perspective of time enabled him to subordinate the inessential to the essential in successive situations. His skill, judgment, and command of the sources prevented him from using that perspective to distort. Halévy would have sympathized with Sir Robert Peel's impatience with critics who insisted that, because a statesman favored a given policy in any one session of parliament, he should necessarily follow the same line in another, whatever the circumstances. In no other historian is the compelling action of circumstances upon political leaders more clearly set out. His *History* describes what course was followed in each situation and also, given the abilities, opinions, prejudices, and political and religious affiliations of the determinant figures, why the course was followed. Past politics are treated in the light of what was possible, and the conduct of public affairs emerges as the uncertain, inexact, and tentative art which a glance at the current newspaper should remind one that politics have always been. Statesmen of varying ability and fallible judgment, subject to all sorts of pressures, deal according to their lights with events of which they (unlike the historian) cannot know the outcome. Halévy's reader comes about as close as is possible to living through history by entering into the minds of the people who, all ignorant of the future, were shaping a still unfinished story and shaping it not in time units of decades to be summed up in a single chapter but month by month, year by year, and election by election, seldom even certain that a given episode was finished.

Halévy was thus able to penetrate beneath the surface of politics to the motives and interests conditioning the public mind and the minds of statesmen. In doing so, he was led to question one of the serious oversimplifications in English history—the view that two-party government was the fundamental vehicle of British politics and that in the two parties are to be found the respective bases of the strains of progressivism and conservatism in British development. Halévy granted this conventional picture a certain superficial validity. But the regular pendulum swing of power between two evenly matched and coherently organized parties was, he thought, largely a fiction, except for the period of William E. Gladstone and Benjamin Disraeli. For the rest of the century, Halévy could not find that the party key unlocked the historical reality. Progress toward the modernized state was forwarded not solely, perhaps not even chiefly, by leaders who led their parties. Rather the determinant influences were exerted by those who sponsored some particular policy which seemed to be demanded by the national interest (whether or not rightly conceived) as against their party's interest and who thereby split and temporarily wrecked their parties—George Canning and Peel, for example, and in later periods Joseph Chamberlain and David Lloyd George. It was because his organization started from events instead of from the superficial continuity of party that Halévy was able to penetrate to the basic current of English history and to find the parties floating on the surface of the issues. Liberal progress in social, political, and economic affairs carried all parties with it.

Halévy's reinterpretation of the significance of parties explains how it was that, a good liberal himself, he nevertheless felt a greater sympathy for Peel than for any other English statesman. The liberalism that Halévy admired in English history was not the monopoly of a political sect. Instead, it was the basic characteristic of the whole society, administered by free and limited institutions in an atmosphere of voluntarily accepted restraint and self-imposed social discipline. The popular demagoguery of the radical agitator, the too obvious class interest of the professional mercantile liberal, the aristocratic dilettantism of the great Whig lord—Melbourne or Russell—an overdose of any one of these might well have unsettled a stability which rested on acquiescence and self-reliance and not on force. The cautious, realistic, sane, and even conservative liberalism of a Peel was, Halévy felt, the approach best calculated to enlarge the area of freedom without disturbing the safety of society.

In considering critically the major themes in Halévy's history, one is struck by the fact that, despite the success of his narrative, his devaluation of the significance of the party system has not exercised anything like so great an influence on later historical writing as has his basic interpretation of the moral importance of evangelical religion. One reason (though this would not necessarily act as a conscious deterrent) may be that historians would find the organization of political narrative a good deal harder if they abandoned the framework of the party system, whereas no similar difficulty would result from the adoption of Halévy's evangelical hypothesis. But also it seems to me that in Volume III and to a much greater extent in Volume IV[23] Halévy tended to be carried away by his respect for Peel. This emphasis does, it is true, serve as a useful corrective to the simple identification of the Whig and then the Liberal party with the progress toward a middle-class society. And it also redresses the tendency of some sentimental writers to ignore the commercial conservative strain running from the younger Pitt through William Huskisson to Peel and to find, instead, the real, sound conservative tradition in the romantic, paternalistic toryism of Robert Southey, Richard Oastler, Young England, and Benjamin Disraeli. Nonetheless, Halévy probably overdoes the conception of a liberalizing conservatism.

Halévy's account of foreign policy, however, is, in my opinion, the weakest aspect of his *History.* Lord Palmerston was the one English statesman whom Halévy treats without either understanding or sympathy. His hostility has been attributed to a good patriotic resentment at Palmerston's cavalier treatment of France, but this criticism is probably unjust. The war had intervened between the publication of his first and second volumes, and Halévy himself explains how, after 1918, he was led to reconsider the foreign policies of Canning and of Palmerston, conventionally regarded as both patriotic and liberal.[24] (The relative superiority of Volume I, incidentally—excellent as the rest of the *History* is—may well be attributable to the fact that it appeared before the war had disturbed the serenity of Halévy's judgment.) Palmerston's conduct and Canning's seemed to Halévy, in the wake of the war, less liberal than irresponsible. He saw in their appeal to patriotic sentiments a sample and a foretaste of the nationalistic, bumptious adventuring which, once it became the prevailing mode in European diplomacy, excited popular suspicions and distrust to the point of war. It was Palmerston's levity that alienated Halévy, not the injuries to French *amour propre*.[25] But even when his aversion for Palmerston is put in the correct light, it remains true that Halévy's innate distaste for power politics growing out of his general dislike for the factor of force in public affairs did prevent him from treating diplomacy with the sympathetic penetration that characterized his discussion of internal problems. In all Halévy's work, one feels that his interests in isolating historical causes (that is to say, the underlying dynamic structure of a social process) stems from his concern for the effects (that is to say, the character of the actual life in a society). He investigated and admired English domestic development because its result in this sense was freedom; he disliked international relations because its result seemed to be war.[26] Once, on the occasion of a lecture in 1934, he explained to a Chatham House audience why he chose not to address them on foreign policy: "Dites ce que vous voudrez, mais affaires extérieures, cela signifie, en général, guerre. J'abandonnai donc ma première idée, et je me mis à travailler la politique intérieure anglaise. Après tout, la structure intérieure de la société a bien aussi son intérêt,—en réalité, un intérêt plus grand que les rapports extérieurs entre les nations."[27] This is, perhaps, not an attitude that Halévy would have found altogether easy to defend in view of his own emphasis (to be discussed later) on the role of the war as the determinant event which directed the domestic polities of all European states into socialistic channels. But even for the earlier nineteenth century it is doubtful whether domestic history can properly be considered as having developed independently of foreign relations. Halévy does not separate the two in his narrative, but he does, it seems to me, tend to think of them in different categories and of the external world as having, somehow, corrupted domestic events.

One further consideration ought to be taken into account before evaluating the central interpretative theme of the *History*. Despite his hostility to Palmerston, Halévy was far from being an anglophobe. On the contrary, he tended at times, though only at times, to be carried away by his admiration not only for Peel but for England in general, an unusual bias for a Frenchman. This, in turn, was because it was the English success in developing a society at once free, tolerant, and stable that first attracted him to the study of the English genius for compromising differences peacefully. He even suggested in his Rhodes memorial lectures that if only members of the League of Nations (including Britain) would adopt British domestic methods in international relations, all would be well with the world.[28] Perhaps he intended this rather as a compliment to his Oxford hosts than as a serious proposal. But in any case, despite occasional touches of this sort, Halévy's work has generally the great merit of looking at England from the outside. His annual visits never developed in him the vague Oxonian vice of regarding England as a norm to which other societies ought, somehow, to have conformed. He regards, for example, the influence of a generalized evangelical morality as the factor which in England provided the voluntary social discipline that permitted a liberal state to be stable, but he does not suggest that evangelicalism was a prerequisite to the success of liberal polity everywhere.

To turn to a criticism of this underlying evangelical hypothesis itself, it must, in my opinion, be qualified in several respects. Halévy probably claimed too much for the moral influence of evangelical religion both by magnifying its importance as a causative factor in social history and by unduly minimizing the strength and coherence of political and economic causation. Not many students of English history would accept unreservedly his view of the insufficiency of the constitution and political institutions in 1815. After all, the old house of commons did prove responsive enough to primarily political pressures to reform itself. Not only that, but in 1815 it had managed to govern England through a quarter-century of war. Halévy's own picture of the unreformed parliament is not so bad as the interpretation requires. And it is commonplace to point out that one of the major reasons for England's immunity to revolution in the nineteenth century was that she had already had her revolution in the seventeenth century and had secured by it a constitutional framework within which the objectives of continental revolutionaries could be realized without violence. In this connection it must be remarked, however, that in a later essay, "Grandeur, décadence, persistance du libéralisme en Angleterre," Halévy did attribute the origin and much of the strength of British liberalism to the Whig tradition.[29]

Neither is it possible fully to agree with his picture of hopeless economic anarchy at the beginning of his period. Economic problems, like political problems, certainly worked themselves out, at least to some extent. Indeed, in his last volume, which, had he lived to carry out his plan, Halévy intended to conclude with another general survey of English society in 1852, he himself remarks that the great expansion of middle-class wealth rendered it no longer necessary to look outside the economic world for an explanation of the stability of English life.[30] He reiterates that this had been necessary in 1815, but he does not draw the conclusion that the prosperous middle class, which, by 1852, he describes as bridging over the extremes of wealth and poverty, had, as a matter of fact, grown out of the 1815 economic situation. The development of a middle-class stake in society cannot very well be considered the product of religious, rather than of economic, causation. Moreover, it was not until sometime *after* 1848 that British stability became so rocklike a phenomenon. Contemporaries did not find society notably stable in the evangelical, but economically uncertain, decades of the twenties and thirties. Halévy, of course, might well reply that evangelical influences were what prevented the uneasy ferment of these years from boiling up into revolution. But even if the point is granted, as it probably should be in a qualified sense, that is not quite the same thing as to ascribe the development of Victorian stability basically to evangelical influences while regarding the roots of economic development as leading in themselves only to chaos.

Finally, Halévy sometimes read evangelical influences into affairs which they do not actually explain very well. For example, he attributed the relative success of pre-1835 local government largely to the humanizing of the justices of the peace by evangelicalism and to the presence of a large number of clergymen on the county benches.[31] But—as the Hammonds amply demonstrated—clerical justices tended to be on the savage rather than the lenient side,[32] and in any case, the clergyman who would be a justice of the peace was far more likely to be an old tory high-and-dry "squarson" than an evangelical.

Nevertheless, whether or not religious factors were a sufficient "cause" of social stability, Halévy's emphasis on the practical role of evangelicalism and nonconformity was an enormous and original contribution to the understanding of English history. After his work, no one could afford to neglect the religious mold of Victorian society or to dismiss it lightly as a faintly ridiculous, but not too important, shell of prudery and cant. Mr. Gladstone's nonconformist conscience was too formidable a fact for such treatment, and no one before Halévy had ever explained the process which implanted such a conscience in a high-church Anglican bosom. Neither has anyone else so clearly described the formation and practical effects of that palpable, but oddly indefinable, fact, the Victorian *Zeitgeist*.

One would like to know Halévy's own opinion of evangelical religion, apart from his undoubted admiration for its results. He never offered an explicit judgment—in French politics he was a good anticlerical—but the reader occasionally has the sense of an envious indulgence, not altogether unmixed with contempt, for the simple-minded religion which produced such a fortunate society as the English, enterprising, dutiful, vigorous, and self-reliant.[33] For Halévy, the main thing about evangelicalism was not that it was a true religion but that it led to individualistic self-restraint. One point, however, Halévy does not raise, and yet it seems that he might well have done so. He attributes the late Victorian decay of liberalism in part to the decline of Protestant conviction, but he does not suggest that so uncritical, so theologically shallow, a religion as evangelicalism could scarcely provide organized Christianity with the sharp intellectual tools required for the problems of the modern world—that, in fact, its emotionalism probably took the edge off whatever tools there were.

The comparison of Leslie Stephen's *The English utilitarians* with *The growth of philosophic radicalism* makes it apparent that one reason for the relative inferiority of Stephen's treatment was its lack of any basic critical or interpretative scheme. Similarly, much of the success of Halévy's *History* derives from the fruitfulness of the hypothesis which he adopted to explain what he conceived to be the central problem of the subject—the combination of stability and freedom in British life. To take this as the central problem was, of course, an arbitrary choice. Nevertheless, the fact that Halévy had a conscious conceptual scheme is what lifts his *History* above the level of an account of miscellaneous events and makes of it at once a history and an analysis of a civilization. Although there are difficulties with the central evangelical hypothesis, as well as with some of the subsidiary interpretations, this does not mean that they were not indispensable in giving form and unity to the work and in providing the author with a basis for selecting his material from the evidence. Halévy had to a remarkable degree the ability not to distort in the process of selection. He made his central hypothesis explicit, and he also described the reasoning which had led to his crucial judgments. The honesty of his portrayal is attested by the fact that his interpretations can be modified out of a study of his own work. This surely is the ultimate triumph of the historical method: to find a meaning which gives significance and unity to one's work and which can be modified out of the work itself. Whether an interpretative hypothesis is "true" or not is less important than whether it is honestly come by and fruitful. Perhaps the fundamental reason for the loss of power which one feels in the two volumes of the *Epilogue* is that there Halévy had outrun the period where his hypothesis was at all applicable.

Opinions differ about whether the *Epilogue* is of an excellence equal to that of the rest of the *History*.[34] In my view it is rather a pity that Halévy interrupted his continuity in order to publish the material that he had already prepared on the Edwardian era. The work delayed him so that, except for the fragmentary *Age of Peel and Cobden,* we have none of what he would otherwise have written on the mid-century. And he was fundamentally out of sympathy with the later period. Not that the thoroughness of his research or the carefulness of his narrative failed. The *Epilogue* is probably the best thing on the period and is a remarkably successful example of the historical method applied to events within the author's memory. But one feels that Halévy had constantly to subdue his impatience with his subject. He had become convinced that for England the nineteenth century ended with the decay of the old, sober, individualist polity, a development symbolized most conveniently by Gladstone's retirement.[35] Now that the Conservative party was fully imperialistic and increasingly jingoistic and now that the Liberal party when it returned to power would be in the hands of advocates of the welfare state, England had, Halévy felt, passed into another age. At bottom he seems to disapprove of the people he is writing about: Chamberlain, Lloyd George, Winston Churchill, and Sidney and Beatrice Webb, all of whose activities he is likely to summarize by bombardments of somewhat querulous rhetorical questions. The Webbs in particular, both in the *Epilogue* and in *L'ère des tyrannies,* have a way of appearing at the most discreditable moments—justifying the Boer War, darkly plotting the expansion of the state, and as "les Webb" assuming in French a sinister quality oddly at variance with the other impressions which one receives of that earnest and humane couple. However inevitable or at least explicable the various evils of the late century—imperialism, statism, and demagoguery—to Halévy they afforded a depressing contrast to the sensible commercial liberalism of Victorian days.

Throughout the preceding volumes of the *History* and *The growth of philosophic radicalism,* what chiefly engaged Halévy's attention were the forces making for liberty. His work is sufficient testimony that he would have described liberty in an orderly society as the highest political good. Stable liberty is possible only, Halévy thought, in a really self-governing country, but he would not have regarded popular sovereignty and representative institutions as simply equivalent to self-government. A really self-governing country—to generalize the interpretation of the *History*—is one that does not require much government. Liberty is freedom from controls outside the individual. Order, in the measure that it is to complement rather than to restrict liberty, must be the result of voluntary restraints and not of external regulation of the individual, no matter by what sort of authority imposed. Not that Halévy was an anarchist or a doctrinaire. He was pre-eminently a man of common sense and would not have permitted a rigorous extension of his position to reduce it to absurdity, nor would he have denied that some kinds of external authority are vastly preferable to others. Neither was he a dogmatic opponent of state intervention in social and economic affairs, provided that the state limited itself to a sort of pragmatic trouble-shooting. What he deplored was the tendency to make statist authority an end in itself.

Halévy's idea of liberty, then, was in the classical liberal tradition. There was no *mystique* about it, however attenuated. Liberty is personal independence. Most collectivist definitions of liberty, on the other hand, are more or less pluralistic and positivistic. You realize liberty *through* something or other, through some group association or some provided opportunity or some guaranteed security. When, therefore, Halévy approached the study of socialism, his own predispositions were antisocialist. Nonetheless, if he was broadly correct in his interpretation of the historical growth and the environmental conditions of liberal polity, the application of his analytical skill to socialism ought to yield interesting results. And it is, of course, always possible that a hostile critic may be worth attending to and even that he may be right.

L'ère des tyrannies and *Histoire du socialisme européen* do not give the impression that Halévy enjoyed being a hostile critic. The two books are in no sense attacks upon socialism. His considered conclusions are deeply pessimistic, but the tone of the discussion is never invidious. He is, perhaps, unfair to the Webbs. With that qualification, however, his criticisms of the socialist tradition, like his criticisms of utilitarianism, do not involve anything in the nature of blame. But, while it was not difficult for him to avoid blaming the utilitarians for the contradictions in their thought, since he was sympathetic to the individualist strain running through the contradictions, it is more noteworthy that he was able to maintain successfully the judicial attitude with which he describes the development of the socialist movement.

Halévy had long been interested in the subject. At the turn of the century he began a course of lectures at the École des Sciences politiques on the history of socialism, and thereafter he alternated this with the course he gave on England. It was on their notes of these lectures that his friends who acted as his editors based *Histoire du socialisme européen*. In spite of the difficulties unavoidable in a volume of lectures reproduced at second hand, the book seems to me the most convenient survey of its subject in any language. It describes the development both of socialist doctrine and of socialist politics, and, as always, Halévy's narrative is informed with analysis. The vein is one of exposition, however, and, like the student at the lectures, the reader is left to draw his own conclusions and to make his own evaluations.

For explicit statements of Halévy's own judgment, one must turn to *L'ère des tyrannies* and particularly to the title chapter, which summarizes Halévy's considered conclusions. Since the book is a collection of essays, to a certain extent it lacks unity. The essays were written at different times and represent different stages in Halévy's thought. One feels a steady deepening of pessimism as the postwar years unfolded and as the despotic state extended its control over one country after another. Halévy's only other major publication on socialism was the lengthy preface which he and M. C. Bouglé provided for their edition of *Doctrine de Saint-Simon*,[36] a portion of which is included in *L'ère des tyrannies*. This chapter on the Saint-Simonians is the most extensive and, simply as a historical account, probably the best in the book. A French version of the separately published *World crisis of 1914–1918*, already mentioned, forms another chapter. An essay on Léonard Sismondi and four shorter discussions of the problems of postwar socialism in England complete the volume. On the whole, *L'ère des tyrannies*, despite the fragmentary nature of its contents, is an impressive and a depressing book—impressive in the display of the historical record, depressing in the conclusions based on it.

L'ère des tyrannies is not simply a collection of occasional essays, however. Halévy's technique of resolving his material into systematic dichotomies gave him a central analytical principle which revealed, in his view, the basic contradiction lying at the root of the entire socialist movement. "Les socialistes croient en deux choses qui sont absolument différentes, et peut-être même contradictoires: liberté—organisation."[37] And further: "Le socialisme, depuis sa naissance, au début du XIXe siècle, souffre d'une contradiction interne. D'une part, il est souvent présenté, par ceux qui sont les adeptes de cette doctrine, comme l'aboutissement et l'achèvement de la Révolution de 1789, qui fut une révolution de la liberté, comme une libération du dernier asservissement qui subsiste après que tous les autres ont été détruits: l'asservissement du travail par le capital. Mais il est aussi, d'autre part, réaction contre l'individualisme et le libéralisme; il nous propose une nouvelle organisation par contrainte à la place des organisations périmées que la Révolution a détruites."[38]

Halévy thought it inevitable that the egalitarian values issuing out of the French Revolution should produce a demand for justice when confronted with the conditions imposed on the working class by industrialization. Almost from the beginning, this urge to emancipation in the socialist movement expressed itself in the demand for the creation of a regulatory authority strong enough to protect the weak against the powerful. "Le socialisme, sous sa forme primitive, n'est ni libéral, ni démocratique, il est organisateur et hiérarchique."[39] The subversion of the end of emancipation by the means of authoritarianism may have been unavoidable; it was, nonetheless, the basic characteristic of the socialist movement. Still, the theoretical inconsistency which Halévy had uncovered between the natural identity and the artificial identification of interests had not turned him against utilitarian liberalism; and, in the abstract, this might be thought just as fatal an irreconcilability as the one he found in socialism, which in a way grew out of the earlier contradiction. More practically, if the working class was materially dissatisfied with its lot in liberal society, then perhaps both the theoretical and the actual failures of liberalism called for something like the socialist movement. But all Halévy's work had the unusual characteristic of combining theoretical analysis with empirical value-judgment. For him, the constructive, emancipating side of socialism was vitiated by the statism which was to realize it, while the shortcomings of liberalism were more than compensated in the freedoms which it made possible.

In spite of divergences and fierce quarrels between different socialist schools and notwithstanding the professed internationalism of most of them, the movement as a whole became increasingly associated with the expanding area of state activity, whether as a means or as an end. In Halévy's view, however—and this is the distinctive feature of his interpretation—what actually impressed socialist practices upon the European states was neither socialist doctrine nor socialist political success but rather the practical necessities of the war of 1914–18. The organization of society for war brought to full fruition the authoritarian implications already latent in socialist blueprints, and the intimate relation between the development of what amounted to socialist polity and the demands of the military state imprinted on contemporary times the character which Halévy uses for his title:[40]

L'ère des tyrannies date du mois d'août 1914, en d'autres termes du moment où les nations belligérantes adoptèrent un régime qu'on peut définir de la façon suivante:

   a) Au point de vue économique, étatisation extrêmement étendue de tous les moyens, de production, de distribution et d'échange;
—et, d'autre part, appel des gouvernements aux chefs des organisations ouvrières pour les aider dans ce travail d'étatisation—donc syndicalisme, corporatisme, en même temps qu'étatisme;

   b) Au point de vue intellectuel, étatisation de la pensée, cette étatisation prenant elle-même deux formes: L'une négative, par le suppression de toutes les expressions d'une opinion jugée défavorable à l'intérêt national; l'autre positive, par ce que nous appelerons l'organisation de l'enthousiasme.

C'est de ce régime de guerre, beaucoup plus de la doctrine marxiste, que dérive tout le socialisme d'après-guerre. Le paradoxe du socialisme d'après-guerre c'est qu'il recrute des adeptes qui viennent à lui par haine et dégoût de la guerre, et qu'il leur propose un programme qui consiste dans la prolongation du régime de guerre en temps de paix.

In Halévy's view this connection between socialism and the militant nation-state—a connection neither simple nor obvious but nevertheless real and profound—might have been discerned even before the first World War fastened it upon Europe. The common nexus between leaders of nationalist politics and leaders of socialist politics was their emphasis upon augmenting the integral power of the state. Their means were much the same, though their ultimate ends may have been very different. But Halévy was not impressed by ends which are too ultimate. "Ce qui m'intéresse," he once impatiently remarked à propos of the Marxist utopia, "c'est le présent et le prochain avenir: au delà il y a ce que Jules Romains appelle l'ultra-futur."[41] Halévy turns to the nineteenth century to point out the complex association between socialist and nationalist developments. In France the reaction against the socialism of 1848 produced the Caesarism of Napoleon III, which quickly exhibited a sort of authoritarian socialist character itself. In Germany it was Ferdinand Lassalle who directly inspired the "monarchie sociale"[42] through which Bismarck consolidated the power of the Reich. At the time only an authoritarian regime could have taken so long a step toward the social-service state as Bismarck's administration did. Lassalle himself, while he hated the liberals, felt a strong admiration for Prussian order. Similarly, in England the Fabian group, while expressing a lively contempt for Gladstonian liberalism, was notably indulgent toward the Tory imperialists. The Webbs themselves were imperialists "avec ostentation,"[43] and they, too, greatly admired the efficient Prussian bureaucracy. After 1918 they were determined to perpetuate the very similar controls which war had imposed on England and to utilize them to cope with the problems of peace and reconstruction. However persuasive this idea, Halévy viewed it with profound suspicion as early as 1922. "Mais quoi?" he asks, "Tout le résultat de la guerre n'aurait été, suivant le voeu secret des Webb et de lord Haldane, que de faire triompher, chez les nations qui venaient de vaincre la Prusse, le militarisme et le bureaucratisme prussiens?"[44]

After 1918, Russian Communists, in possession of a national state, did not jettison the instruments of power they had seized but perfected and greatly strengthened them. Italian fascism issued directly out of a socialist background perverted by the fear of anarchy: naziism in Germany came out of a decade of semisocialist politics. Both modeled themselves directly upon Russian methods of government. By the mid-thirties the Communist and the fascist states differed only in ideologies, one nationalist, the other socialist; in techniques of government they were identical, and necessarily so because each had to adopt so much of the character and program of the other as the real conditions of maintaining power required. "Bref, d'un côté, en partant du socialisme intégral, on tend vers une sorte de nationalisme. De l'autre côté, en partant du nationalisme intégral, on tend vers une sorte de socialisme."[45] Basically this was because "ces 'frères ennemis' ... ont un père commun, qui est l'état de guerre."[46] But the war did more than give birth to both movements; it also furnished them their techniques of control: "Il appartenait à la guerre mondiale de 1914 de révéler aux hommes de révolution [the Bolsheviks] et d'action [the fascists] que la structure moderne de l'état met à leur disposition des pouvoirs presque illimités."[47]

As for the future of the limited-state democracies, Halévy was increasingly pessimistic as the postwar years wore on. For a time he was cautiously sanguine about the prospect that England might show the way out. Possibly, he thought in the twenties, the English, having in their time submerged so many religious and political conflicts in compromises which averted violence by tolerating peaceable political rivalries—possibly the English would achieve the same tour de force with the class struggle, which must be domesticated, since it could not be denied. But he had little hope for the Labor party. The problems of socialism in a parliamentary regime he thought insuperable. Moderate socialists, to achieve their complicated program, the technicalities of which were beyond the comprehension and remote from the immediate interests of working-class voters, would have to enforce a regimen of discipline and sacrifice. This would deny their followers exactly the material improvements promised them by socialism. Labor, Halévy thought, was fortunate never to have had a clear parliamentary majority.

In any case, the darkening international scene cast all such problems into shadow, both for England and for France: "Les tyrannies qui nous touchent de plus près—celle de Berlin, celle de Rome—sont étroitement nationalistes. Elles ne nous promettent que la guerre. Si elle éclate, la situation des démocraties sera tragique. Pourront-elles rester des démocraties parlementaires et libérales si elles veulent faire la guerre avec efficacité? Ma thèse . . . c'est qu'elles ne le pourront pas. Et le recommencement de la guerre consolidera l'idée 'tyrannique' en Europe."[48]

Halévy thus ended his career on a note of blackest pessimism. It is difficult to think that events of the last decade have proved him entirely mistaken. By weaving so closely together the long backgrounds of socialism and of modern war—socialist organization, on which in its moderate sense most enlightened people seem now to rely for a solution to Europe's problems, and war which produced those problems—he lights no very cheerful path into the future.

The issues raised by Halévy's valedictory work are far too immediate for a definitive judgment on it to be possible, though several obvious lines of criticism suggest themselves. It is too early to tell, of course, but it begins to look as if he might have been too hasty in writing off the English capacity for adjusting parliamentary institutions to new situations. Nor were democratic practices entirely submerged in war, at least in the western powers. More generally, it might be inquired what Halévy's solution would have been to the problems to which socialists addressed themselves, inasmuch as he admitted the reality of the problems? What he seems to have hoped for was the sort of restraint and spirit of compromise which, recognizing the reality of differing class interests, would permit of their being adjusted by voluntary *ad hoc* agreements, refereed, perhaps, by public agents, but not imposed by a state asserting its authority over both—indeed, over all conceivable—interests. But that he was never very explicit along these lines does not in itself vitiate the grounds for Halévy's alarm at the statist solution.

It might also be objected that Halévy's discussion is too abstract, too intellectual, and that it does not really illuminate the great variety of concrete circumstances and actual events. This objection would certainly apply to some of his earlier work. But in the abstract contradiction of assumptions around which Halévy organized his treatment of utilitarianism, he certainly caught up an epitome of the real political success of individualist liberalism as compared to its relative social and economic shortcomings. Halévy himself met this criticism in answer to a friend who remonstrated with him on this account, and his reply is a good defense of the utility of intellectual history. He admitted that in his early career he had treated history too abstractly, but he had modified his approach radically until what he now offered was a "cours d'histoire tout court." "Ce qui ne veut pas dire que je n'aie pas été heureux, que je ne sois pas heureux encore, d'avoir abordé l'histoire du socialisme par le biais de l'histoire des doctrines. Car ... les doctrines stylisent, schématisent les faits. Et rien ne me paraît plus utile, pour la connaissance des faits, que cette schématisation."[49]

Halévy's emphasis upon the concrete role of the war rather than the influence of Marxist or other collectivist doctrines in the progress of socialism might be regarded as calling for a serious qualification of his generally idealist interpretation of historical causation. To be sure, he held that the war itself was the product of men's feelings, stirred by the long chain of causes to a pitch of patriotic emotion, rather than of their interests. It would also be difficult to reconcile the view that history is formed fundamentally by beliefs and ideas with the contention that, however different the ideologies, Communist and fascist regimes are identical in practice and hence in fact. But these considerations only indicate that what Halévy offers is not a totally consistent philosophy of politics of history. He himself was primarily an idealist, but from his own work, dispassionate and objective as it is, one can argue that the qualifications on the idealist interpretation are so far reaching as to destroy the position. Whatever opinion one adopts on this issue, there is a more fundamental truth to be appreciated in Halévy's work. In the fact that a great French scholar produced the most distinguished work on modern English history, a work in which the central interpretation is the supreme value of liberty, one has a reminder that between the best of the English tradition and the best of the French tradition there is more in common than is always remembered; that, in fact, the West probably disputes fewer and more trivial values than the ones it shares.

For the historian there is a particularly interesting passage at the end of Halévy's work. He is describing the state of his opinion when he first began to lecture on the history of socialism:[50]

Je n'étais pas socialiste. J'étais 'libéral' en ce sens que j'étais anticlérical, démocrate, républicain, disons d'un seul mot qui était alors lourd de sens: un 'dreyfusard.' Mais je n'étais pas socialiste. Et pourquoi? C'est, j'en suis persuadé, pour un motif dont je n'ai aucune raison d'être fier. C'est que je suis né cinq ou six ans trop tôt. Mes années d'École Normale vont de l'automne 1889, juste après l'effondrement du boulangisme, à l'été de 1892, juste avant le début de la crise du Panama. Années de calme plat: au cours de ces trois années, je n'ai pas connu à l'École Normale un seul socialiste. Si j'avais eu cinq ans de moins, si j'avais été à l'École Normale au cours des années qui vont des environs de 1895 aux environs de 1900; si j'avais été le camarade de Mathiez, de Péguy, d'Albert Thomas, il est extrêmement probable qu'à vingt-et-un ans j'aurais été socialiste, quitte à évoluer ensuite, il m'est impossible de deviner en quel sens. Lorsque, appliquant à nous-mêmes les méthodes de la recherche historique, nous sommes amenés à découvrir les raisons de nos convictions, nous constatons souvent qu'elles sont accidentelles, qu'elles tiennent à des circonstances dont nous n'avons pas été les maîtres. Et peut-être y a-t-il là une leçon de tolérance. Si on a bien compris cela, on est conduit à se demander s'il vaut la peine de se massacrer les uns les autres pour les convictions dont l'origine est si fragile.

Historians generally may find sobering the profound modesty and perfect candor with which so painstaking a member of their profession, one whose achievements are as impressive as Halévy's, examines and leaves open the possibility that the interpretations of his entire life work may be simply the result of a circumstantial personal accident. But also this same modesty can cast some light both on why history is worth writing and why it must be rewritten and reinterpreted as the perspective changes. And it is consoling to reflect that in a self-questioning attitude like Halévy's lies too, perhaps, the soundest antidote to so pessimistic a view as his own of the prospects for the future.

Notes and References

1. (*Journal*, XXII [1950], 155). Although not named on the title-page, M. C. Bouglé, Raymond Aron, J.-M. Jeanneney, Pierre Laroque, Étienne Mantoux, and Robert Marjolin edited the book. Its plan and content are Halévy's. The writing of most of it, unfortunately, is not but is the result of the editorial collaboration of his friends and former students at the École des Sciences politiques, working from his papers and from notes taken during his course of lectures there. Only two chapters on Marxism and four "annexes" (on Friedrich List, German historicism and political economy, Thomas Carlyle and John Ruskin, and Marxism and syndicalism) are from Halévy's own pen, and these were written around 1903 in his early, highly abstract manner. Since they were completed, the editors thought it better to include them than to attempt a revision.

2. The other two are *L'ère des tyrannies: études sur le socialisme et la guerre*, ed. M. C. Bouglé (Paris, 1938), a collection of lectures and previously printed articles; and Vol. IV of *Histoire du peuple anglais au XIXᵉ siècle*, ed. Paul Vaucher (*Journal*, XX [1948], 259–61).

3. For biographical information and previous critical discussion see Sir Ernest Barker, "Élie Halévy," *English historical review*, LIII (1938), 79–87; J. B. Brebner, "Halévy: diagnostician of modern Britain," *Thought*, XXIII (1948), 101–13; Léon Brunschvicg, "Élie Halévy," *Revue de métaphysique et de morale*, XLIV (1937), 679–91; C. H. Smith, "Élie Halévy," in *Some historians of modern Europe*, ed. B. E. Schmitt (Chicago, 1942), pp. 152–67; and the collection of interesting, if somewhat redundant, obituary notices published by the École libre des Sciences politiques under the title *Élie Halévy* (Paris, 1938).

4. Halévy retained his interest in philosophy. He was co-founder of *Revue de métaphysique et de morale* and remained an editor of it throughout his life.

5. Paris, 1896.

6. *La formation du radicalisme philosophique* (Paris, 1901–4). Citations herein refer to the English edition, trans. Mary Morris (London, 1928), which includes a preface by Sir Ernest Barker and a bibliography of Jeremy Bentham by C. W. Everett.

7. Under the general title *Histoire du peuple anglais au XIXᵉ siècle,* the original editions, separately subtitled, appeared as follows: Vol. I, *L'angleterre en 1815* (Paris, 1912); Vol. II, *Du lendemain de Waterloo à la veille du Reform Bill, 1815–1830* (Paris, 1923); Vol. III, *De la crise du Reform Bill à l'avenement de Sir Robert Peel, 1830–1841* (Paris, 1923); Vol. IV, *Le milieu du siècle, 1841–1852* (Paris, 1946). Footnotes herein refer to the English translation (hereafter cited as *"History"*). Vol. I was translated by E. I. Watkin and D. A. Barker and the remaining volumes by E. I. Watkin (London, 1924–47). The English edition of Vol. IV has the separate title, *The age of Peel and Cobden* (London, 1947).

8. After Vol. III of the *History,* Halévy interrupted his chronology to treat the period 1895–1914. The general title is still *Histoire du peuple anglais au XIXᵉ siècle* and the subtitles as follows: *Épilogue, 1895–1914; I: les impérialistes au pouvoir, 1895–1905* (Paris, 1926); and *Épilogue, 1895–1914; II: vers la démocratie sociale et vers la guerre, 1905–1914* (Paris, 1932). Footnotes herein refer to the English translations by E. I. Watkin (London, 1929–34) (hereafter cited as *"Epilogue"*). Some catalogues list the two volumes of the *Epilogue* as Vols. IV and V of the original title. Since the writing of this article the first two volumes of what is to be a six-volume revised edition of the entire *History,* including the *Epilogue,* have been published (London and New York: Peter Smith, 1949——). Apparently, the new edition is to follow the general plan of the original edition except that subtitles will be different and that Vol. IV will include an extensive essay by R. B. McCallum on the period 1852–95.

9. In this article I shall not discuss Halévy's use of the sources from a technical standpoint. An estimate of Halévy's success in the handling of source materials may be found in Graham Wallas' introduction to Halévy's *History,* I, v–viii.

10. *La théorie platonicienne des sciences,* p. xvi.

11. *Ibid.,* p. 378.

12. One feels Halévy to have been a little overenthusiastic about his own argument here. Was Burke's emphasis on consequences and his appeal to irrational sentiments and loyalties really utilitarian in the mechanistic sense of the greatest-happiness principle?

13. *The growth of philosophic radicalism,* p. 154.

14. London, 1900.

15. *The growth of philosophic radicalism,* pp. 485–86.

16. *History,* introd., I, vi.

17. *Thomas Hodgskin* (Paris, 1903).

18. *History,* I, 176.

19. *Ibid.,* pp. 334–35.

20. Halévy had already published an account of the foundation of the Methodist movement itself, "La naissance du méthodisme en Angleterre," *Revue de Paris,* IV (1906), 519–39 and 841–67. This article, seldom mentioned in bibliographies, seems to me the best available account of early methodism and its appeal in the general economic, political, and religious background of the period. It is the first statement of Halévy's thesis that the Methodist revival was a strong and continuing influence for social conformity, that it largely accounts for the quality of voluntary moral discipline which differentiated English from continental society, and that it explains the persistence of a generalized Puritan psychology in England long after continental Protestantism had been dissipated in rationalism or indifference.

21. *History,* III, 166.

22. *Ibid.,* II, vi.

23. Since Vol. IV was published only in 1946, it cannot yet have exercised any great influence. It should be emphasized, too, that Halévy would probably have revised much of it.

24. *History,* II, vii–x.

25. One does not find patriotism creeping into Halévy's remarks on the conclusion of the Napoleonic wars in *England in 1815* or into a later discussion of English misunderstanding of French affairs in his rather slight essay, "English public opinion and the French revolutions of the nineteenth century," in *Studies in Anglo-French history,* ed. Alfred Coville and Harold Temperley (Cambridge, 1935), pp. 51–60.

26. It should, however, be remarked that he served on the commission for editing the French diplomatic documents, but I suspect—though without authority, and I may be quite wrong—that he did so out of a sense of duty rather than out of real enthusiasm for the project.

27. *L'ère des tyrannies,* p. 200.

28. *The world crisis of 1914–1918* (Oxford, 1930), p. 56.

29. É. Halévy, R. Aron, M. C. Bouglé, and Others, *Inventaires, la crise sociale et les idéologies nationales* (Paris, 1936), pp. 5–23. Here, however, Halévy rather overdoes the limited character of the Tudor monarchy.

30. *History,* IV, 292–93.

31. "Before 1835," in *A century of municipal progress,* ed. H. J. Laski, W. I. Jennings, and W. A. Robson (London, 1935), pp. 28 and 35–36.

32. J. L. and B. Hammond, *The town laborer* (London, 1920), pp. 268–78, and *The age of the Chartists* (London, 1930), pp. 217–20.

33. See particularly *History,* IV, 301–2.

34. Compare, e.g., Sir Ernest Barker's review of Vol. II of the *Épilogue* (*English historical review,* XLVIII [1933], 674–76) to the enthusiasm of Robert Dreyfus, who regarded the *Épilogue* as Halévy's masterpiece ("L'ami," in *Élie Halévy,* p. 39).

35. *Épilogue,* I, ix–xiii. Halévy describes his state of mind explicitly both in the introduction just cited and in an interesting essay on John Morley, whom he describes

as the last of the Benthamites ("Les souvenirs de Lord Morley," *Revue de métaphysique et de morale*, XXV [1918], 83–97).

36. Paris, 1924.

37. *L'ère des tyrannies*, p. 208.

38. *Ibid.*, p. 213.

39. *Ibid.*

40. *Ibid.*, p. 214.

41. *Ibid.*, p. 225.

42. *Ibid.*, p. 213.

43. *Ibid.*, p. 217.

44. *Ibid.*, p. 155.

45. *Ibid.*, p. 227.

46. *Ibid.*, p. 235.

47. *Ibid.*, p. 249.

48. *Ibid.*, p. 222.

49. *Ibid.*, p. 220.

50. *Ibid.*, pp. 216–17.

—☞— 13 —☜—

Alexandre Koyré

NO SENIOR SCHOLAR was nearly so important to my formation as a historian of science as was Alexandre Koyré. The same in greater or less degree was true of Marshall Clagett, Bernard Cohen, René Taton, Pierre Costabel, Alistair Crombie, Tom Kuhn, Erwin Hiebert, and John Murdoch, to mention no others among our generation. My wife and I first met Koyré and Madame Koyré in 1954 when I was on sabbatical beginning research on science in the French Revolution. The occasion was a soirée given by René and Juliette Taton. Gaston Bachelard and other notables were of the company. Koyré did seem a little daunting when I was introduced. "Vous êtes professeur de quoi, Monsieur?" was the first thing he asked me. Since the only answer at that juncture was English history, I rather think I babbled a bit. The austerity was nothing but a seemly European reserve, however, and Madame Koyré and he were cordiality itself during the rest of our year in Paris.

It was thus an extraordinarily lucky coincidence that the opportunity to teach a course in history of science in 1956 should have presented itself at just the time when Koyré became a member of the Institute for Advanced Study. Not that I ever discussed the details of teaching with him. Undergraduate pedagogy formed no part of his purview. Nevertheless I do not know how I would have conceived the thematics of that first course if I had not already read, and often reread, *Études galiléennes* (of which more is in the essay that follows). Koyré spent half of every academic year in Princeton from 1956 to 1962. Until then no one had ever made much use of the fine collection of first editions in history of science that Lessing Rosenwald had presented to the Institute many years previously. That was one reason, though certainly not the main one, that led Robert Oppenheimer as Director to take the initiative in inviting Koyré to the Institute. Oppenheimer could be very gracious. Introducing his guest on the occasion of a public lecture, he said, "Professor Koyré does us the honor of speaking very softly." Koyré did himself no less honor. The topic was the now-famous "Hypothesis and Experiment in Newton."[1] Never have I been part of an audience that listened more intently, applauding at the end by a virtually reverent silence.

Koyré was deeply touched by the offer of a professorship, which followed shortly. He felt he must decline, however, and accepted instead a permanent membership. The latter allowed him, as it did Otto Neugebauer, to take up residence whenever he liked. He thus did not need to abandon the École Pratique des Hautes Études, to which institution he felt a loyalty stemming from the days when it had afforded a post to the unknown young Russian refugee scholar he was in the early 1920s. What I learned in his presence was nothing concrete or specific. It had to do not only with himself but with the company to which his friendship gave access, that of Erwin Panofsky, of Harold Cherniss, of Ernst Kantorowics, of Robert Oppenheimer. Theirs was a quality of urbanity and cultivation that is somehow no longer attainable. Not by me, at any rate. Still, whatever my thinking may be, it was both elevated and deepened by close association with Alexandre Koyré.[2]

That the strongest influences on my early work, Halévy in history proper and Koyré in history of science, should both have been French Platonists must say something about my temperament. I am not aware of having written Platonic history, except perhaps to some degree in *The Edge of Objectivity.* I have always found, however, that reading Plato is a pleasure while reading Aristotle is a duty.

Notes

1. An extended version forms Chapter 2 of Koyré's *Newtonian Studies* (Cambridge: Harvard University Press, 1965).

2. The above is a close paraphrase of the passage that appeared in my autobiographical essay, "Apologia pro Vita Sua," which I contributed to the issue of *Isis* marking the 75th anniversary of the History of Science Society, *Catching up with the Vision, Isis* 90 (1999), pp. S84–S94.

THIRTEEN

Alexandre Koyré*

—ɯ—

(*b.* Taganrog, Russia, 29 August 1892; *d.* Paris, France, 28 April 1964), *history of science, of philosophy, and of ideas.*

Koyré's work was threefold. First, he exercised a formative influence upon an entire generation of historians of science, and especially in the United States. In France, secondly, where his circle was mainly philosophical, he also initiated the revival of Hegelian studies in the 1930's and published important studies of other pure philosophers, most notably Spinoza [6]. Thirdly, his essays on Russian thought and philosophical sensibility were important contributions to the intellectual history of his native country [4, 11]. A strong vein of philosophical idealism inspired all his writings, which proceeded from the assumption that the object of philosophical reasoning is reality, even when the subject is religious. A remark in the preface to his study of Jacob Boehme might equally well be applied to any of his books: "We believe ... that the system of a great philosopher is inexhaustible, like the very reality of which it is an expression, like the master intuition that dominates it."[1]

For Koyré was ever a Platonist. Indeed, the best introduction to the unity of view and value inspiring the whole body of his work is his beautiful essay *Discovering Plato* [9], published in 1945 in French and English editions in New York, and originally composed in the form of lectures given in Beirut after the fall of France in 1940. Koyré never despaired of European civilization, however Hellenic its apparent disintegration. It was always his inner belief that mind might yet prevail. The contempla-

* Reprinted from the *Dictionary of Scientific Biography* 7 (1973), pp. 482–490.

tive tone disarms resistance to the hortatory discourse, which, mingling jest with seriousness in true Platonic style, opens to the reader the implications of philosophy for personality and of personality for politics, those being the themes that invest the dialogues with dramatic tension.

Koyré said little here of Platonism in the development of science, but the relation of intellect to character and of personal excellence to civic responsibility that this essay brings out explains his sympathy for the Platonic inspiration that he detected (and in other writings perhaps exaggerated) in the motivations of the founders of modern science, particularly Galileo.

Koyré began his secondary education at Tiflis and completed it at the age of sixteen at Rostov-on-Don. His father, Vladimir, was a prosperous importer of colonial products and successful investor in the Baku oil fields. Husserl was the idol of Koyré's schooldays, and in 1908 he went to Göttingen, where, besides the master of phenomenology he had come to follow, he also met Hilbert and attended his lectures in higher mathematics. In 1911 he moved on to Paris and the Sorbonne, where he listened to Bergson, Victor Delbos, André Lalande, and Léon Brunschvicg. Although he did not become as familiar with any of his teachers in Paris as he had with Husserl and his family (Frau Husserl had mothered him a bit), he felt at ease in the cooler climate of French civilization.

Before the war he had already begun work on a thesis on Saint Anselm under the direction of François Picavet, then teaching at the École Pratique des Hautes Études. In 1914 Koyré, though not yet a citizen, enlisted in the French army and fought in France for two years. Then he transferred his service to a Russian regiment when a call came for volunteers and went back to Russia, where he continued to fight on the southwestern front until the collapse in October 1917. During the civil war that followed, Koyré found himself among opposition groups which can be best compared to resistance forces, fighting against both Reds and Whites. After a time, he decided to disengage himself from the melee, and, the war being over, he made his way back to Paris. There he was married with great happiness to Dora Rèybermann, daughter of an Odessa family. Her sister also married his elder brother. In Paris he resumed a life of scholarship and philosophy, finding to his astonishment that the proprietor of the hotel where he had lodged in his student days had faithfully preserved the manuscript of his thesis on Anselm throughout the war.

Always a philosopher in his own sense of professional identity, Koyré began his career in the study of religious thought, though it was in the history of science that he later did his deepest work. His first books were theological: *Essai sur l'idée de Dieu et les preuves de son existence chez Des-*

cartes (1922), *L'idée de Dieu dans la philosophie de St. Anselme* (1923), and *La philosophie de Jacob Boehme* (1929). Completion of the first qualified him for the diploma of the École Pratique and won him election as *chargé de conférences,* or lecturer, in that institution, with which he remained associated throughout his life. The work on Anselm, completed earlier, was published later and satisfied the Sorbonne's requirements for the university doctorate, a degree elevated into the *doctorat d'État* by virtue of the Boehme thesis.

Students of Koyré's later writings on the history of science will recognize characteristic motifs and methods in the analysis he gave these early subjects. The theological tradition that appealed to him was that most highly intellectualized of apologetic strategies, the ontological argument for the existence of God. In the versions given both by its originator, Anselm, and by Descartes, mind rather than religious experience made the connection between personal existence apprehended subjectively and external reality, of which the important aspect in this context was God—though it could as easily be nature when Koyré turned his interest to the natural philosophers. His central proposition in regard to Descartes was that the philosopher of modernity owed much to medieval predecessors. It is one that would no longer need to be argued. Neither would his more interesting, supporting assertion of the philosophic value of scholastic reasoning, "subtleties" being a word that Koyré never thought pejorative.

To historians of science the most interesting feature of the discussion is the use that Koyré found Descartes making of the concepts of perfection and infinity. In handling the latter, he showed how the mathematician in Descartes had fortified the philosopher and invested the ontological argument with a sophistication unattainable by the reasoning of Anselm. Occasional asides presaged the direction in which Koyré's own interests would afterwards develop: for example, "we consider that the most notable achievement of Descartes the mathematician was to recognize the continuity of number. In assimilating discrete number to lines and extended magnitudes, he introduced continuity and the infinite into the domain of finite number."[2] In this book, however, Koyré had his attention on the *Meditations* and on Descartes the theologian and metaphysicist. Only later, in the beautiful and lucid *Entretiens sur Descartes* [8], did Koyré handle instead the *Discourse on Method,* emphasizing that it was the preface to Descartes's treatises of geometry, optics, and meteorology. Koyré would then no longer have agreed with his own youthful statement to the effect that, although Descartes altered the whole course of philosophy, the history of science would have been little different if he had never lived.[3]

Indeed, Koyré's own natural predilections emerge from the contrast in tone between his two major writings on Descartes. The *Entretiens* is an enthusiastic book, sympathetic and almost affectionate in its treatment of Descartes. Not so the thesis, a little stilted in its quality, wherein the author does not seem quite at ease with his subject. The constraint comes out overtly in passages concerning Descartes's want of candor, but the reader is left with a more general feeling of artificiality about the very enterprise of treating Descartes theologically. Koyré's having been a candidate in the division of the École Pratique concerned with "sciences réligieuses," the Vᵉ Section, may quite naturally have affected his choice of a subject. What is surprising, however, is that Koyré remained associated with that section throughout a life devoted largely to the history of science, there being no appropriate provision for the latter subject in the academic structures of Paris. It was a circumstance bespeaking both the rigidity of institutions in the French capital and the flexibility of their administrators, despite whose generosity Koyré felt some difficulty over his commitments in his later years.

No ambiguities beclouded the simplicity and serenity of Anselm's commitments, and though Koyré's monograph on the founder of the ontological proof of God's being is less suggestive of his later interests in its thematics than the thesis on Descartes, it is more so in its treatment, specifically on the score of sympathy and penetration of the man through the texts.

Perhaps Koyré's most characteristic gift as a scholar (it was the manifestation in scholarship of his personal quality) was his ability to enter into the world of his subject and evoke for the reader the way in which things were then seen: in this case, the spiritual and intellectual reality in which Anselm perceived both beatifically and logically the necessity of God's being; in other instances, Aristotle's world of physical objects apprehended by common sense and ranged into an orderly philosophy; Jacob Boehme's tissue of signatures and correspondences between man and nature; the Copernican globes spinning and revolving for the simple and sufficient reason that they are round; Kepler's vision of numerical form and Pythagorean solidity; Galileo's abstract reality of quantifiable bodies kinematically related in geometrical space; and finally Newton's open universe, with consciousness situated in infinite space instead of in the cosmos of ancient Greek philosophy.

It was through meticulous analysis of essential texts, however, and not through general summary or paraphrase that Koyré thus opened spacious implications out of the intellectual constructions of his subjects. He liked to print extensive passages from the text to accompany his analysis in order that the reader might see what he was about. Indeed, his writing adapted the French instructional technique of *explication de texte* to the highest purposes of scholarship. Most of his later works derived from courses, often from individual lectures, given in the many institutions in France, Egypt, and the United States, where he taught regularly or was a guest. In later years his knowledge sometimes made him seem severe to younger scholars unsure of their own. The effect was altogether unintended. Fundamentally his was a deeply humane intellectual temperament, critical in the analytical and never in the denigrating or destructive sense. He wished to bring out the value in the subjects that he studied, not to expose what might be found of hollowness or falsity in them. Easy targets never tempted him. His own self-assurance was thus compatible with the most serious humility, for he subordinated his gifts to enhancing the merits of those who by mind, daring, imagination, and taste had contributed to civilizing our culture, and who had thereby aroused his admiration.

Such qualities of empathy animated the important studies he made of Hegelian philosophy and of the intellectual culture of 19th-century Russia. Neither of those concerns bore directly on the history of science, but perhaps a word may be said. His knowledge of Hegel derived from youthful immersion in Husserl's phenomenology. In the early 1930's he thought to convey the interest it held to his circle of philosophical friends in Paris—formed for the most part in the École Normale Supérieure—to whom it was largely alien, not to say terra incognita. These papers were well received,[4] and readers whose case is similar may find particularly illuminating his "Note sur la langue et la terminologie hégéliennes."[5] Similarly, the papers in his two volumes on Russian intellectual history developed for a French learned public a subject for which Koyré had special competence: the dilemma of Russian writers torn between the necessity for assimilating European culture if their country were to become civilized, and resisting it if Russia were to establish its own national identity [4, 11]. Admirers of Koyré's writings in the history of science would do well to read the most considerable of those studies, a monograph on Tchaadaev.[6] Although it has nothing to do with their subject, it is one of the finest, most sympathetic and revealing pieces that he wrote.

By contrast, Koyré's work on the German mystics did have an important if somewhat enigmatic bearing on his historiography of science, for although he was never more earnest than in mediating between this inaccessible tradition and his modern reader's sensibility, his own reaction to it was to turn from theological subjects back to the scientific interests of his student days at Göttingen. His major doctoral thesis remains the most considerable and reliable study of Boehme, a lucid book on an obscure writer. Koyré also gathered into a little book four short pieces on Schwenkfeld, Sebastian Franck, Paracelsus, and Valentin Weigel, Boehme's most important sources [12]. Its reissue in 1971 coincided with a revival of the occult that the author would have deplored. True, it might be held that Boehme took an interest in the natural world even as did Galileo, Descartes, and Kepler, his contemporaries. Any resemblance is only apparent, however, for Boehme's sense of nature was altogether symbolic, the reality of phenomena residing for him in the signatures they bear of the divine. It is true that Koyré's awareness of how the world had impinged on consciousness before modern science destroyed these symbolic meanings sensitized his later writings on the scientific revolution. But he came to feel a certain futility in the enterprise of exploring the experiences of mystics, which by definition could be known only by him to whom they happened. Boehme was consistent in always seeking to read the correspondence between man and the world out of what he often called the book of himself, whereas the Koyré of *Études galiléennes* observed in the opening lines that only the history of science invests the idea of progress with meaning since it records the conquests won by the human mind at grips with reality.[7]

However that may be, the leitmotif of Koyré's work in history of science was the problem of motion; and he first identified it in a philosophical essay, "Bemerkungen zu den Zenonischen Paradoxen," published in 1922, prior to these theological writings.[8] In this, his first substantial publication,[9] Koyré argued that understanding Zeno's puzzles required analysis not merely of motion but of the manner in which its conceptualization in parameters of time and space involved ideas of infinity and continuity. After reviewing the Zenonian contributions of Brochard, Noël, Evelyn, and Bergson, Koyré (no doubt thinking back to his studies with Hilbert) invoked the findings of Bolzano and Cantor on the infinite and the nature of limits, and distinguished between motion as a process involving bodies in their nature and motion as a relation to which they are indifferent in themselves. A footnote anticipated Koyré's lifework in a single sentence: "All the disagreement between ancient and modern physics may be reduced to this: whereas for Aristotle, motion is necessarily an action, or more precisely an actualization (*actus entis in potentia in quantum est in potentia*), it became for Galileo as for Descartes a state."[10] Towards the end of his life, Koyré was sometimes asked how he happened to turn from theology to science, and once said, "I returned to my first love."[11]

His own career was full of movement. He had prepared his materials on Russian intellectual history in the first instance for a course at the Institut d'Études Slaves of the University of Paris. In 1929, the year *Jacob Boehme* appeared, he was appointed to a post in the Faculty of Letters at Montpellier and taught there from September 1930 until December 1931, enjoying the climate and quality of life in the Midi while regretting the inaccessibility of libraries. In January 1932 he was elected a *directeur d'études* at the École Pratique and returned to Paris, where his course treated of science and faith in the sixteenth century. Having read Copernicus for that purpose, and found how little was really known of his epochal accomplishment, Koyré prepared a translation of book I of *De revolutionibus,* its theoretical and cosmological part, together with a historical and interpretative introduction [5]. It was his initial contribution to history of science proper. In it Copernicus stands forth a thinker about the universe and no mere manipulator of epicycles, a thinker at once archaic and revolutionary. He was archaic in his addiction to the Platonic aesthetic of circularity, making it into a cosmic kinematics. He was revolutionary in his conviction that geometric form must comport with physical reality, and that no hypothesis joining the two was too daring to adopt, let the consequences be what they might for tradition and common sense. By implication, form itself became geometric, instead of substantial, and down that road lay modern science.

When Koyré published his Copernicus edition in 1934, he was teaching on a visiting basis at the University of Cairo. Finding his colleagues and students most congenial, he returned there in 1936–1937 and again in 1937–1938. For that audience he prepared lectures later developed into the *Entretiens sur Descartes.* Having turned from Copernicus to Galileo, it was also in Cairo that he settled down with the great Favaro edition of the latter's works, a set of which he had brought to Egypt, and there composed his masterpiece, *Études galiléennes.*[12] The title page gives 1939 for the date of publication. Actually it appeared in Paris in April 1940, just prior to the German invasion. Koyré and his wife were once again in Cairo. He wished to serve amid the disasters, and they hurried back to France, reaching Paris just as the city was surrendered. Thereupon they turned about, making their way first to Montpellier, and then by way of Beirut back to Cairo. Koyré had already determined to rally to the Free French and offered his services to De Gaulle when the General came to Cairo. Since Koyré held an American visa, De Gaulle felt that the Free French cause might benefit from the presence in the United States of a man of intellectual prominence able to express the Gaullist point of view in a country where government policy was favorable to Pétain. Somehow, the Koyrés found transportation by way of India, the Pacific crossing, and San Francisco to

New York. There he joined a group of French and Belgian scientists and scholars in creating the École Libre des Hautes Études, and he taught there as well as in the New School for Social Research throughout the war, making one trip to London in 1942 to report to De Gaulle. In New York he developed the familiarity with American life that made it natural for him to spend in his later years something like half of his professional life in the United States.

It was in the United States in the immediate postwar years that *Études galiléennes,* not much noticed amid the distraction of scholarship by war, found its widest and most enthusiastic public, a case of the right book becoming known at the right time. A new generation of historians of science, the first to conceive of the subject in a fully professional way, was just then finding an opportunity in the expanding American university system, which more than made up in flexibility and enthusiasm for science whatever it may have lacked in scholarly sophistication and philosophical depth. Casting about through the literature in search of materials, they came upon *Études galiléennes* as upon a revelation of what exciting intellectual interest their newly found subject might hold, a book which was no arid tally of discoveries and obsolete technicalities, nor a sentimental glorification of the wonders of the scientific spirit, nor yet (despite the author's Platonism) a stalking horse for some philosophical system, whether referring to science like the positivist outlook or to history like the Marxist.

Instead, they found a patient, analytical, and still a tremendously exciting history of the battle of ideas waged by the great protagonists, Galileo and Descartes, in their struggle to win through to the most fundamental concepts of classical physics, formulations that later seemed so simple that schoolchildren could learn them with ease and without thought. It was a struggle waged not against religion, nor superstition, nor ignorance, as the received folklore of science would have it, but against habit, against common sense, against the capacity of the greatest of minds to commit error amid the press of their own commitments. Koyré sometimes observed, indeed, that the history of error is as instructive as that of correct theory, and in some ways more so, for although nothing to be celebrated—he was no irrationalist—it does exhibit the force and nature of the constraints amid which intellect needs must strive in order to create knowledge. (The more strictly philosophical problem of the false was one that he developed intensively in its classical context in a charmingly ironic essay, *Epiménide le menteur* [10].)

Koyré's technique was to study problems both intensively and broadly, intensively for themselves and broadly in the awareness of their widest

significance. *Études galiléennes* consists of three essays published in separate fascicles. The first is entitled *À l'aube de la science classique,* the latter phrase meaning classical physics. The theme that unites all three is the
emergence of that science (without which the rest of modern science is
unthinkable) from the effort to formulate the law of falling bodies and the
law of inertia, the subjects respectively of the second and third fascicles.
The subtitle of the first fascicle, "La jeunesse de Galilée," implies that
Galileo's early education and first researches recapitulated the main
stages in the history of physics from its origins in antiquity. Koyré's sympathetic summary of Aristotelian physics emphasizes the anomaly of the
cause attributed to motion in projectiles and explicates the reasoning of
Benedetti and Bonamico, from whom Galileo learned physics and who developed the 14th-century impetus theory into a scheme for explaining
the flight of missiles and fall of heavy bodies. Only when Galileo abandoned the idea of causal impetus, however, did he begin to lead the way
from a physics of quality to a physics of quantity. He first attempted that
step in the analysis in his youthful *De motu,* left in manuscript. There he
substituted Archimedean for Aristotelian methods and formulated the
relation between a body and its surrounding medium in terms of relative density.

In Koyré's view, geometrization of physical quantity in the Archimedean sense was the crux of the scientific revolution. The intellectual
drama, becoming at times a comedy of errors in *Études galiléennes,* is
made to consist of a counterpoint between Galileo and Descartes striving
to disengage the law of falling bodies and the law of inertia, respectively
the earliest and the most general laws of modern dynamics, from con-
cealment by the gross behavior of ordinary bodies throughout the everyday world. In the end Galileo achieved the law of fall and Descartes the
concept of inertia. Galileo began in 1604 in private correspondence with
a correct statement of the former law—that the distance traversed in free
fall from rest is proportional to the square of the elapsed time—and simultaneously attributed it to an erroneous principle—that the velocity
acquired at any point is proportional to the distance fallen.

In fact, velocity is proportional to time in constant acceleration, and
the irony that reveals the depth of the mistake is that Descartes independently repeated these same confusions fifteen years later in his correspondence with Beeckman. The specific trouble lay in the mutual un-
familiarity of mathematics and dynamics. However clearly Galileo saw
the need for formulating the latter in terms of the former, his only tools
for mathematicizing motion were arithmetic and geometry. Analytical
though his mind was, proportion had to do the work of functional inter-

dependence, and it was not intuitively clear to him at the outset that lapse of time could naturally be expressed in geometric magnitudes. His instinct having been eminently that of a physicist, Galileo eventually worked through to a resolution of his error. The *Discorsi* incorporates a fully mathematical derivation of the law from the principle of uniform acceleration, followed by the famous experimental verification on the inclined plane (which Koyré in his own excessive skepticism about the experimental component of early physics dismissed as a thought experiment).

Less fortunate with this problem was Descartes. Committed to identifying physics with geometry, he never did perceive that his formulation of fall was inconsistent with the physical description of the phenomenon. But if this tendency to "géometrisation à outrance" concealed the elements of the physical problem from Descartes, it was on the other hand just such mathematical radicalism that led him to the law of inertia, unconcerned to say where motion would stop and what could hold the world together if bodies tended to move in straight lines to infinity. Before this physical problem, Galileo finally drew back into the traditional conception that on the cosmic scale motion endures in circles, and left it to Descartes to enunciate the more general, the universal law of motion. Attributing the law of inertia to Descartes was certainly one of the most original and surprising of Koyré's findings in *Études galiléennes,* and it is central to the argument. In consequence of that principle, the ancient notion of a finite cosmos centered around man and ordered conformably to his purposes disappeared into the comfortless expanse of infinite space. In Koyré's view, the scientific revolution entailed a more decisive mutation in man's sense of himself in the world than any intellectual event since the beginnings of civilization in ancient Greece, and it came about because of the change that solving the basic problems of motion required in conceiving their widest boundaries and parameters.

In the postwar years Koyré resumed his post in Paris while lecturing from time to time at Harvard, Yale, Johns Hopkins, Chicago, and the University of Wisconsin. Western Reserve University awarded him its honorary doctorate of L.H.D. in 1964. In 1955 he came to the Institute for Advanced Study in Princeton, where he was appointed to permanent membership in the following year. From then until his health began failing in 1962, he spent six months of the year in Princeton, returning to Paris each spring to give his annual course at the École Pratique. The tranquillity of the Institute, and specifically its Rosenwald collection of first editions in the history of science, were essential to the completion of his further works. He was greatly stimulated and encouraged in their composition by his association with Harold Cherniss and Erwin Panofsky, and also by the acumen and criticism of Robert Oppenheimer, then the director of the Institute, in whose bracing company Koyré was one of the very few people with the intellectual self-possession to feel at ease.

Those works carry further the main themes that Koyré discerned in the scientific revolution, its history and philosophical aspects. *La révolution astronomique* was the last book he left in finished form, and consists of a very substantial treatise on Kepler's transformation of astronomy, preceded by a resume of Koyré's earlier discussion of Copernicus and followed by an essay on the celestial mechanics of Borelli. This last was one of his most original contributions to the literature, for although Borelli has been well known to scholars for his mechanistic physiology, the intricate rationalities of his world machine had been very little studied in modern times. As for the main part of the book, Kepler was always one of Koyré's favorite figures, appreciated for his boldness, for his imagination, for his Platonism, finally for his accuracy. Koyré distinguished his touch from that of Copernicus by making him out an astrophysicist needing a physical explanation of the planetary motions, in search of which he came upon his mathematical laws. In no way did Koyré underplay the fantastic and Pythagorean aspects of Kepler's thought. Indeed, it might be said that Koyré's earlier interest in German mysticism met his later commitment to science in his study of Kepler. Ultimately, however, we have Kepler making his mark through the fertility of an imagination controlled by fidelity to physical fact.

The themes that interested Koyré reached their dénouement in the Newtonian synthesis, and his essay on its significance is one of the most lucid, serene, and comprehensive of his writings. It opens the volume of *Newtonian Studies* published after his death. Perhaps it is a pity that he did not see fit to include "A Documentary History of the Problem of Fall From Kepler to Newton" [13], for that meticulous monograph exhibits at his scholarly best his gift for treating the ramifications of a single problem in detail and in generality as they appeared to the succession of analytical minds that handled it. For the rest, Koyré was not given the time to establish the same degree of coherence among his several studies of Newton that he did in *Études galiléennes.* In the last years of his life, he was collaborating with I. Bernard Cohen on the preparation of a variorum edition of the *Principia,* currently in press. The essay in *Newtonian Studies* on "Hypothesis and Experiment in Newton" translates the famous "hypotheses non fingo" to mean "feign" not "frame," and takes issue with the attribution of a positivistic philosophy to Newton himself. The most substantial essay in the volume contrasts Newtonian with Cartesian doctrines of

space, and carefully explores the theological implications of the difference, a theme worked out more fully in Koyré's *From the Closed World to
the Infinite Universe.*

Completed earlier than *Newtonian Studies,* this important work follows the metaphysical course of the transition epitomized in the title, beginning with the cosmology of Nicolas of Cusa and culminating in the
Newtonian assertion of the absoluteness of infinite space and the omni-
potence of a personal God distinct from nature. Theologically, the critical
issue throughout was the relation of God to the world, for it appeared, and
most subtly so to Henry More, that Cartesian science escaped atheism
only by falling into pantheism. In regard to these issues Koyré's discussion may seem a little bodiless to readers whose sensibilities are less finely
attuned to the metaphysical and theological implications of the old ontologies. The problems will come alive, however, if they are transposed
from a metaphysical into a psychological key. It is the sort of reading that
would be consonant with his own admiration for the writings of Émile
Meyerson, to whose memory he dedicated *Études galiléennes,*[13] and that
would place *From the Closed World* alongside that work as its more philosophical complement or companion, concerned with what Koyré now
calls "world-feelings"[14] in contrast to world views.

The central theme is that of alienation, the alienation of consciousness
from nature by its own creation of science. Put in those terms the metaphysical anxieties about God and the world will take on reality in modern eyes, and that is precisely what the destruction of the Greek cosmos
entailed:

> The substitution for the conception of the world as a finite and well-
> ordered whole, in which the spatial structure embodied a hierarchy
> of perfection and value, that of an indefinite and even infinite uni-
> verse no longer united by natural subordination, but unified only
> by the identity of its ultimate and basic components and laws;
> and the replacement of the Aristotelian conception of space—a dif-
> ferentiated set of innerworldly places—by that of Euclidean geom-
> etry—an essentially infinite and homogeneous extension—from
> now on considered as identical with the real space of the world.[15]

Yet if this emphasis in Koyré might give aid to the current fashion for
deploring science as something set against humanity, his treatment gives
protagonists of antiscientism no comfort. It is significant that of all the
great minds of the seventeenth century, the only one apart from Bacon
with whom Koyré felt little sympathy was Pascal [18q]. For he always held

the creations of intelligence to be triumphs in the long battle between mind and disorder, not burdens to be lamented.

Notes

1. *La philosophie de Jacob Boehme,* p. viii.
2. *L'idée de Dieu et les preuves de son existence chez Descartes,* p. 128.
3. *La philosophie de Jacob Boehme,* p. vi.
4. Jean Wahl, "Le rôle de A. Koyré dans le développement des études hégéliennes en France," in *Archives de philosophie,* 28 (July–Sept. 1965), 323–336.
5. *Études d'histoire de la pensée philosophique;* originally published in *Revue philosophique,* 112 (1931), 409–439.
6. *Études sur l'histoire des idées philosophiques,* pp. 19–102.
7. *Études galiléennes,* p. 6.
8. *Jahrbuch für Philosophie und phänomenologische Forschung,* 5 (1922), 603–628; published in French in [17a].
9. He had published one small note prior to World War I, "Remarques sur les nombres de M. B. Russell," in *Revue de metaphysique et de morale,* 20 (1912), 722–724.
10. *Études d'histoire de la pensée philosophique,* p. 30, n. 1.
11. Koyré left among his papers a curriculum vitae of 1951 which sets out his own sense of the interconnectedness of the work that he had accomplished and that he then projected; see *Études d'histoire de la pensée scientifique,* pp. 1–5.
12. Two articles containing parts of the work had already appeared: "Galilée et l'-expérience de Pise," in *Annales de l'Université de Paris,* 12 (1937), 441–453; "Galilée et Descartes," in *Travaux du IXᵉ Congrès international de Philosophie,* 2 (1937), 41–47.
13. Cf. Koyré's "Die Philosophie Émile Meyersons," in *Deutsch-Französische Rundschau,* 4 (1931), 197–217, and his "Les essais d'Émile Meyerson," in *Journal de psychologie normale et pathologique* (1946), 124–128.
14. *From the Closed World to the Infinite Universe,* p. 43.
15. *Ibid.,* p. viii.

Bibliography

A Festschrift entitled *Mélanges Alexandre Koyré,* 2 vols. (Paris, 1964), was organized on the occasion of Koyré's seventieth birthday. The second volume opens with the list of his principal publications, comprising some seventy-five titles. We limit the present article to identifying his books, together with the more important of his articles, those mentioned in the footnotes above and under items [17], [18], and [19] below. It is a testimonial to the continuing interest in Koyré's specialized studies that in his later years and after his death, associates and publishers thought it important to collect and reissue these writings in book form. Readers may find it helpful to know the contents of those collections.

[1]   *L'idée de Dieu et les preuves de son existence chez Descartes* (Paris, 1922; German trans., Bonn, 1923).

[2] *L'idée de Dieu dans la philosophie de S. Anselme* (Paris, 1923).

[3] *La philosophie de Jacob Boehme; Étude sur les origines de la métaphysique alle-mande* (Paris, 1929).

[4] *La philosophie et le mouvement national en Russie au début du XIXᵉ siècle* (Paris, 1929).

[5] *N. Copernic: Des Révolutions des orbes célestes, liv. 1, introduction, traduction et notes* (Paris, 1934; repub. 1970).

[6] *Spinoza: De Intellectus Emendatione, introduction, texte, traduction, notes* (Paris, 1936).

[7] *Études galiléennes* (Paris, 1939): I, *À l'aube de la science classique*; II, *La loi de la chute des corps, Descartes et Galilée*; III, *Galilée et la loi d'inertie*.

[8] *Entretiens sur Descartes* (New York, 1944); repub. with [9] (Paris, 1962).

[9] *Introduction à la lecture de Platon* (New York, 1945); English trans., *Discovering Plato* (New York, 1945); Spanish trans. (Mexico City, 1946); Italian trans. (Florence, 1956); repub. in combination with [8] (Paris, 1962).

[10] *Epiménide le menteur* (Paris, 1947).

[11] *Études sur l'histoire des idées philosophiques en Russie* (Paris, 1950).

[12] *Mystiques, spirituels, alchimistes du XVIᵉ siècle allemand: Schwenkfeld, Seb. Franck, Weigel, Paracelse* (Paris, 1955; repub. 1971).

[13] "A Documentary History of the Problem of Fall From Kepler to Newton: De motu gravium naturaliter cadentium in hypothesi terrae motae," in *Transactions of the American Philosophical Society*, 45, pt. 4 (1955), 329–395. A French translation is in press (Vrin) under the title *Chute des corps et mouvement de la terre de Kepler à Newton: Histoire et documents du problème*.

[14] *From the Closed World to the Infinite Universe* (Baltimore, 1957; repub. New York, 1958); French trans. (Paris, 1961).

[15] *La révolution astronomique: Copernic, Kepler, Borelli* (Paris, 1961).

[16] *Newtonian Studies* (Cambridge, Mass., 1965); French trans. (Paris, 1966).

[17] *Études d'histoire de la pensée philosophique* (Paris, 1961).
    (*a*) "Remarques sur les paradoxes de Zénon" (1922).
    (*b*) "Le vide et l'espace infini au XIVᵉ siècle" (1949).
    (*c*) "Le chien, constellation céleste, et le chien, animal aboyant" (1950).
    (*d*) "Condorcet" (1948).
    (*e*) "Louis de Bonald" (1946).
    (*f*) "Hegel à Iena" (1934).
    (*g*) "Note sur la langue et la terminologie hégéliennes" (1934).
    (*h*) "Rapport sur l'état des études hégéliennes en France" (1930).
    (*i*) "De l'influence des conceptions scientifiques sur l'évolution des théories scientifiques" (1955).
    (*j*) "L'évolution philosophique de Martin Heidegger" (1946).
    (*k*) "Les philosophes et la machine" (1948).
    (*l*) "Du monde de l''à-peu-près' à l'univers de précision" (1948).

[18] *Études d'histoire de la pensée scientifique* (Paris, 1966).
    (*a*) "La pensée moderne" (1930).
    (*b*) "Aristotélisme et platonisme dans la philosophie du Moyen Age" (1944).
    (*c*) "L'apport scientifique de la Renaissance" (1951).
    (*d*) "Les origines de la science moderne" (1956).
    (*e*) "Les étapes de la cosmologie scientifique" (1952).

(*f*) "Léonard de Vinci 500 ans après" (1953).

(*g*) "La dynamique de Nicolo Tartaglia" (1960).

(*h*) "Jean-Baptiste Benedetti, critique d'Aristote" (1959).

(*i*) "Galilée et Platon" (1943).*

(*j*) "Galilée et la révolution scientifique du XVIIᵉ siècle" (1955).*

(*k*) "Galilée et l'expérience de Pise: à propos d'une légende" (1937).

(*l*) "Le 'De motu gravium' de Galilée: de l'expérience imaginaire et de son abus" (1960).**

(*m*) "'Traduttore-traditore,' à propos de Copernic et de Galilée" (1943).

(*n*) "Une expérience de mesure" (1953).*

(*o*) "Gassendi et la science de son temps" (1957).**

(*p*) "Bonaventura Cavalieri et la géométrie des continus" (1954).

(*q*) "Pascal savant" (1956).**

(*r*) "Perspectives sur l'histoire des sciences" (1963).

  *English original republished in [19].

  **English translation published in [19].

[19] *Metaphysics and Measurement* (London, 1968). English versions of [18] *i, j, l, n, o,* and *q.*

SECONDARY LITERATURE. Accounts of Koyré and his work have appeared as follows: Yvon Belaval, *Critique,* nos. 207–208 (1964), 675–704; Pierre Costabel and Charles C. Gillispie, *Archives internationales d'histoire des sciences,* no. 67 (1964), 149–156; Suzanne Delorme, Paul Vignaux, René Taton, and Pierre Costabel in *Revue d'histoire des sciences,* 18 (1965), 129–159; T. S. Kuhn, "Alexander Koyré and the History of Science," in *Encounter,* 34 (1970), 67–69; René Taton, *Revue de synthèse,* 88 (1967), 7–20.

E. J. Dijksterhuis: Mechanization of the World Picture

THE AUTHORS OF THE BOOKS reviewed in the three essays that follow, E. J. Dijksterhuis, Shmuel Sambursky, and Salomon Bochner, had in common with Halévy and Koyré that they belonged to the last generation of European scientists and scholars to receive a disciplined humanistic education in a pre-1914 lycée, Gymnasium, or the Dutch or Russian equivalent, before going on to a university to specialize in history, philosophy, physics, or mathematics, as the case might be. The last four wrote on history of science in ways which, now that the subject has become a professional subdiscipline, may be considered to combine learning with idiosyncrasy. This is an appropriate place to note, therefore, that many informative works either long antedate or accompany the literature produced by those who professionalized the field in the last half century. To mention only writers further back in time whom I have found invaluable, I think of Henri Daudin on 18th- and 19th-century taxonomy, J. T. Merz on European thought in the nineteenth century, Auguste Comte in the historical passages of *Cours de philosophie positive,* J. B. Delambre on the history of astronomy, and J. B. Montucla on the history of mathematics.

There is no one now, it is safe to say, who would undertake the history of a single discipline from antiquity to modernity as Dijksterhuis does, and no one known to me who would be capable of so nuanced a treatment of the evolution of the mechanistic outlook and treatment of phenomena. I had the privilege of meeting him twice, once at the Madison meeting in 1957, and again in 1962 when the History of Science Society awarded him the Sarton Medal at the close of the International Congress of the History of Science held in Ithaca and Philadelphia. That prompted the following essay, which appeared in the series of "Perspectives" published in *American Scientist* during the editorship of Sir Hugh Taylor.

FOURTEEN

E. J. Dijksterhuis:
*Mechanization of the World Picture**

—⚏—

T he immediate occasion of this Perspectives is the availability of Dijksterhuis's most comprehensive book, *The Mechanization of the World Picture,* first published in Amsterdam in 1950 and now most skillfully translated into English by C. Dikshoorn and published by Oxford University Press.[1] Among those (very few they are) who may properly be called great historians of science, Dijksterhuis has so far been perhaps the least known in the English-reading world. It is difficult to understand why this should be so. He is Professor of the History of Science in the University of Utrecht and has published much in his own country and his own language. His is the admirable tradition of Dutch science and scholarship, however, and his work seems equally at ease in French, in German, and in English. Indeed, he has brought unobtrusively into print within the space of a few years what only a lifetime of the finest scholarship could yield in modest and in full knowledge of a great subject. In 1951, Dijksterhuis followed the first edition of his *Mechanization* with a documented set of studies on Cartesianism in his native country, Descartes having returned the hospitality of Holland by the philosophy which served in effect as the nursery of its science. In 1955 appeared the first volume of an edition of *The Principal Works of Simon Stevin,* to whom Dijksterhuis has devoted one of his deepest and most considered studies. In the same year appeared *The First Book of Euclid's Elementa* edited with a glossary. In his renderings of the mathematics of antiquity, Dijksterhuis follows a policy of his own devising. The Greeks used no symbols in writing mathematics. Their geometry is in words, and modern translations, notably those of T. L. Heath, devoted admirer though he was of the corpus

* Reprinted from "Perspectives", *American Scientist, 50,* no. 4 (December 1962, pp. 626–639).

of ancient mathematics, convey rather the results than the style and quality of the reasonings. Dijksterhuis does not go to the other extreme of
literal translation, which would be of an intolerable verbosity, and he
has perfected instead what may be called a free rendering. He treats his
sources as one might in translating poetry. Through his own command
both of language and of the author's intent, he conveys the latter in the
former, and thus by art admits the reader more closely into the thinking
of Euclid or of Archimedes than other scholars have succeeded in doing.
On Archimedes Dijksterhuis published two volumes in 1956, one an edition of the *Arenarius* with glossary, the other a small but general study of
inestimable value in expounding what scholarship does in fact know of
the greatest scientific mind in all antiquity.

This constitutes a notable body of work, and it presents the reviewer
with a peculiar though not unwelcome problem. Everything in it seems
right. Everything in it seems necessary. Dijksterhuis offers no handle for
argument as between the scholar and the one who would discuss what he
tells us. And perhaps this is why he is not yet so famous in the subject as
have been more programmatic or more eristic scholars—his work makes
no noise. He rides no hobby. He does not cast about in his knowledge of
the past of science for the bits that seem to foretell and hence to justify
what he wants it to be or do or mean today. He has no formula for the history of science. He simply tells us how it came about. Gently the opening
pages of *The Mechanization of the World Picture* deny the reader who is
too eager to come to grips with the subject either definition or judgment. It is clear enough what the mechanistic conception has done. It has
brought forth physical science: "experiment as the source of knowledge,
mathematical formulation as the descriptive medium, mathematical deduction as the guiding principle in the search for new phenomena to
be verified by experimentation."[2] It has rendered rational and even possible the technology by which the modern world has necessarily to live.
It has penetrated and polarized philosophy. "Owing to all these factors the
mechanization of physical science has become much more than an internal question of method in natural science; it is a matter that affects the history of culture as a whole, and on this account it deserves the attention of
students outside the scientific world."[3]

Less evident than what mechanization did, is what we are to think of
it and what it was. Reluctantly but unavoidably Dijksterhuis introduces
the term "mechanicism" in order to distinguish the movement as a whole
from the science of mechanics. Its champions within science and apologists without have advocated mechanicism on pragmatic grounds as the
mode of handling experience which admits of clarification and verifica-

tion, which replaces errors and figments with reliable and reproducible results. Its critics, unable to deny the evidence of its gathering success, find this but a brutal justification for what they take as the encroaching sands of a philosophical desert, a wasteland of material plenty and spiritual discontents. Not that Dijksterhuis thinks choice or judgment unimportant, but we cannot well proceed to make them until we know how mechanicism in fact evolved and what the reasons were for its hold upon the thinking of scientists, until we appreciate why it was that Huygens (to take a distinguished example from among many) laid down as a policy that we must necessarily adopt the mechanical mode of reasoning or else renounce the hope of comprehending anything in physics.

Ultimately the reasons for appreciating mechanicism have to be historical, for the subject itself has changed in scope and content. What do we mean by mechanistic? There is no easy answer: "In one sense the whole of this book is an attempt to discover to what extent it is possible to speak of a mechanistic world picture."[4] Do we mean that conception which would consider the universe as a machine, such a notion as would end with making mind itself an implement? Or do we mean that nature may best be described in its operations by means of formulations drawn by extension or analogy from the science of motion? This is quite a different matter. The one implies a science of machines and the other of matter in motion. The one usually pre-supposes the notion of a designer-creator of the world, and the other that reality is bits and pieces in random whirl. How, then, did a word drawn from the former come to designate the latter? And if (however inconsistently) we do mean by mechanics primarily the science of motion, what kind of kinematics was at the heart of mechanization? That of the classical world, most notably of Aristotle, which is inconsistent with the later world picture of corpuscular mechanics? That called confusingly enough "classical" by physicists who mean by it Newtonian rather than antique (though as we shall see Newton had not reached it)? Or finally do we mean by mechanistic the concepts of quantum physics? These are not questions to be answered *a priori*, and though Dijksterhuis puts them searchingly, he attempts no complete resolution. His account is of the genesis of classical science, and that culminates in the work of Isaac Newton.

The scope of the book thus delimited, let it be said at once that this is no narrow monograph. In a quite extraordinary way the treatment combines comprehensiveness with discrimination in the selection of material. Dijksterhuis is not one who would make all science emerge from some egg in the Archimedean lever fertilized by the Democritan atom. His method is to survey in turn each of the ancient schools and each succes-

sive period and distinguish therein the elements which in historic fact
and regardless of initial philosophical incompatibilities did come together in the evolution of mechanism. He addresses himself, moreover,
to a general (though certainly an intelligent) public of whom he requires
no prior knowledge of the subject. Here again, however, he makes the best
of both worlds, never compromising the scruples of the specialist in the
interest of a specious clarity. His is a true clarity in which there is no over-
simplification and which bespeaks a serene and comprehensive mastery.
The reader is bound to feel secure in his sense of deep reserves of scholarship behind every statement and every judgment. One's only complaint
must be of the author's disclaimer of intent to write a book from which
professional historians of science too will learn. They can and do. There is
scarcely any topic touched, on which some observation will not strike
them, occasionally by its novelty, more often by its economy, justice, and
accuracy. And so widespreading are the roots of this subject, so central the
stem, that the book may be recommended without qualification to those
unfamiliar with the literature of the history of science, both as the best
treatment extant of physicomathematical science from the beginnings
through Newton, and with no less confidence as an exemplar of what the
historical approach at its most conscientious may accomplish in illuminating the scientific enterprise in the context of general culture.

No summary would be useful of a book which is itself a summary, indeed a *Summa,* and an idea of the contents will be better conveyed by an
account of the anatomy of the work, for its structure offers several features which, if not exactly novel—in what sense can sound history offer
novelty?—nevertheless throw the subject into a distinctive relief. The historian orders his information in the first instance by appropriate choice
of periods, and Dijksterhuis disposes his material into four—Antiquity,
which he carries through Patristic times; the Middle Ages, for our purposes the twelfth through the fourteenth centuries (he appreciates but
does not develop the Arabic transmission); that which he calls a prelude
to classical science and which essentially consists in the Renaissance continuing into the sixteenth century; and finally the rapid inauguration in
the seventeenth century of classical physics itself. It has not been usual
in the historiography of science to bring forward the Renaissance phase
as a period in its own right. Thus to promote it appears a most valuable
device for it eliminates the false and sometimes angrily disputed question
whether the Renaissance as a movement of culture was favorable or inimical to the scientific enterprise. Dijksterhuis never enters as a partisan
into such disputes, nor does he take issue in point of method with the concept of historiography whence they arise, by which a unique course led

with right and wrong turnings from the Greeks to Newton and on to the present. This is not his practice. He treats things in succession and in their own context and not only relative to some datum, the seventeenth century or the twentieth, taken as a goal. Rather than judge the men of the Renaissance as if their job had been to convey 14th-century mechanics into 17th-century minds, Dijksterhuis arranges his exposition so as to make it appear that they contemplated nature in their own way and with their own interests. Their science was different from that of the scholastics and also from that of Galileo and Newton. It was in historical fact a prelude, not to be assessed either as a regression or advance.

The historian must identify themes as well as periods, the strong threads that run longitudinally along the historical fabric, and Dijksterhuis distinguishes three central topics which appear in evolving content and proportions of varying importance in the successive historical environments. He tells first of the main schools of philosophy in each period with respect to their contributions to science. In the ancient world the Aristotelian philosophy outweighed all the others. It would be difficult to better in brief or even in extended compass the account which Dijksterhuis gives of its salient features, especially of the relation of its metaphysics to its physics and cosmology. He has no illusions about its inhospitality to mathematical physics, yet one feels that his respect goes out to Aristotle for the fullness and fidelity of his constructions, and that he admires these qualities more profoundly than flashes of brilliance in less careful but seemingly more auspicious thinkers: ". . . a fuller acquaintance with Aristotle's scientific works leads to a much greater appreciation. If his works are read with an unbiased mind, it is obvious that here is a student of nature who possesses an extensive knowledge of physical phenomena, is vividly interested in explaining them, and bent on giving such an explanation on purely physical grounds. A comparison with Plato shows that the tendency—known from the *Timaeus*—of constructing an imaginary nature by reasoning from preconceived principles and forcing reality more or less to adapt itself to this construction had now been replaced by a purely empirical attitude, based on the recognition that true knowledge of nature can only be gained from carefully collected observational data."[5]

In the medieval period Thomism merging into nominalist scholasticism formed the expectations held of science; and in the Renaissance, humanism set the style, primarily under the inspiration of Plato. Dijksterhuis's discussion of the influence of humanism on science is a small masterpiece of nuance. Humanism was, he observes, as verbal and as respectful of antiquity as ever Aristotelianism had become. At most we are to regard it as a restriction upon literal adherence to Aristotle. Yet neither may one overlook its service in the discovery and purification of ancient scientific texts, or in its occasionally mystical summons to mathematicization in a Pythagorean vein. In the final period, that of the seventeenth century, philosophies seem attenuated in importance by comparison to the youthful strength, the growing bodies of the sciences themselves. There was coming into its own what Dijksterhuis in a slightly earlier connection calls the doctrine of a double truth, by which "an idea may appear useful in practical scientific work and may be applied quite thoughtlessly, although it may not satisfy a man's critical and philosophical faculties. On the other hand, conceptions which are acceptable to the latter may fail to have any influence on positive science."[6]

Secondly within each period, Dijksterhuis turns to the development of those parts of what we would regard as positive science that are relevant to the theme of mechanization. The account of ancient astronomy explains the principle of the Ptolemaic computational devices, and displays the issue as between them and the spherical models imagined for physical reality. In the Middle Ages, kinematics and formalization of motion as a quality of variable intensity occupy pride of place. Nothing in the whole historiography of science has been more revolutionary in this century than the rehabilitation of the claims to scientific importance of 14th-century kinematics and philosophy of science. Dijksterhuis does full justice to the magnificent researches of Pierre Duhem which wrought (and sometimes overwrought) this change and to the meticulous studies of Anneliese Maier, who has corrected some of Duhem's enthusiasms and expounded the natural philosophy of the later scholastics in relation to their own outlook and metaphysical commitments rather than in the light of later mechanics. Writing in 1950, Dijksterhuis could not reflect the further studies of A. C. Crombie, notably on Grosseteste and his methodology of experimental science, and of Marshall Clagett, who has summarized much recent work in *The Science of Mechanics in the Middle Ages,* a book which is the indispensable starting point for anyone who would make an acquaintance with medieval achievements in physics.[7] The section on the seventeenth century, finally, occupies itself equally with cosmology and mechanics. Dijksterhuis puts very openly the dilemma which the historian must confront in organizing his material. Shall he follow straight through the lines of development of each of the sciences? If he does, the connections are slighted and so too are the scientific personalities of the great men. But if he follows a biographical approach, he will cut up the story of each of the sciences and lose its internal logic of development. Convergence of all the lines in Newton gives Dijksterhuis a goal, toward which he works by choosing with great discernment those parts of the work of the great figures—Copernicus, Kepler, Stevin, Galileo, Pascal, Descartes, Boyle—which reveal their qualities in the parallel lines of astronomy, cosmology, statics, hydrostatics, mechanics, and the structure of matter.

The third major theme in this, the vertical aspect of Dijksterhuis' organization, opens to his treatment prospects comparable to those deriving from his horizontal particularization of Renaissance science as preliminary and distinct. He tells in each of the four periods of the prevailing theories of matter and its structure and discusses it as a subject equal in dignity and distinction to astronomy and the mathematicizable parts of physics, i.e., optics, statics, and ultimately kinematics. This permits him to work into the evolving fabric of mechanistic science the antecedents of those subjects which certainly have come to belong in it since the seventeenth century, that is chemistry and the physics of matter and force as well as of motion and the stars. Dijksterhuis treats alchemy and hermetic thought as considerately as his other topics. Alchemists there were, we are assured, who according to their lights truly sought to conform their thinking to what they learned in their experiments. Philosophers too in the Middle Ages concerned themselves with the problem of mixture, whether in a true *mixtum* the elements are still present, and if so then how this is to be understood.

By no means did that question disappear in modern chemistry, and the logicians of the fourteenth century worked out its alternatives with scrupulous subtlety, holding on the whole to the Aristotelian doctrine of natural *minima* (not to be confused with Democritan atoms since they undergo a qualitative change in mixture). Below that level bodies could not be divided without ceasing to exist, though whether qualities too had minimal intensities—motion, for example—was a vexatious puzzle. In all this pre-chemistry, however, Dijksterhuis attaches greater significance on the whole to "the voices of those who not only reflected upon the nature of a chemical compound, but also handled substances in practice and actually produced compounds or decomposed them into their components. . . ." We shall therefore have to turn to the alchemists, not because they were the only people who were in a position to handle substances, but because the altogether practically minded technical workers, who also did so by profession, did not as a rule commit their findings to writing. The fact that the aim of the alchemist was to make gold is of no account in the present context. Indeed, alchemy can only be distinguished from chemistry by the objective which it set itself, and where this is irrelevant there is no longer any reason to maintain the difference of name."[8] As to combination, moreover, the working chemists seem in practice not to have taken the Aristotelian view of combination as mixture occurring with transformation of the elements: they simply supposed themselves to be making some substances out of others and decomposing compounds into their original constituents. There, moreover, where Dijksterhuis' two unexpected themes cross, in the discussion of theories of matter in the Renaissance, he is at his most felicitous with what others have found to be either an embarrassment or a cause to espouse. The juncture enables him briefly but naturally to place Paracelsus in the history of science, not as some Faustian *magus* posturing and declaiming (though no doubt he did), but as the transformer of chemistry who in making it the science of the "principles" of mercury, salt, and sulfur, instead of the four ancient elements of merely philosophical fame, thus brought it to closer grips with actual materials, and who, further, in turning the subject to medical account rather than to making gold enhanced its dignity and started its popularity as the science specially apt for giving access to the real workings of nature in body and in depth. In this wise all the hermetic tradition, the correspondences of macrocosm and microcosm, the sense of cosmic sympathy, the parallelism of stars and metals and their empire over fortune, the trinities on every hand—all that becomes background for the history of science and not (as this reviewer confesses he has always regarded it) simply as magic or superstition to be exorcised by reason and experience.

One must not leave the impression of excessive schematization, however, admirably adapted though the scheme is to the material, for it is not there that the excellence of the book resides, but rather in the reconciliation of breadth of theme with economy of statement, of urbanity in judgment with fullness in criticism, that most rare and welcome genre of criticism which values without blaming. On reading Dijksterhuis the historian of science will again and again enjoy the satisfaction of hearing struck the note one knows is right and could not have played oneself. Consider, for example, the observation on the blending of what might otherwise seem the antitheses of rationalism and mysticism in the Pythagoreans and by extension in Platonism and much highly abstract science since, how the ideal of pure knowledge can become also an escape from the imperfections of a spotted world, and how this tradition, albeit a stimulus to mathematicization, might also disparage the empirical study of nature, particularly in times when religion should happen to inculcate contempt for the material world. Nor will Dijksterhuis adopt drastic alternatives and categorically rule Aristotelianism out of any relation to the genesis of mathematical physics. He agrees, and there can be no dispute, that Aristotle generally stands for the a-mathematical aspects of natural science, but he also observes that the very structure of his axiomatization and demonstration is derived from mathematics, and that Euclidean geometry, Archimedean statics, and much of the best of 17th-century physics were written in the form of just such a demonstrative science.

Dijksterhuis emphasizes a general criticism of Greek science which, though one has not found it given such importance elsewhere, again arrests attention by its justice. This is a practice of valuing by which the philosopher or the educator is forever posing choices between polar alternatives on grounds of relative nobility, perfection, or excellence—mind over matter, the finite over the infinite, the square over the rectangle, the male over the female, circular motion over rectilinear, the celestial spheres over the sublunary. Carried over into the Renaissance, this became the convention for the philosophic or aesthetic dialogue—painting over sculpture, poetry over prose. Does this habit of false and unreal choice persist in our own world, even if mainly in the realm of intellectual snobberies, as perhaps the least altered by the centuries of our legacies from Greece, so that the famous two cultures are only the latest installment of the issue between a liberal and a practical education, or so that the superiority complex of the theoretical physicist vis-à-vis his experimental colleague still expresses *artes liberales* (for such was geometry) lording it over *artes mechanicae* the which "do not befit the free Hellene"?[9] So pervasive is what he calls "axiology" that Dijksterhuis gives it an importance equivalent to the more familiar teleological point of view as a fundamental and detrimental characteristic of Greek thinking.

For though this is a measured book, it is never bland, never permissive of wishful scholarship. How moderately and yet how firmly Dijksterhuis discounts the fashion, prevalent not so long ago and by no means among Marxists only, which would make science one of the intellectual products of a process rising in the social relations of classes and in the methods of getting a living. According to Borkenau, for example, mechanistic science was a projection upon nature of the practice of standardized procedures in the early factory system. According to Simmel and others, experience with a monetary economy created the prospect of an exact and quantitative treatment of the universe. Alfred von Martin, sociologist rather than economic historian, has held that the ideal of pure theoretical knowledge was a function of the emancipation of the moneyed classes from feudalism and a special case of their pleasure in business for its own sake. "It is doubtful," writes Dijksterhuis mildly, "whether the history of science is enriched by such constructions."[10] These and similar speculations fail to convince because, on the one hand, they do not demonstrate the causal relation they would establish, and on the other, their authors needs must pretend that no mathematical account of nature had ever been conceived prior to the influence of Renaissance capitalism. Not that Dijksterhuis wishes to neglect the demonstrable role of technology itself as distinct from its supposed social setting, and one may quote as an instance of incisiveness and restraint his paragraph on the practical relation to science of the instrument-makers, the cartographers, and the artist-engineers of the type Leonardo, Brunelleschi, and Dürer:

> A good deal of the knowledge and skill displayed by these men was still purely empirical, but the constant handling of matter, which is always refractory, could not fail to stimulate the desire for a causal explanation and to induce efforts to devise a more rational working-method. It thus becomes understandable that the first branch of science in which the revival was to take place was mechanics (at first still in the sense of science concerned with tools and implements). In this case empirical knowledge did not have to be sought deliberately, but arose naturally from the pursuit of technical trades; the waiting was only for theoretical reflection, which, however, was helped by the fact that there is no single department of physics which calls more urgently for mathematical treatment and lends itself to it more naturally than mechanics. The first essential element of classical physics, the mathematical approach, thus came into its own as spontaneously as the second, the empirical foundation.[11]

Dijksterhuis is no more indulgent when it is the scientists who habitually abuse the facts. His discussion of Galileo in no way diminishes that extraordinary and provocative man's stature there at the beginning of the battle for classical physics, but he gives the real Galileo, indicating the Aristotelian moorings from which his thought broke loose, the persistence even in *Discourses on Two New Sciences* of fragments from the earlier essays, and the inconsistencies and occasional failures in strict frankness. Nothing is more difficult than to say just what his greatness consisted in, just what his contribution was, and this is because he was so embedded in a past which he vehemently denounced in all his works:

> The matter is complicated even further by the fact that besides the genuine, though often one-sided, portraits of Galileo based on the study of his works there exists a spurious picture presented by what may be called the Galileo myth, the current popular representation. This latter is the picture that has been produced and is perpetuated

by writers on modern physical science who feel the need of an historical introduction, but who omit to perform the simple duty of exactness, which consists in verifying their statements in the original
sources. This picture is totally false, but it is much more vivid than
any of the genuine ones, and consequently many readers will be inclined to be content with it. Moreover, it tends to simplify matters
greatly: its splendour eclipses all secondary figures. It also furnishes
a simple terminology because, whenever a term is needed to characterize the special features of classical science, the adjective Galilean at once suggests itself. It is, therefore, comprehensible that any
criticism of this idealized picture tends to cause annoyance. . . .[12]

The entire book is a preparation for the chapter on Isaac Newton which
concludes the narrative. Dijksterhuis has no thought to question the consensus by which Newton's work is taken to complete the genesis of classical physics and to initiate its career. If any position is secure in the history of science, surely it is that. Nevertheless, this is a remarkable chapter.
The felicity with which Dijksterhuis combines the veins of exposition
and criticism here finds its worthiest occasion. Whatever one's standpoint in regard to Newtonian physics, neither the historian of science nor
the physicist who reads attentively will again say quite the same simple
things about Newton's own physics. What, to take only the most notable
example, are we to understand in the literal sense of Newton's Laws after
Dijksterhuis' dissection of the axiomatization of the *Principia*?

In the *Principia* what are usually taken as enunciations of the classical
principle of inertia, of force in terms of mass and acceleration, and of the
equivalence in opposite directions of action and reaction appear as Axioms
or Laws of Motion. Of the first of them, the principle of inertia, Dijksterhuis observes that Newton had already stated its content in his Definition
III, which reads, "*The* vis insita, *or innate force of matter, is a power of resisting by which every body, as much as in it lies, continues in its present
state, whether it be of rest or of moving uniformly forwards in a right line.
This force differs nothing from the inactivity of the mass, but in our manner
of conceiving it.*" In Dijksterhuis' view (very carefully worked out) this
means that Newton did not in fact hold the concept of inertia universally
attributed to him, i.e., that rectilinear motion is a persistent state of matter requiring no cause or explanation. What he says is rather Aristotelian,
and it is that every motion requires a motor, the *vis inertiae*, identical in
function with the "impetus" of the 14th-century schoolmen and with
what Galileo in his turn had called *Vis Impressa.*

What is to be appreciated is that Newton like every creative scientist wrote in full ignorance of the future of his science and in full knowledge only of its legacy. Even more startling than Dijksterhuis' assertion that he did not think of classical inertia is the further analysis which finds no warrant in the *Principia* for what became the expression of the second law, the relation that equates force to the product of mass into acceleration. Again Dijksterhuis goes back from the axioms to the Definitions, where Newton says what he means by his terms. Axiom II lays down: *"The change of motion* [i.e., of the quantity of motion or momentum] *is proportional to the motive force impressed, and is made in the direction of the right line in which that force is impressed."* Definition VIII says in effect that a force produces what we call momentum in some certain time and that two forces are to each other as the momenta produced in like times.

The question is, do these statements taken together assert the same thing as the relation $\underline{F} = m \times \underline{a}$? And it is Dijksterhuis' view that they do not. He puts it thus, to paraphrase him very closely: begin with Definition VIII, in the case in which a constant force F acts for time t upon a mass-point m starting at rest. Then from

$$F = m \times a \quad \text{and} \quad v(t) = a \times t$$

it follows that

$$F \times t = m \times v(t)$$

or that force is proportional to the momentum developed in a given time. But one may not reverse the conclusion. For example, suppose that a constant force F produced an acceleration proportional to time, such that these relations applied

$$F = m \times c \quad \text{and} \quad a = c \times t;$$

then in motion beginning from rest, velocity would vary with time thus

$$v(t) = \frac{1}{2}ct^2$$

so that

$$F \times t^2 = 2 \times mv(t),$$

and again force is proportional to momentum developed in a given time. From this it appears that for Newton's Definition VIII and Axiom II to be valid, the condition that $\underline{F} = m \times \underline{a}$ is sufficient but not necessary. We may not, therefore, take Newton's language to mean what it did not say, i.e., that it is a law of motion that force should equal mass times acceleration.[13]

Indeed, Dijksterhuis describes reading the classical force law, $\underline{F} = m \times \underline{a}$, into Newton as a case of the Emperor's clothes. Thus to construe Axiom II one has to suppose that Newton means "rate of change of motion" when what he says is that "change of motion is proportional to the motive force impressed. . . ." Nor need that be imagined if one abandon the notion, for which again there is no warrant in the language, that in speaking here of forces Newton has in mind continuously acting forces, notably gravity. In Dijksterhuis' view he is not at all thinking of gravitational or centrally attractive forces in the second axiom, but only of instantaneous forces and particularly those of impact. This is why there is no reference to time in the statement of the second law. Both the history and the corollaries of the laws of motion and impact bear out the validity of this restriction. It is a very serious one since it means that Newton's own Second Axiom or Law of Motion says nothing about the proportionality of force and acceleration. And if it be urged, as properly it may be, that Newton in all his discussion of gravity knows very well that a constant force creates uniformly accelerated motion, Dijksterhuis answers that like Huygens and indeed Galileo before him, Newton nowhere argues this or provides for it in axiomatization but considers the fact simply as self-evident. Only later when his successors did him the service—or disservice?—of supposing that for change he must have meant rate of change, only then was his Second Axiom taken as asserting their Second Newtonian Law of Motion. Truly, Dijksterhuis observes of the *Principia,* "like other fundamental works in the history of science (one thinks of the writings of Copernicus and Kepler) it lacks the qualities required of a good textbook. If anyone attempted to learn Newtonian mechanics from it, he would be faced with grave difficulties."[14]

This by no means exhausts Dijksterhuis' criticisms either of Newton's axiomatization or of other features of the Bible of classical physics. In the midst of it he seems to remember as it were that he is writing for those unaccustomed to the give and take of scientific dialogue, and his explanation to them may, perhaps, be taken as an apology for all that realm of discourse which is so foreign to the ways in which men normally exchange their thoughts and in which, nevertheless, it seems likely that a rapidly, perhaps a logarithmically increasing volume of the world's affairs are being and will be carried on:

> This critical discussion of the axiomatic system on which Newton founded classical mechanics may well have appeared a little astonishing to readers without a mathematical training. First Newton is pictured as the universal genius who inaugurated a new era in the history of thought, as the founder of a science which was to lead men to a previously unsuspected insight into nature, and to enable them to achieve technical wonders surpassing the boldest dreams of the past. And then the manner in which he dealt with the principles of the science on which all this is based is severely criticized. If the first statement is true, is not the second devoid of importance?

Does it not amount to petty cavilling to point out imperfections in the introductory part of a work every page of which testifies to the brilliant way in which the author handles the new conceptions?

The mathematically trained reader, however, will understand that what counts in mathematics is not only the efficacy of the developed methods, but also the exactness of their motivation, and that the unremitting work of strengthening the foundations on which the edifice of mathematics rests is of no less significance than the erection of the superstructure. Such a reader does not merely want to be able to speak and act, but also wants to know what he is saying and on what grounds his actions are based.[15]

Nor does Dijksterhuis evade the question he posed in the beginning for this last observation brings out what mechanization has been about. It is in the connotations of the word "mechanics" that the confusion and the disputes arise, and not in the content of the science itself. Like the other words which physics draws from common experience, it must be stripped of its etymological associations if its bearing is to be seized. We have no longer to do with mechanical models nor with ontologies of motion, nor with soul or mind or with their absence: "classical mechanics is mathematical not only in the sense that it makes use of mathematical terms and methods for abbreviating arguments which might, if necessary, also be expressed in the language of everyday speech; it is so also in the much more stringent sense that its basic concepts are mathematical concepts, that mechanics itself is a mathematics. . . . We can then give a positive answer to the question we set ourselves in the introduction to this book. The mechanization of the world picture during the transition from ancient to classical science meant the introduction of a description of nature with the aid of the mathematical concepts of classical mechanics; it marks the beginning of the mathematization of science, which continues at an ever-increasing pace in the twentieth century."[16]

Notes and References

1. Dijksterhuis' major writings are:
   (a) *Val en Worp. Een bijdrage tot de geschiedenis der mechanica van Aristoteles tot Newton.* Groningen, 1924.
   (b) *De Elementen van Euclides.* 2 vols. Groningen, 1929–1930.
   (c) *Het getal in de griecksche wiskunde.* Groningen, 1930.
   (d) *Descartes als wiskundige.* Groningen, 1930.
   (e) *Archimedes I.* Groningen, 1938.

(f) *Archimedes II.* In Euclides XV–XVII, XX. Groningen, 1938–1944. The English translation is published as Volume 12, *Acta Historica Scientiarum naturalium et medicinalium* of Bibliotheca Universatitis Hauniensis. Copenhagen, 1956.

(g) *Hellenistische kosmologie* and *Van Coppernicus tot Newton* in *Antieke en moderne kosmologie.* Arnhem, 1941.

(h) *Simon Stevin.* The Hague, 1943.

(i) *Descartes et le cartésianisme hollandais,* ed. with Cornelis Serrurier *et al.* Paris, 1951.

(j) *De betekenis van de wis- en naturkunde voor het leven en denken van Blaise Pascal.* Amsterdam, 1952.

(k) *The Principal Works of Simon Stevin.* Vol. I. Amsterdam, 1955.

(l) *The First Book of Euclidis Elementa.* Leiden, 1955.

(m) *Renaissance en natuur wetenschap.* Amsterdam, 1956.

(n) *The Arenarius of Archimedes.* Leiden, 1956.

(o) *Geminus Rhodius: Gemini elementorum astronomiae capita* I, III–VI, VIII–XVI. Leiden, 1957.

2. Dijksterhuis, *Mechanization of the World Picture,* p. 3.

3. *Ibid.,* p. 3.

4. *Ibid.,* p. 4.

5. *Ibid.,* p. 69.

6. *Ibid.,* p. 282.

7. A. C. Crombie, *Robert Grosseteste and the Origins of Experimental Science, 1100–1700.* Oxford, 1953.

A. C. Crombie,*Medieval and Early Modern Science.* 2 vols. New York, 1959.

Marshall Clagett, *The Science of Mechanics in the Middle Ages.* Madison, 1959.

Of the writings of A. Maier, one may cite particularly (a) *Die Vorlaüfer von Galileis im 14. Jahrhundert,* Rome, 1949, and (b) *Metaphysische Hintergrunde der spätscholastischen Naturphilosophie.* Rome, 1955. For her other monographs and for bibliographical detail on those of Pierre Duhem, see Dijksterhuis, pp. 519 and 523.

8. Dijksterhuis, pp. 206–207.

9. *Ibid.,* p. 74.

10. *Ibid.,* p. 241.

11. *Ibid.,* p. 243.

12. *Ibid.,* p. 334.

13. This discussion on pp. 466–471.

14. *Ibid.,* p. 474.

15. *Ibid.,* pp. 476–477.

16. *Ibid.,* pp. 499–501.

—ᴥ— 15 —ᴥ—

Shmuel Sambursky: A Physicist Looks at Greek Science

IT IS WITH SOME trepidation that I include the following review essay in this collection. Current fashion contemns the merest whiff of Whig history. Past science is to be viewed strictly and uniquely in the context of its own time. In my opinion, try as one may, it is impossible to erase hindsight from one's mind in the writing of history. One need only restrain it. Nevertheless, there are extremes. Sambursky exercized no such restraint. Nothing of the sort was enjoined upon historians of science in his generation, however, and in any case he was a physicist by profession and a classicist and historian by avocation.

Reading over his book, I still find it intelligent, provocative, and in important respects illuminating. To be sure, his deprecation of Plato is questionable and of Aristotle unacceptable. (Dijksterhuis appreciates the latter properly.) As for the atomists, their importance in the history of science is well known, though subject to different interpretations, of which Sambursky's is not the least interesting. His discussion of the Stoics in relation to history of science was altogether new to me, however. It may be that I bought it too uncritically in later writings. I would no longer insist on the subjective versus objective distinction between Stoic physics and modern science nor suggest a filiation between the former and romantic attitudes to nature in the eighteenth century. I do think that the patterns are comparable but doubt the validity of a historical connection.

Sambursky clearly did believe in filiation. The aspect of his book that I would now see as most misleading, however, is that he selects out of the corpus of ancient Greek perceptions of nature only those that in his interpretation resemble features of modern physics. That a modern physicist learned in the classics should find them there is intriguing but not historically persuasive. Beyond that Sambursky considers that the nature studied by science is unchanging while knowledge of it increases. This proposition is acceptable to me, but not to many or perhaps most current historians of science, particularly those of a constructivist persuasion. In their eyes, science does not give an account of natural reality but is constructed by a sequence of experiments and theories imposed on nature by individual scientists or, more generally, by research schools.

Having been born in the region of Poland ruled by Germany before 1918, Sambursky studied in the University of Berlin. He made his career, however, in the Hebrew University of Jerusalem, beginning while Palestine was under the British mandate. We became friends after this review appeared, and he was foremost among those inviting me to give the Balfour lectures at the Weizmann Institute in Rehovoth in 1971.

FIFTEEN

Shmuel Sambursky:
*A Physicist Looks at Greek Science**

—ɯ—

I t is not often that a book appears which by sheer force of originality compels one to essay a review article in a period as far from one's own as is Greek antiquity from the eighteenth century. Such, however, is the irresistibility of *The Physical World of the Greeks*[1] by Professor S. Sambursky of the physics faculty in the Hebrew University of Jerusalem. His book is bound to hold very great interest for everyone who has any concern in the history of science, whether as scientist, philosopher, or historian. And I take the liberty of discussing it because its primary interest lies in the light which it throws on the whole history of scientific ideas rather than simply in its exposition of the positive content of Greek science. This is not to imply that it is unusable in the latter respect. On the contrary, it succeeds very well with my own students. But the reader in search of introductory information will do better, perhaps, to begin with the well-known Cohen-Drabkin *Source Book*[2] in conjunction with Marshall Clagett's recent *Greek Science in Antiquity*.[3] For it is no reflection on Sambursky's fine classical scholarship to say that his perspective is that of the physicist he is, rather than of the classicist or humanist who studies antiquity for its own sake, or for the sake of philosophy.

No other work on Greek science is so explicitly concerned with the parentage of modern science in Greece. (In order to avoid the confusion which would arise in this article from the term "classical physics," I shall use "modern" to mean the seventeenth century and after, and "contemporary" to mean the twentieth century.) In our own world as in Greece, science is conceptual thought mediating between consciousness and nature.

* Reprinted from review of Shmuel Sambursky, *A Physicist Looks at Greek Science* (New York, 1956) in *American Scientist* 46, No. 1 (March 1958), pp. 62–74. Copyrighted 1958, by The Society of the Sigma Xi and reprinted by permission of the copyright owner.

But an intellectual revolution has reversed the direction in which information flows. Modern science takes its starting points outside the mind and reduces events to concepts. Greek science started inside the mind whence concepts were projected out onto the cosmos to explain phenomena. This transposition from subjectivity to objectivity accompanied the shift from qualitative to metric formulation in which Galileo stands cleanly at the divide. A final essential point of difference lies in the role of technology, which has associated itself ever more closely with science as a consequence and continuation of the Renaissance thrust toward mastery of the world. In reciprocal fashion, science itself has become increasingly instrumental until experiment has displaced imagination and reason, not as creator of science, but as mode of inquiry and arbiter of theory.

But though science may change, nature, its object, does not, and for this reason any true science will by definition exhibit certain constant features: "Similar basic scientific conceptions necessarily forge for themselves the same moulds of thought and means of expression, regardless of the technical resources of the age."[4] This assertion of the morphological continuity of science since its invention in Greece is Sambursky's central proposition. To exemplify it he fixes attention on those aspects of Greek science which may be taken as structural anticipations of the idea patterns and great dilemmas of contemporary physics. This is an approach which shines an unexpected beam upon the antique facade of thought. It throws into shadow certain well-known features, brings into prominence others less noticed in the ordinary view, and reveals the whole in a perspective at once arresting and provocative.

In this perspective, science began, not as some scholars (most notably Neugebauer)[5] would have it, with the empirical or numerical techniques worked out by Babylonians or Egyptians for handling astronomical and practical problems, but with the Milesians, who proposed the very idea of nature in their rational conception of the cosmos as an orderly whole working by laws discoverable in thought. They were the first to set science its great task of explaining nature. Moreover, Sambursky sees in their formulations the adumbration of basic modes of scientific reasoning. The search for a fundamental substance served the principle of economy of hypotheses. The definition of all changes as transformations in that substance introduced conservation laws. Anaximander's cosmology presupposed the principle of symmetry (by which Sambursky seems to mean sufficient reason, though he does not call it that). The attempt to construct a physical picture of the cosmos inaugurated the use of models in scientific reasoning. Finally, in the assertion that, "Motion is from eternity"— i.e., that it calls for no explanation—the Milesians appreciated that, "Quality can be reduced to quantity. . . . For in this brief sentence is comprised the very essence of science from the time of Anaximenes to the present day."[6] But only after Galileo described motion dimensionally would the breach into objectivity glimpsed by the Milesians be opened so that all the forces of modern physics might pour through it.

Dimensional thinking or measurement moves attention to the Pythagoreans, who introduced into science this most crucial of all aspects in their attempt to comprise nature in number. Sambursky attaches the utmost importance to the emergence of the geometrical continuum from the Pythagorean consideration of irrational numbers. He describes enunciation of the laws of musical harmony as the first application of mathematical description to physical quantity. Indeed, he goes so far as to compare in imagination the feelings of the Pythagoreans at this discovery to the awe with which the contemporary physicist may properly contemplate the velocity of light or Planck's constant of action. Here are universal and fundamental quantities which science discerns in nature. "We may say, then, that from the standpoint of modern science, the scientific method of the Pythagoreans was correct."[7] And with this, one begins to suspect the direction in which Sambursky's originality will lead. For he assigns to Greek achievements in handling the continuum by physico-geometrical reasoning that importance which it is more usual to attribute to the concept of statical moment. In Sambursky the harmonic ratios play the role which Koyré in his discussion of the debt of modern science to Greece assigns to Archimedean statics.[8] The laws of harmony were the germ or matrix from which a measuring physics has unfolded. These laws expressed the application of metrics to nature, and this before ever Plato and Aristotle had separated mathematics from physics in opposing but equally divisive and defeatist ontologies.

Sambursky sees Greek physics in a far more dynamical light than one would have believed possible. From what has been said, it is obvious that he will drastically diminish the credit allowed Platonic influences in scientific development. He admires Archimedes, of course, but his interpretation leaves nothing really crucial for Archimedes to do. Instead of emerging as *the* Greek precursor of Galileo and modern science, Archimedes appears rather as an honorable exception to Platonic idealism which sterilizingly says that truth is not in the world of things. (Sambursky himself, however, takes a highly intellectual view of science, and it is pleasant to be able to record—rather as a compliment to his sensitivity of mind than as a complaint over inconsistency—that Plato has cast his spell over one moving passage in which contemporary physics is presented as bearing the battle in the continuing conflict between an ordering Reason and a chaotic Necessity.[9])

In deprecating Plato, Sambursky deprecates Aristotle too, since implicitly at least, he follows Jaeger on their relationship.[10] Indeed, "In the sphere of natural science, there was no essential difference of opinion between the two philosophies," and this is the next of the surprises afforded the observer who looks on the image of Greek science through the diffraction grating of current physics. Put so baldly, the surprise amounts almost to shock. Yet it is a point. If mathematics and nature do not fit, it matters little to the possibility of science (though much to philosophy) which is at fault. Extending the breach to cosmology, Plato worked a separation of heaven from earth in metaphysics. But "Once Aristotle had accepted it and given it a broader physical base, the fate of Greek science was sealed."[11] For the uniform cosmos of the Milesians was fatally dichotomized.

Indeed, Platonic-Aristotelian philosophy misled in quite general ways. Its very conception of science was humanistic, not naturalistic. To resort to teleology for understanding was to turn the scientist into an art critic forever discovering and explaining the technique or purpose of the ultimate artist. Teleology intruded systematically into science that conception of causality which from Socrates through Darwin bedevilled and embittered discussion between science and humanities. What offense might have been spared had the protagonists been able to recognize that they were talking of different things.

As to method, science in Aristotle moves, not from observation or experiment through induction and abstraction to predictive verification, but from common sense through categorization and deduction to definition. Such a route could never lead to a comprehensive dynamics. It is the measure of Aristotle's genius that, though astray, he succeeded with local motion as well as he did. (Again, Sambursky startles his reader into an appreciation of the unconventionality of his focus when he writes: "The study of this question is in itself Aristotle's chief claim to fame."[12]) Aristotle did develop a quasi-quantitative theory employing mathematical reasoning. But in principle he could never attain objectivity, which would have required considering motion apart from the missile. His success could be only partial and provisional within the confines of his fundamentally unscientific metaphysics of cause, change, place, and plenum. On the whole, therefore, Aristotle's influence in science, despite some helpfulness in biology, was "more negative than positive." Not that Sambursky would deny the extent of that influence, however regressive. On the contrary, "The chief interpreters of the three great religions finally blended the main principles of Aristotle's philosophy with their religious conception of the universe, thus turning the whole of that philosophy, including its physical and cosmic aspects, into unquestionable dogma."[13]

So far as science is concerned, then, the antithesis between Plato and Aristotle was unreal. And having practically dismissed from the history of science the two philosophies in which historians and philosophers have generally looked for the Greek contribution, Sambursky turns for its culmination and portent rather to the opposing but complementary resolutions of the great problem of "the unity of the cosmos and the plurality of its phenomena" given by the two Hellenistic schools, the Stoics and the Epicureans.[14] Never before have their theories of the universe been taken so seriously in the historiography of science: "There is no question here of comparing the two theories in terms of absolute scientific achievement: such a comparison would obviously be both pointless and unfair. The main purpose of the comparison is rather to estimate the validity of a method as shown in its internal logic and the extent to which it succeeded in developing to the full its basic premises at a time when scientific evidence was in the main qualitative. For we must bear in mind that the Ancient Greeks hardly knew of experimentation and mathematical deduction as means of applying scientific intuition to reality: for them analogy and the scientific model were the only connecting-links between the between the invisible and the visible."[15]

Sambursky makes atomism rise out of the necessity to save the conservation of matter from Zeno's paradoxes by establishing in principle a lower term to physical division. The atomists, therefore, reversed Aristotle on infinity by excluding it from subdivision and postulating its extension. As Aristotle observed, they, too, in their fashion were saying that the world is made of number. By the sound instinct which led them to follow the Milesians in accepting motion as a fundamental fact, they evaded the metaphysical difficulties created not by nature but Aristotelian reasoning. And, finally, by a stroke of imaginative genius, they situated their infinite extension in that void which rendered motion possible in a monistic science.

Thus in a conceptual and qualitative sense, the atomists identified the elements which were to go into post-Galilean physics—all except one. The idea of statistical order eluded them. They never decided how the stability is possible of large bodies which are in fact a swarm of atoms motion. How do they congeal to the appearance of fixity? But this difficulty was less serious than having no science at all. So they wisely learned to live with it, and went on to imagine lattice structures for matter and the molecular formation of compounds on the analogy by which language is made up of letters through forms which follow the rules of spelling and syntax. Their alphabet of nature was composed of atoms, and Sambursky is very persuasive in claiming for this empiricism of the imagination, this mode of inference from the visible to the invisible, concepts which are adopted because they make objective reasoning possible, the highest place among the scientific achievements of Greece.

One of the most ingratiating features of this civilized book is the felicity with which the author chooses aphorisms from the Old Testament as mottoes for his chapters. On Stoic physics ("'And the breath came into them, and they lived'—Ezekiel, 37.10") he is at his most provocative. For if atoms-and-the-void ("'Here a little, there a little'—Isaiah, 28.10") presaged Newtonian mechanics, the Stoic *pneuma* addressed itself dynamically to field phenomena and thereby established the pattern in which the geometric continuum, the mathematical expression of the real tension and unity in nature, has always retorted upon atomizing kinetics. The Stoic answer to the atomists, therefore, joined that dialogue which science has conducted between the unity of nature and the variety of phenomena for over two millenia and which continues so movingly in Niels Bohr's "Discussion with Einstein":[16] "The corner-stone of Stoic physics is the concept of a continuum in all its aspects—space, matter, and continuity in the propagation and sequence of physical phenomena."[17] For the Stoics it is activity which has ontological significance, rather than matter. And the carrier of unity, the principle of activity in the cosmos is the breath of the spirit—if I may take the liberty of combining the two translations which Sambursky suggests for the *pneuma* which binds the world into one dynamic whole, which is to the world what life is to the animal, and which, therefore, is the life of the world.

Lest the reader object that it is a mystic rather than a physicist who should be fascinated by the energetic unity which the Stoics seat in *pneuma*, he may be reminded that out of the mouths of many a physicist, a mystic has been known to speak in the language of geometry. Not that the *pneuma* can be expected to satisfy the modern mind as can the conceptions of the atomists, for in principle it is ineffable. It cannot be defined, since—as Sambursky acutely says—it creates unity in nature by abandoning simplicity for itself. But "Today we know that every self-consistent scientific system can only be achieved by compromises of this kind."[18]

Therefore, we can never say what the *pneuma* is. We can only say what it does. And what it does, is first of all to bind together substances which would otherwise be passive, undifferentiated stuff to provide them with permanence and stability. It is the erethritic agent in matter,

which it permeates as a stimulus penetrates erectile tissue in animals—
the analogy suggests itself in reflecting on the function of the quality of
tautness which Lamarck calls "orgasm" in his vegetable physiology, of
which Stoic physics are remarkably reminiscent (or rather prophetic).[19]
But *pneuma* is also the agent of differentiation between entities. Those
characteristics of bodies which the Epicureans attribute to the nature and
arrangement of atoms, the Stoics read back to its state of permeation by
pneuma. For them there are no boundaries in nature. Combination can
never arise from mixture or aggregation, but only from total blending of
principles, and the significance of this, too, will be familiar to students of
the modern period who remember the assertion before Lavoisier in the
qualitative chemistry of the seventeenth and eighteenth centuries that
principles are what combine, not masses.

The *pneuma,* finally, in its most comprehensive function is the principle of harmony throughout the whole cosmos, which it unifies through
the sympathies it creates between terrestrial and celestial things. The
Stoic "conception of the cosmos as a single unit, a homogeneous organism, rules out the existence of a void by giving a new revolutionary significance to the continuum concept. Instead of being represented as a
mathematical and topological 'juxtaposition,' the continuum now appears
as an 'interrelation'—a physical field of activities and influences passing
from place to place and from substance to substance and transforming
the whole mass of entities into a structure which acts and is acted upon
through the harmonious interpenetration of its parts."[20]

Sambursky argues very daringly that these Stoic hypostases were actually physics. If the Pythagoreans brought metrics into science, the Stoics introduced dynamics by the sophistication with which they built on
old Milesian notions of air and fire. This was "a first tentative approach to
the conception of thermodynamic processes in the inorganic world."[21]
And the Stoics are credited with real scientific achievement. Their theory
of the tides actually related the movements in the sea to the position of the
moon. But their most impressive formulation was of wave phenomena
and the propagation of vibrational motion through continuous media.
This aspect of motion had escaped both Aristotle and the atomists. On it
the Stoics based a theory of perception not unlike that of Descartes, their
toniké kinesis playing the part of pressure in his plenum.

Perhaps the most interesting of Sambursky's claims for the Stoics relates to their mathematics. The problem of infinite subdivision posed by
Zeno's paradoxes was even more pressing physically for them than for the
atomists, since they could not fall back on matter to escape it. Neither did
they have the infinitesimal calculus to introduce into mathematics dy-

namical tools suited to the requirements of their physical theories. Nevertheless, they made a greater step than any other school. Plutarch explicitly formulated the method of convergence on a limit for purely spatial quantities: "'There is no extreme body in nature, neither first nor last, with which the size of a body comes to an end. But every given body contains something beyond itself and the substratum is inserted infinitely and without end.'"[22] Sambursky describes this as the application to physics of the purely geometric method of convergence in Archimedes' calculation of *pi* or Eudoxos' proof that the cone is one-third the volume of the circumscribing cylinder.

The tragedy of the Stoics' otherwise dynamic physics was that they were unable to imagine the analogy which would have permitted them to treat time dimensionally. If they had, they could have conceived the abstract quantity, velocity, which embraces the functional relationship of time and distance and resolves Zeno's paradox in a ratio. But the subjectivism and biological emphasis of all Greek science were peculiarly difficult psychologically to transcend in the treatment of time. Discussing time, the Stoics abandoned the dimensional for the organic continuum. This forced them to adopt an atomic solution—a concept of intervals—for the physical aspects of time and prevented them, too, from breaking out of subjectivism to the Galilean definition of motion.

But with all this, it was in their causalism that the Stoics assumed most impressively the stance of modern physics. Democritus had already taken over the rule of law from the Milesians, for whom it was the assumption that the necessity in things simply is their causality, and he introduced it comprehensively at all levels of science. Unfortunately, Epicurus and the later atomists abandoned strict causality out of concern for free will. It was, therefore, not atomism, but rather the Stoic continuum which anticipated the standpoint of Laplace and modern science on the uniformity and determinism of nature. Strict causality is implicit in the unity of nature. In Stoic doctrine it assumes the guise of destiny or fate, which permits the reconciliation, or rather identification, of causality with Providence—not a capricious Providence, but a lofty Providence which knows its own mind utterly. "The absolute rule of causality thus becomes an integral part of the cosmos conceived as a continuum, and the Stoic conception of cosmos is completed by including the causal relation in it."[23] But the Stoics believed man to be a responsible moral agent. They, too, had to redeem free will, and Sambursky regards their solution, though ultimately unacceptable, as similar to the isolation of the problem practised by modern physics. They would divide causes into primary and secondary—the man dropping a stone down a slope, its rolling "'because that is the nature of the stone and of the roundness of its shape.'"[24] Just so will the physicist who would describe the total motion of a system distinguish the initial conditions from the particular impulse given some projectile.

So much will suffice to indicate the main features of Greek science in Sambursky's reading of its ancient concepts. Many other excellent things are in the book. The account of cosmogonies is clear and sensitive. The geometrical models devised by astronomy are well handled. There is an interesting discussion of Plutarch's *Face in the Moon.* The inability of Greek science to transcend the qualitative, the static, the speculative, and the rational, is explained on the grounds of technological quietism, disinclination for manual work, and failure to move beyond the formation of schools to institutions which might embody a consensus about nature. All this has been said before,[25] but Sambursky resays it well and adds two further elements. He attaches great weight to the Greek failure to think scientifically about the possible event, or to perceive order in statistical regularity. And he feels that the association of science with philosophy was distracting and ultimately disastrous to science—in the seduction, for example, of the Pythagoreans by Plato or of the atomists by Epicurus.

These last, of course, are judgments which come with great authority from a physicist. They contribute to the summons posed by the book as a whole to re-examine our structure of interpretation of Greek science. Since this is not a project to be undertaken lightly, it may perhaps be well to restate in a sentence what I take to be the main argument. It is this: that the confrontation of Plato with Aristotle is not what gives the history of science its dialectic, for there is no serious issue between philosophers both of whom see nature as the creation of artful mind; the fruitful antithesis is between the atomists, who see nature as a problem in mechanics and multiplicity, and the Stoics, who see it as a problem in dynamic force and unity—nature as whirl versus nature as fate.

This is an extremely interesting proposition. It is impossible to express too keen an enthusiasm for its suggestiveness. Only very rarely is so well traversed a field of scholarship stimulated by so novel an approach. It makes one wish that more physicists would illuminate with scholarly erudition the path their science has traveled. But it would not be natural if a few little clouds did not remain in the mind—remnants, perhaps, of its penetration by the *pneuma.* Sambursky believes firmly (as do I) in the reality of the scientific revolution, in the new science of Galileo established in generality and once for all in Newton. From physics, the cutting edge of this science, he looks back on its Greek progenitor to establish correspondences. Now, if the march of physics since Newton has indeed been forward, then the choice of a point along its way from which to survey

Greek science ought to change, not the arrangement, but only the distance of the picture. I cannot criticize Sambursky's interpretation as a contemporary physicist. Nevertheless, it must be confessed that when I, a student of 18-century physical science, read his book, I see in it different things.

Sambursky has, indeed, taught me more about certain features of the history of science in the Enlightenment than any book on the eighteenth century. But he has taught me to understand, not Newtonian science, but the resistance to it, the effort to take refuge in humane alternatives. For the Enlightenment saw a moral revolt against physics, expressed in moving, sad, and angry attempts to defend a qualitative science, in which nature can be congruent with man, against a quantitative, numbering science which alienates him by total objectification of nature. And this qualitative science, as urged for example by Goethe and Hegel, did not stem, as Sambursky says it did, from Aristotle.[26] (Teleological or providential science does, but that is quite another story.) Goethe's science derived instead, from that Stoic tradition to which Sambursky would trace mechanistic dynamics. So, too, did the ferociously vulgar science of Marat, the emanationist evolutionary science of Lamarck, and even the naturalistic and moralizing science of Diderot (as I have recently had occasion to argue, in consequence, partly of the stimulus afforded by Sambursky's book).[27]

All these are of Stoic inspiration. But the dynamical formulations of Newton and Maxwell are not. Here one must take direct issue with Sambursky's attribution of correspondences. He discerns in the *pneuma* the shape of the 18th-century mechanical aether developing into 19th-century fields of force. But can that be shape "which shape hath none?" For like the apparition with which Lucifer confronts us, the *pneuma* conducts one, not into the heart of science, but into the 18th-century underworld of nature. Its true heirs were the subtle permeating fluids of the Enlightenment, the principles which merge quality into consciousness in a subjective science—fire, caloric, Mesmer's magnetic current, Nollet's electric fluid. These were bearers of that activity which is fundamental as matter in motion is not. They were attempts to embrace phenomena discovered by the new experimental science in Stoic concepts. But the inspiration is not scientific, nor even proto-scientific, so long as the entities remain in principle indefinable. And that they had to do, because the ultimate function of such imponderable fluids, their *raison d'être,* was to penetrate, to permeate, to blend everything into everything, not—as Sambursky says himself of the *pneuma*—by topological juxtaposition however intricate, but in the intimacy of that perfect union which goes beyond junction to identity.

But the Newtonian aether was different, fundamentally different in nature and in function. In the first place, the aether is definable. It is, so Newton says, an elastic medium "vibrating like air, only the vibrations far more swift and minute." And again, it is "much of the same constitution with air but far rarer, subtiler, and more strongly elastic." Nor does it permeate matter to unite it with space, or rather to dissolve it into activity. Quite the contrary, "the universal Impenetrability of matter" is one of the cornerstones of Newtonian doctrine. The aether stands freely and most densely in empty space, and in bodies it "pervades the pores," that is to say it fills the interstices between the corpuscles which themselves "are void of Pores." Neither is the influence of aether on matter the vehicle of change as permeation by *pneuma* is for the Stoics. Rather, "That Nature may be lasting, the Changes of corporeal Things are to be placed only in the various Separations and new Associations and Motions of these permanent Particles."[28] And in Query 21 it is hinted that the aether itself must be particulate—a suggestion which precludes its assimilation to the world of the *pneuma* and the qualitative continuum.

The Newtonian aether, therefore, was no escape from atomistic mechanism. Schematically, it played the role, not of *pneuma* in Stoic physics, but of the void in the physics of Democritus. The classic objection to this atomic tradition has always been that it postulates the existence of the nothing. But the difficulty is only semantic. For what is really postulated is the existence of that which motion occurs *in*—translational motion in the case of the void, vibrational or harmonic motion in the case of the aether. The aether, therefore, belongs to the expanding and progressive history of objective science and not to the contracting and self-defeating history of subjective science. Sambursky might, perhaps—and properly—object that in distinguishing the aether from the *pneuma,* I have taken no account of its claims in energetics, as the first conceptualization of the phenomena embraced by modern thermodynamics. But even if one were to admit the force of the analogy, the Stoic *pneuma* presages neither Einstein's triumph over the multiplicity of nature, nor the magnificence of his ultimate failure, but only the somewhat pathetic irrelevance of Duhem.[29]

A further objection to accepting the Stoics as progenitors of our physics suggests itself from the appeal exerted by their principles in the Enlightenment. These were seized on by Diderot and d'Holbach, and after them by Goethe and the *Naturphilosophen* of German romanticism, because they permit the moralizing of nature. Historically speaking, the only other way in which it has ever proved possible to moralize nature is to have recourse to the government of an intervening Providence. But this gives up the uniformity of nature. It was an escape which seemed simply childish to all but the English natural theologians and a few French nature writers of no philosophical importance. Moreover, if one reads Bréhier, it appears that moralizing nature was the object of Stoic physics from the outset.[30] Its concepts had descended, of course, from the prehistoric and primitive representations of natural forces in myth and legend. It was an attempt to elevate this legacy out of the collective subconscious into science and philosophy. But Stoicism preserved the belief that the common understanding holds the truth and that virtue rises out of nature to be found by science. Sambursky's perspective does not encompass this aspect of Stoic physics. But that its object was moral must surely be reckoned into any estimate of its relevance for modern physics. And this is a consideration, to come to the essential point, which would tend to restore Archimedes to the place of progenitor, the Archimedes of the tradition of Platonic mathematical realism. For there is Platonic beauty and truth in statics. But there is no morality.

An additional reservation may further repair the formal Platonic filiation of the scientific revolution back through Galileo to Archimedes. It has to do with the incapacity to see order in probability as an allegedly limiting barrier in Greek science. In Sambursky's emphasis this is one factor dividing Greek science from that which has supervened since the scientific revolution. But so far as I know, neither is there in Galileo any significant conception of chance in nature. No more is it to be found in Kepler or Descartes or Newton. Indeed, the notion that an order of chance reigns in things (and not just in games or affairs) was systematically and analytically introduced into science only in Bernoulli's kinetic consideration of gases. Perhaps, therefore, this ought to be taken as the beginning of a second scientific revolution (though this is only a suggestion, for it is a subject to which I propose to address myself in a future investigation).

Sambursky offers two or three more handles to dissent—smaller ones and easier to grasp though providing less leverage. One sees what he means when he describes the subordination of science to philosophy as a disaster for science. (Is it exacting its retribution in our own century?) There is no need to reply except in pointing out that historically and culturally it could not have been otherwise, or perhaps by saying that qualified scholars have equally described the post-Newtonian separation of science and philosophy as a cultural disaster for conceptual science. Indeed, if one may be pardoned a rejoinder not meant to be malicious, failure to consider the philosophical implications of Stoic physics is what permits Sambursky's altogether admirable enthusiasm. Other interpretations are vulnerable to similar criticisms. For example, the Stoic causal distinctions are less analogous to the division of the problem in modern physics than to the Aristotelian classification of causes.

More generally, the historian is likely to find startling the readiness with which Sambursky attributes to ancient Greek thinkers wrestling with the first problems of philosophy distinctions like those forced on the theoretical physicist by the dilemmas posed him in his science. One is doubtful whether the pedigree of ideas is not read back from their arrangement in modern physics rather than out of the texts. So it is that the Aristotelian doctrine of place is compared to the concept of space in general relativity, the thermic processes of the Stoics to heat death, the separation of the opposites in early cosmogonies to gradients in potential. Many similar comparisons are adduced to illustrate the thesis "that the inner logic of scientific patterns of thought has remained unchanged by the passage of centuries and the coming and going of civilizations: the same models and associations recur, only in new forms suited to the more advanced stage reached by physical knowledge."[31] Nevertheless, if the historian's somewhat priggish sense of the fitting is at first dismayed at the ease with which Sambursky moves back and forth across twenty odd centuries, his surprise soon becomes a pleasant one and waxes into admiration. For no writer before Sambursky has made so explicit and so clear how Greek science was a marvelous complex of rational speculation, how it was a truly new thing in the world, and how whatever its answers, it discovered the problems. It is interesting to differ on emphasis and significance. But there can be only applause for the closing sentence of this really splendid book: "Whoever makes a close study of the scientific world of Ancient Greece cannot but be filled with veneration and his veneration will but increase, the more he realizes that, beyond all differences and changes, the cosmos of the Greeks is still the rock from which our own cosmos has been hewn."[32]

Notes and References

1. New York, 1956.
2. Morris R. Cohen and I. E. Drabkin, A Source Book in Greek Science (New York, 1948).
3. New York, 1956.
4. Sambursky, p. 140.
5. Otto Neugebauer, The Exact Sciences in Antiquity (Princeton, 1952).
6. Sambursky, p. 11.
7. Ibid., p. 42.

8. Alexandre Koyré, "A l'aube de la science classique," fascicule I of *Études galiléennes* (Paris, 1939).

9. Sambursky, p. 49.

10. Werner Jaeger, Aristotle (Oxford, 1948).

11. Sambursky, p. 55.

12. *Ibid.,* p. 92.

13. *Ibid.,* p. 80.

14. *Ibid.,* p. 105.

15. *Ibid.,* p. 106.

16. Paul Schilpp, ed., Einstein, Philosopher-Scientist (New York, 1949), pp. 201–241.

17. Sambursky, p. 132.

18. *Ibid.,* p. 136.

19. Philosophie zoologique (Paris, 1809), Part II. For further discussion, see my article, "The Formation of Lamarck's Evolutionary Theory" (see Chapter 3 of this work), *Archives internationales d'histoire des sciences,* Oct.–Dec., 1956, pp. 323–338.

20. *Ibid.,* p. 142.

21. *Ibid.,* p. 133.

22. *Ibid.,* p. 155.

23. *Ibid.,* p. 170.

24. *Ibid.,* p. 172.

25. Particularly by Ludwig Edelstein, "Recent Trends in the Interpretation of Greek Science," *Journal of the History of Ideas,* XIII, pp. 573–604.

26. Sambursky, p. 92.

27. "The *Encyclopédie* and the Jacobin Philosophy of Science," *Papers of the Madison Institute of the History of Science,* ed. Marshall Clagett (Madison, Wis., in press).

28. Isaac Newton, Opticks; see Dover Publications edition (New York, 1952) with preface by I. B. Cohen and introduction by Sir Edmund Whittaker, lxx (quoting Newton's paper of 1675), and queries 18–24, 28, 31 (esp. pp. 349–352, 364–365, 400).

29. Pierre Duhem, La théorie physique (Paris, 1906).

30. E. Bréhier, Chrysippe et l'ancienne stoicisme (Paris, 1951).

31. Sambursky, p. 203.

32. *Ibid.,* p. 244.

SIXTEEN

Salomon Bochner as Historian of Mathematics and Science*

—⚏—

I t is impossible for me to convey an appreciation of Bochner and his accomplishment in history of mathematics and science otherwise than in personal terms or to imagine how his writings would read in the eyes of people who did not know him. His judgments were alive with idiosyncrasy. When something that was being presented in seminar pleased him, he would fairly tingle in his place at the table, straining forward, eyes alight, urging the point onward with "Beautiful, very beautiful," half under his breath and perhaps not consciously uttered. Even at those moments we could never feel quite comfortable with our success, for it remained unclear whether we were the ones exciting the response or whether it was Aristotle or Fermat or Leibniz refracted through the glass of our optically uneven sensibilities, now mercifully translucent and too soon again opaque. The opaque interludes were terrible. Bochner's expression would collapse through phases marking disappointment, incomprehension, boredom, and despair until at the worst moments he rescued his sanity, his slightly puckish sanity, by falling asleep. On such occasions he might murmur something at the beginning of the coffee break and go off into the stacks, ostensibly in search of a book, never to return that day. Whether he taught us anything explicitly, I should be hard

* Reprinted from *Historia Mathematica* 16 (1989), pp. 316–323. This essay was written at the request of mathematical colleagues in 1983, shortly after the death of Salomon Bochner (August 20, 1899–May 2, 1982). It was to have been published with several other memorials concerning the main aspects of his life and career. That project did not go forward. I am grateful to *Historia Mathematica* for allowing me thus to make known the admiration and enjoyment experienced by my students and myself throughout Bochner's participation in our studies during the last 5 years that he served on the Princeton faculty. He retired in 1968 and moved to Rice University, where he was Chairman of the Mathematics Department from 1969 until 1976.

pressed to say, but certainly his presence, and the dread of provoking his absence, put us on our mettle. That was not his intention, and I think he was unaware of his effect.

His intention was to participate in discourse about the history of science, his violon d'Ingres, which he played to the strains of an inner melody. It happened that a doctoral program in history and philosophy of science got under way at Princeton in 1960. Apparently we did not disgrace ourselves in the first year or two, for Albert Tucker, then Chairman of the Mathematics Department, took me (a fairly junior faculty member) aside one day and intimated that Bochner might be receptive to an invitation to associate himself with our work, but that he was too shy to say so. Bochner, he went on, was not only a leading but a learned mathematician. The former fact had already been born in on me by the subliminal vibrations that transmit the general consent of a university in matters of reputation. The latter information, such is the frailty of scholarly natures, only increased my apprehensiveness. Would that all unworthy fears turned out to be as groundless, for Bochner's joining the staff of our program in its early years was the making of it.

Bochner had long since been giving himself the pleasure of studying (he might rather have said "savoring," a favorite word) the classics in the history of mathematics and mechanics beginning in antiquity. Since he had been educated in the classical Gymnasium, he knew Greek and Latin no less than modern languages, and he always sought out the earliest edition to be found of any text. His taste in the secondary literature was more haphazard, not to say quixotic. There were penetrating remarks on occasion, but Bochner's judgment of the scholarship of others was not, in my view, always illuminating. On the other hand, he proved to be an excellent judge of the qualities of our students.

A word first about his vision of the subject—I do not want to call it anything so mundane as a point of view. Emphatically, his was not the attitude that historians of science recognize and tend to deplore as characteristic of scientists and mathematicians: a taste for anecdote, often spiced by a touch of malice; a nose for scandal, particularly in the matter of priorities; an interest in substance only insofar as pieces of past science appear to be approximations, more or less awkward, to what is known now.

Not so Bochner. He was that rare, perhaps that unique mathematician whose historical sensibility was formed, not at bottom by the mathematics he practiced, but by the philological tradition out of which he explored its reaches in civilization. Central European rather than strictly Polish, Bochner was of the last generation of scholars who still incarnated the reality that German universities made of the ideal of culture down to the catastrophes of the twentieth century. He sometimes exhibited the classicist's or the humanist's instinct, the reverse of the scientist's, that the ancients, an Aristotle, an Archimedes, a Euclid, were and had to be of greater stature than the moderns. These are only half heard overtones, however, not symptoms of adulation. More important was the complement. Nowhere in his writings is there the faintest trace of condescension toward scientists or science in the past. When criticism is in order, it is the kind that might be visited on a contemporary.

And yet, we cannot make of Bochner an historian's scientist–historian, treating science in relation to its own time and context rather than as a function of its future. The difference is not only that he betrayed no interest in effects of social and political conditioning, a preoccupation that in the 1970s might have gone to an extreme from which it shows signs of receding. There was more to Bochner's distinctiveness than that. Historians, whatever their persuasion about fundamental factors in the historical process, are interested primarily in development: not just in what the event or, in the case of science, the discovery, theorem, or treatise was in itself, though of course they need to know that, but in how one event followed from and led to others in time and circumstance. In Bochner's perspective all of mathematics wears the appearance not altogether of intellectual contemporaneity, as I was about to say, but of a deeper contemporaneity residing in intuition. His writings, therefore, convey nothing of the process of development of exact science even though they do identify stages through which it has passed. These stages or phases are seen as conditions, however, as a series of states of knowledge with no nexus running between them.

A certain tension, not to say ambivalence, in his outlook is evident in a lengthy essay "How History of Science Differs from Other History" (1966, Chap. 2). Bochner was widely read in the great historians—Herodotus, Thucydides, Tacitus, Gibbon, Ranke, Mommsen. There are hints that he really thought general history a richer and more rewarding body of subject matter than history of science. He once said so to me. That is not just because general history concerns the whole life of mankind, but because the strict criteria of science impose limits on what can be said about its history. Archimedes was a judge of his own accomplishments in a way that Alexander the Great was not. Science internalizes its own history and in some degree represses it. The historian of science must be circumspect and respect the taboos that are inseparable from its creation and also the success of the outcome that has prevailed. General history is free of such constraints and is inexhaustible. At the same time, Bochner's sense of general history was by no means that of a professional historian. The distinctions on this side are of a different order, however. His was the historiography of Hegel, of the Germanic personification of abstractions, of—not to put too fine a point upon it—Spengler, and no practitioner of history as a succession of mind-sets of the Zeitgeist could have been more out of favor than Spengler among historians for the last half-century, for both professional and political reasons. In Bochner's essays, too, the great historical periods become actors in the drama: the Middle Ages, the Renaissance, the seventeenth, eighteenth, and nineteenth centuries. Each is endowed with aspects of collective intellect and will; thus: "The 17th century was an age of revelation: the 18th century was an age of patristic organization; and the 19th century was an age of canonical legislation. If we dared to continue we might suggest that the 20th century is an age of reformation. . . ." Periods and even sciences become subjects of verbs in the active voice, thus: "The Renaissance did little for physics but much for mathematics," and "Rational Mechanics also gradually introduced the concept of a purely mathematical space which is multi-dimensional . . ." (1966, 180, 220, 246).

Although Bochner conveyed no notion of strictly historical development, in the case of mathematics and mechanics he nevertheless did see growth, a kind of intrinsic maturing of collective powers. These actualizations of a potentiality in the life of the discipline are virtually Aristotelian. They are not explained. They are recognized. They come about "somehow," another favorite word, which Bochner somehow managed to employ with particular precision, and not as a vague gesture pointing away from ignorance. The organic quality of their emergence is further and pervasive witness to the chrysalis of Germanic historical sensibility.

A deeper ambivalence underlies Bochner's uncertainty over the value to be placed upon history of science in relation to general history. What is to be thought of mathematics itself, its very possibility, its claim upon culture, its penetration of other and perhaps one day all formal knowledge? In many passages he refers to mathematics as an esoteric activity, almost as a kind of pastime that by some destiny akin to that of myth imposes its rules upon ever wider sectors of reality. The figure of a game helps resolve a feature of his treatment that might otherwise appear inconsistent with the notion of an intellectual contemporaneity and comparability of exact knowledge. For Bochner does feel free to take the Greeks to task for the ultimate sterility of their absorption in geometry, and specifically for their having missed developing a system of spatial and temporal coordinates. They were not inferior thinkers in his eyes. No, they were athletes who refused to modify their strategy and gain what was within their powers. Their failure appeared to Bochner almost "inexcusable" (1966, 52). A game is to be won.

Apart from that, it is by no means obvious that other elements in the surrounding community have reason to welcome extrapolation of the rules of any such game. Their own pleasure may be spoiled, their autonomy infringed. In fact, however, expansion of mathematics has transpired more by invitation from recipients than by imposition from practitioners. At all events, there is nothing to be done about it. Bochner emphatically did not believe in the possibility, let alone the desirability, of channeling or guiding mathematical investigations. When he spoke of younger generations and the directions they were taking, not all of which were to his taste, he would say, "You cannot prevent them," without even a hint that he wished you could.

The first of Bochner's historical papers, "The Role of Mathematics in the Rise of Mechanics," appeared in 1962. Substituting the word "science" for "mechanics," he also made that the title of his first book, a collection of this and other essays published in the next 4 years. Another volume, *Eclosion and Synthesis, Perspectives on the History of Knowledge* (1969), was written in part out of his participation in the seminars of our Program in History and Philosophy of Science. In his last years at Rice University, Bochner composed further articles and monographs, several of them in the nature of reminiscence. In most instances, the titles of his papers are less indicative of the contents than is customary in historical writing. Instead, each of the ostensible topics serves as the occasion to embark on discussion of certain favorite themes to which he recurs from these various points of departure. It will be more pertinent, therefore, to identify the themes themselves than to attempt a summary of discrete contributions.

The underlying preoccupation, and perhaps the most signal motivation of the entire *oeuvre*, is indicated by the above title as modified to cover the contents of the earlier of the two books. How does it happen that mathematics applies to nature and even to society in such wise that it has become the most powerful strain in science? Ultimately, Bochner regards that, the dominant fact of Western intellectuality, as a mystery which he proposes to bring into relief in various ways but not to dispel. The Greeks, for their part, never developed a mathematical physics. Lacking, in Bochner's view, were an idea of quantity expressible in real numbers, the practice of analysis, and the capacity for abstraction. Only in 17th-century mechanics did the concept of a moment become possible, the real-number product of unlike magnitudes.

At this overt level, these findings of Bochner, and others like them, contain few surprises. The merit and originality of his discussion lie instead in many shrewd remarks in passing on such matters as functionality in Newton's thinking and in general, 18th-century principles of mechanics, the absence of any need for energy considerations before the nineteenth century, the emergence of distinctions between scalar and vector quantities, the "complexification" of physics through the introduction of complex numbers, and so on. There are many variations on two recurrent themes, first that there is a reciprocity between mathematics and physics, but second that it does not consist in deliberate steps, however successful, to mathematicize physics for specific purposes. Much of the mathematics that has proved most important to physics has originated without any thought of application in the course of purely mathematical investigations. Bochner was particularly intrigued by the frequent recourse in physics to pieces of "pre-fabricated mathematics," resources like the tensor calculus of Ricci and Levi-Civita lying there ready to Einstein's hand.

In my view, the best sustained of Bochner's substantive discussions is the essay "Aristotle's Physics and Today's Physics" (1964) together with the remarks on physical and mathematical space that figure in that paper and are then elaborated in several others. Although his views on the historical process are redolent of 19th-century philosophy, his views on science tend to come fundamentally down to earth. He resonates to Aristotle, not to Plato; to quantum mechanics, not to the mystique of relativity (his essay on Einstein [1979a] borders on the iconoclastic); to Kepler and Newton and Euler, not to Descartes and Leibniz and Kant. Here and there he will indulge philosophers who say things about scientific matters. Their "philosophemes" were another kind of pastime, however, that unlike mathematics came to nothing beyond embroidery. Essentially, he considers that Aristotle's *Physics* is to be read as physics, and not as philosophy. The notion of time as a determinant of motion or change; the correspondence of its oneness to that of the universe; the study of cosmology as an extension of physics; the relation of chance to necessity; *topos* or place involving a notion of spatiality as the setting for a physical system; the explanation of motion by antiperistasis, or action of a medium, as akin to thermodynamical processes; certain of the yes-or-no signals that constitute computerized information—in these and other matters, Bochner finds comparisons between Aristotelian and modern modes of thinking that illuminate the one by the other, without his wanting to say that it is a question of anticipations or reversions. The thought patterns are comparable, that is all.

Eclosion and Synthesis is a rather high-flown title for the second of Bochner's historical volumes (1969). Perhaps the book may best be described as a tentative sketch for a comparative morphology of two great stages in the evolution of systematic knowledge. Bochner designates by "eclosion" the characteristics of the half-century from 1776 to 1825 and by "synthesis" those of the twentieth century, mostly since the 1920s. Clearly, these terms would bear an enormous weight of generalization if they were to be taken literally and exhaustively. I do not think Bochner meant them that way. I think he meant them impressionistically as labels permitting him to gather into two loose bundles reflections upon his reading and experience. The opinions are a good deal more then desultory and a good deal less than systematic. More often than not they amount to insights. For example, the period he calls eclosion had no identity in the historiography of science at the time of his writing. What he ascribes to it is not so much cognitive discovery or conceptual innovation as it is the concretization of the modern learned disciplines in the form of professional entities— physics, chemistry, economics, and others, with emphasis on the very creation of mathematical and theoretical physics. Now, this observation, which Bochner does little more than assert, was right on target. The movement is now called the second scientific revolution, although I doubt that any of the many people working with its problems owe their start to his small, aphoristic book.

The notion of synthesis in the twentieth century is less fully circumscribed by dates and less tellingly illustrated by cross-disciplinary comparisons. By it, he means the momentum acquired by modern bodies of knowledge, their tendencies to internalize differences and disputes rather than to allow themselves to be transformed, the accelerating displacement of intellectuality by factuality, and the widening and tightening of the grasp of mathematics. I think the most interesting passages, however, are those conveying Bochner's skepticism, expressed in relation to modern cosmology, about the commitment of Newton and classical physics to the infinity of the universe, and those concerning Cézanne's liberation of artistic perspective from the straitjacket of three-dimensional Euclidean space.

In general, it should perhaps be said in conclusion, Bochner's historical writings are more rewarding for their asides, their irreverences, their glancing observations, than for their arguments, which are too fragmentary, introspective, and elliptical to be often persuasive. I do not mean that as anything but a high compliment. Professional life is stuffed with colleagues ever at our elbows trying to convince us. How much more agreeable on occasion, and how rare, to be intrigued, to be amused, to be startled, to see something differently, however fleetingly. I close with a few examples. Of Euclid and his immunity to human and historical influences: "In short, it is almost impossible to refute an assertion that the *Elements* is the work of an unsufferable pedant and martinet" (1966, 35) Of rational mechanics versus the rest of physics: "Thus Newton composed not only his formidable *Principia* but also an insinuating *Opticks,* at first not in Latin but in English, so that even poets might read it, as some did" (1966, 221). And, finally, to the complaint that people missed the chance Leibniz had provided to found mathematical logic in the eighteenth century:

> To this I wish to say, from hindsight, that, as developments went, everything turned out as well as one could wish for. The 18th century was extremely wise to give priority to constructing a thick basic layer of mathematics and erecting an edifice of rational mechanics; the 19th century was then the readier to initiate a mathematization of physics and of logic and the theoretization of other science. If this kind of rationalization of mine is too crass a case of "being wise after the event," then I wish to observe that, for my part, I have never felt dismay over savoring history, any history, backward through time, in addition to viewing it forward through time. (1966, 101)

SALOMON BOCHNER:
WRITINGS ON HISTORY OF SCIENCE AND MATHEMATICS

1. 1962. The role of mathematics in the rise of mechanics. *American Scientist* 50, 294–311.
2. 1963a. Revolutions in physics and crises in mathematics. *Science* 141, 408–411.
3. 1963b. The significance of some basic mathematical concepts for physics. *Isis* 54, 179–205.
4. 1964. Aristotle's physics and today's physics. *International Philosophical Quarterly* 4, 217–244.
5. 1965a. Aristotle's notion of place (topos) in physics. *Acts of the Tenth International Congress of the History of Science* (Ithaca, 1962) 8, 471–474.
6. 1965b. Why mathematics grows. *Journal of the History of Ideas* 26, 3–24.
7. 1966. *The role of mathematics in the rise of science.* Princeton Univ. Press. Japanese edition, 1970.
8. 1967. Plato and the one-ness of knowledge. *University Magazine* (Princeton University publication, Spring) 32, 18.

9. 1968. The size of the universe in Greek thought. *Scientia* 103, 511–531.

10. 1969. *Eclosion and synthesis, perspectives on the history of knowledge.* New York: Benjamin.

11. 1973. Five contributions (Continuity and discontinuity in nature and knowledge; Infinity; Mathematics in cultural history; Space; and Symmetry and asymmetry). In *Dictionary of scientific ideas.* New York: Scribners.

12. 1974a. Henry Bateman. In *Dictionary of American biography, Supplement four, 1949–50,* pp. 57–58. New York: Scribners.

13. 1974b. Mathematical reflections. *The American Mathematical Monthly* 81(8), 57–58.

14. 1975a. Note on Kepler's contribution to mathematical analysis: Kepler—Four hundred years. *Vistas in Astronomy* 18.

15. 1975b. Mathematical background space in astronomy and cosmology. *Vistas in Astronomy* 19, Part 2, 133–161.

16. 1975c. Commentary on Curtis A. Wilson, "Rheticus, Ravetz, and the 'necessity' of Copernicus' innovation." In *The Copernican achievement,* Robert S. Westman, Ed., pp. 40–48. Berkeley: Univ. of California Press.

17. 1977. Review of Genevieve Guitel, *Histoire comparée des numérations écrites* (Paris: Flammarion, 1975). *American Scientist* 65, 105.

18. 1978a. The emergence of analysis in the Renaissance and after. *Rice University Studies* 65(2 and 3), 11–56.

19. 1978b. Kepler: A personal footnote. *Vistas in Astronomy* 22, 19–20.

20. 1979a. Einstein between centuries. *Rice University Studies* 65(3), 1–54.

21. 1979b. Fourier series came first. *The American Mathematical Monthly* 86, 197–199.

Thomas S. Kuhn: The Nature of Science

THOMAS S. KUHN's *The Structure of Scientific Revolutions* (1962) has been and still is more widely known than anything written by a historian or philosopher of science in our generation. When asked to review it for *Science,* I was in Paris on leave and accepted eagerly. Tom Kuhn and I had been friends since the academic year 1946–47 when we were resident tutors at Harvard in Kirkland House. We there discovered our mutual, or rather reciprocal, interest in history of science, he from the side of physics and I from history.

What was my dismay, then, when on first reading *Structure* my reaction was puzzled irritation. What I had expected I do not know, but this was not it. The style is not inviting. There is constant repetition of key words and terms such as paradigm, puzzle, Gestalt shifts, incommensurability, anomaly, normal science, and so on, to each of which a special meaning is attached for the sake of the argument. I was at a loss what to do and almost decided to return *Structure* to *Science* on the grounds that I was not the one to review it. Then I thought of Herbert Butterfield on the importance of picking up the other end of the stick when confronted with an intractable puzzle. I put the book aside, and decided to try on a different thinking cap after a couple of weeks. In the meantime I was reading Auguste Comte's *Cours de philosophie positive* in the Bibliothèque Nationale.

When I came back to *Structure,* I saw what it was about (understanding a new paradigm?) and wrote the following review. It was among the first to appear. Kuhn was in Copenhagen at the time, interviewing for his oral history of the origin of quantum mechanics. He wrote me how pleased he was. All this came about at an important juncture in our careers. Kuhn had begun teaching history of science in one of the General Education courses inaugurated by James B. Conant at Harvard in the late 1940s. His first book, *The Copernican Revolution,* was the outgrowth from that course. Noel Swerdlow has recently shown how, all unbeknownst to the author at the time, the seeds of *Structure* were planted in that slim volume.[1] Such was Kuhn's technical acumen that, unversed in Latin though he was, he saw deeply into the problem and, unlike many a historian of early astronomy with impeccable latinity, he got the issues right.

On leaving Harvard, Kuhn moved to Berkeley, where he had a joint appointment in the History and Philosophy Departments. There he got on well with his colleagues in history, but was unable to develop a relationship of mutual confidence with the philosophers. For my part I had initiated an undergraduate course in history of science at Princeton in 1956 and a graduate program in History and Philosophy of Science in 1960. Both went well enough that Nassau Hall authorized enlarging the staff in 1962–63. My first thought was to try to persuade Kuhn to move to Princeton. Although *Structure* had just appeared, the range of its importance was not immediately recognized and certainly not by me. That he accepted strengthened our Program beyond measure. I have had occasion to note elsewhere that ours was a wonderful collaboration during the sixteen years prior to his acceptance in 1979 of what amounted to a research professorship at MIT.[2] It might be said of us that, in Isaiah Berlin's contrast, he was the hedgehog concentrating on one deep thing while I was the fox sniffing around on the surface at many tempting things. Different though Tom Kuhn and I were in background and temperament, we saw eye to eye on all academic, human, and practical aspects of the Program and of university affairs in general. Our only difference was of negligible importance personally.

It concerned merely the fundamental nature of science.

Notes

1. T. S. Kuhn, *The Copernican Revolution: Planetary Astronomy in the History of Western Thought* (Cambridge: Harvard University Press, 1966); N. M. Swerdlow, "An Essay on Thomas S. Kuhn's First Scientific Revolution, *The Copernican Revolution,*" *Proceedings of the American Philosophical Society* (Vol. 148, #1, March 2004), pp. 64–120.

2. "Apologia pro Vita Sua," *Catching up with the Vision: Essays on the Occasion of the 75th Anniversary of the History of Science Society.* Supplement to *Isis,* vol. 90 (1999).

SEVENTEEN

Thomas S. Kuhn: The Nature of Science*

—ɯ—

Normal science is succeeded by a creative phase of revolution out of which new concepts emerge.

—CHARLES C. GILLISPIE

This is a very bold venture, this essay, *The Structure of Scientific Revolutions* by Thomas S. Kuhn (University of Chicago Press, Chicago, 1962. 187 pp.). Kuhn has been turning it over in his mind and developing it in his work ever since the beginning of his career as a physicist-turned-historian. That was some 15 years ago during the course of James B. Conant's program for imparting science through historical examples in the service of general education. Now Kuhn would thrust more deeply into science than pedagogy will reach. His opening sentence is unequivocal: "History, if viewed as a repository for more than anecdote or chronology, could produce a decisive transformation in the image of science by which we are now possessed" (p. 1). For he is not writing history of science proper. His essay is an argument about the nature of science, drawn in large part from its history but also, in certain essential elements, from considerations of psychology, sociology, philosophy, and physics. The reader is not to expect philosophy of science in the usual Anglo-American sense of a study of logical problems found in scientific proceedings or systems. Rather is this a sketch for a genetic philosophy of science, presented in earnest of a fully developed study promised for the future.

* Reprinted from Review of Thomas S. Kuhn, *The Structure of Scientific Revolutions* (University of Chicago Press: Chicago, 1962) in *Science* (14 December 1962), pp. 1251–1253.

The author starts in the conviction that what he calls the accepted concept of science is misleading. This he describes as a view that science consists in an aggregate of facts, observations, laws, theories, and techniques
for getting more such results—in short, that science is a body of information which has accumulated in a linear series of discrete discoveries
about how the world is made. Kuhn's critique of the very notion of scientific discovery may, indeed, be the strongest part of his argument, and is
certainly at the heart of it. It was not, of course, difficult for him (though
it is still very useful) to demolish the notion that inventions of theory—
Maxwell's laws of the field, say, or Newton's laws of motion, or general relativity—were found like hidden treasure or a misplaced hat, there in the
logic or structure of things, wanting mainly to be revealed. Neither, however, will he allow novelties of fact—the indentification of oxygen, for example, or of radioactivity, or even of Uranus—to be counted as discoveries, if we mean thereby elementary increments of information by which
science grows in volume and, in growing, has its history and progress. He
tends to annul the distinction between finding fact and inventing theory.
A fact is itself only by virtue of theory and vice versa.

In what, for instance, did the discovery of oxygen consist? The gas
appears to have been first isolated by Scheele, who did not publish and
had no influence in the matter. Priestley prepared an impure sample from
the red oxide of mercury in 1774, mistook it for laughing gas, and when
he came to recognize it rather as a new species, dubbed it dephlogisticated
air. Lavoisier, for his part, instantly recognized Priestley's gas (though not
Priestley's priority) as what combines in combustion and calcination,
initially took it for pure air, and when he did see it as a fraction of the atmosphere, always considered its essence to be the principle of acidity requiring to be combined with caloric to assume the gaseous form. Valence,
diatomicity, specific heats—knowledge or definition of those and other
properties lay in the future, and who may say at what moment oxygen
took on the identity that it has since attained? This depends on what one
means by oxygen. One could mean simply what we breathe, and what was
fundamental for science was less the specification of the gas than the
chemical revolution wherein study of its combinations played a forcing
role.

That revolution consisted not just in a rush of new reagents, new reactions, new techniques, new methods of analysis, though these there were
in plenty, but also in a new way of seeing these materials of a science. Very
important to Kuhn's argument are findings of the modern psychology of
perception which make the literal notion of seeing as ambiguous as that

of scientific discovery. He refers to the well-known shifts of gestalt in which an observer switches back and forth between seeing a rabbit or a duck in some appropriate design, and he also calls attention to more elaborate experiments which make seeing a function of habitual expectation as well as of optics and physiology. A man may, for example, be conditioned by his expectations of the world to overcome the initial malaise inspired by inverting lenses and see things right side up, and only on repeated exposure will one recognize for what it is the anomaly of a black five of hearts planted in the deck of cards. Kuhn would extend the consequences to the sense of seeing in which the verb means conceiving.

A science, then, is how its practitioners as a highly articulated group see the ensemble of its phenomena, and this proposition takes Kuhn into considerations of the psychology and sociology of scientific communities and of what factors lead them to respond to, or to resist, innovation. They are the keepers of what the author calls *paradigms*. For, although science does not develop as a deposit of empirical discoveries economized now and again by theory, it nevertheless does have its evolutionary pattern. To a phase of "normal science." when scientists are agreed upon their paradigm and seek mainly to perfect it, succeeds, as a consequence of anomalies, a creative phase of revolution out of which emerge new paradigms to replace the old and run their course. What is more, the shift from one to another occurs rather as the conversion of a community than as the persuasion of persons by bits of new evidence or shorthands in theory.

Paradigms of Normal Science

This is an interesting schema. Let us develop and exemplify its elements a little more fully. Kuhn gives to the word *paradigms* a special significance and an importance all his own. One sees what he means, although he gives no precise definition. *Paradigms* are what give coherence to modes, or better, perhaps, to schools of science: to Ptolemaic or, contrariwise, to Copernican astronomy, to Aristotelian dynamics or to statistical mechanics, to phlogiston chemistry or to uniformitarian geology. Usually they are born in achievements like Newton's *Principia* or Lavoisier's *Traité élémentaire*, works exemplary enough to impart a tradition to a train of scientific research. They may be sufficiently comprehensive to contain a whole science like classical physics. Or—paradigms within paradigms—they may govern special and often quite restricted domains, such as the corpuscular picture of light which was displaced in the revolutionary way by the undulatory.

For Kuhn the concept has an importance going beyond a physical model, though it contains that. He particularly-wishes to emphasize that paradigms are what lay down the law to neophytes: "The study of paradigms is what mainly prepares the student for membership in the particular scientific community in which he will later practice. Because he there joins men who learned the bases of their field from the same concrete models, his subsequent practice will seldom evoke overt disagreement over fundamentals. Men whose research is based on shared paradigms are committed to the same rules and standards for scientific practice. That commitment and the apparent consensus it produces are prerequisites for normal science, that is for the genesis and continuation of a particular research tradition" (p. 11).

Nevertheless, a paradigm in Kuhn's sense is no closed set of propositions and practices, not quite an object for replication as it is in grammatical usage where one verb is the pattern for an entire conjugation. "Instead, like an accepted judicial decision in the common law, it is an object for further articulation and specification under new or more stringent conditions" (p. 23). Of such is the business of "normal science," each era of which is lived in service to some paradigm. Gathering fact, choosing that which is relevant, building and using appropriate instruments, developing applications, determining constants, formulating theory in more economical or more general expression—amid such "puzzles" do all scientists lead most of their lives and most scientists all of their lives. Normal science, moreover, is cumulative in the fashion which has been mistakenly attributed to the whole course of scientific progress. It solves its puzzles, extends its range, and refines its measurements. Not here, however, not in this work-a-day activity, does anything original happen. For normal science never innovates. Kuhn is consistent with his argument. He represents most scientists as rather a hidebound lot, not at all eager for fundamental innovation, as men who like to know where they are and where their work fits, who even tend to resist novelties which unsettle the paradigm and with it the intellectual security of the community.

ANOMALIES, INNOVATIONS, AND SCIENTIFIC REVOLUTIONS

Novelty will out, however, and inevitably. Anomalies do occur—the problem of what moves the mobile in Aristotelian dynamics, gathering complexity in Ptolemaic astronomy, augmented weight after combustion in phlogiston chemistry, contradictions required of the ether first after Fresnel and more generally after Maxwell. Logically anomalies are not to be distinguished from puzzles which are the business of normal science. Both may be described as counter-instances to some theory not quite adequate to the matter in hand. Anomalies, however, have the property of persistently suggesting, not simply that we do not quite know how to work the paradigm, but more seriously that at certain points the fit with nature fails. When these points begin to seem crucial or numerous enough, the failure becomes a scandal; the affair reaches the state of physics in Einstein's time or of astronomy in that of Copernicus. Then it is that a revolution occurs, a shift of gestalt into a new mode of seeing things which breaks with the established order, that of the old paradigm, and by first converting and then commanding the allegiance of a (usually new) set of practitioners becomes in its turn the paradigm of a new phase of normal science. Kuhn is very severe with positivists, however. These events are never brought about by the simple sort of methodological precept which has it that a theory is abandoned or modified should some instance of it fail: "The act of judgment that leads scientists to reject a previously accepted theory is always based upon more than a comparison of that theory with the world. The decision to reject one paradigm is always simultaneously the decision to accept another, and the judgment leading to that decision involves the comparison of both paradigms with nature *and* with each other" (p. 77).

Thus, Kuhn sets great store by the necessity for scientific revolutions. He will have none of the sort of reconciliation which makes classical physics a special case of relativistic physics. Newton's laws are derivable from Einstein's work only by presupposing the latter, which created a quite incompatible system of physical referents. "What had previously been meant by space was necessarily flat, homogeneous, isotropic, and unaffected by the presence of matter. If it had not been, Newtonian physics would not have worked. To make the transition to Einstein's universe, the whole conceptual web whose strands are space, time, matter, force, and so on, had to be shifted and laid down again on nature whole. Only men who had undergone or failed to undergo that transformation would be able to discover precisely what they agreed or disagreed about. Communication across the revolutionary divide is inevitably partial" (p. 148).

I do not think that it is overstating Kuhn's argument to say that he regards the revolutions of which he writes as the only creative episodes in science. They shape the research by which the emergent paradigm is perfected throughout the course of its usefulness. Even more fundamentally, the work of imagination and discipline which the men who make the revolutions bring to bear on phenomena are constitutive of nature itself. Afterwards science transpires in a different world, heliocentric as opposed to geocentric, sequential and curvilinear rather than enduring and rectilinear.

Summary

It is not for a historian of science to pass judgment on the central critique of Kuhn's essay, since that is directed to science itself. A few reservations may be ventured, however, of a sort which the author of so searching a discussion will certainly expect, the more so since, in his own terms, he proposes nothing less than a revolution in our concept of science, if not of nature. But it is not clear to me that anyone really holds the view of science which he would demolish. I for one find a great deal more in this book to agree with than might be expected in an exponent of a counterrevolutionary school. The argument depends very heavily on the viability of the terms—*paradigm, normal science, revolution, anomaly, crisis,* and the like. So it has been with many a philosophy of history from Comte to Toynbee. So it has been with many a chapter in the history of science—phlogiston, calorie, ether—and the student of either of these genres (which Kuhn, like Comte, combines) will have learned to be wary of mistaking the terms he gives his subject for its elements, the definitions for the happenings. The argument sometimes comes perilously close to circularity: that is, normal science does not aim at novelty, ergo what is novel is not normal science but an anomaly. On strictly historical grounds, moreover, strong cases might be made for considering books like Newton's *Principia* and Lavoisier's *Traité élémentaire* as summaries of a heritage rather than as models shaping the future. The reader may be referred, for example, to E. J. Dijksterhuis's treatment of Newton in his recently translated *Mechanization of the World Picture,* where it appears that Newton himself did not adumbrate the laws of motion in the sense in which they were fundamental to classical physics. For example, the proportionality of force to the product of mass into acceleration was imported into the second law in the development of analytical mechanics, not forced upon a school by a revolutionary law-giver. Newton was thinking of impact.

Still, there are not many books which find one making eager jottings in the margin, nor fortunately need one act on these; one may instead, and indeed in candor must, await the full development that Kuhn intends to provide. Meanwhile there can be only admiration for the erudition, the scholarship, the fidelity, and the seriousness that the enterprise reflects on every page. One is safe in predicting that whatever the final success, there will be no petty faults to find. Every historian, moreover, will surely applaud one recurrent and fundamental emphasis, which is that the development of science must be set into the context of a Darwinian historiography and treated as a circumstantial evolution from primitive beginnings rather than the ever closer approach to the telos of a right and perfect science. It is odd, and Kuhn is absolutely right about this, that by instinct scientists tend to see it the latter way. At least their students do, and who else could be responsible for that?

PART V

Science and Society

18

Mertonian Theses

IF MEMORY SERVES, Robert K. Merton and I first met at the Madison conference in 1957. I had, however, read and reread his monograph *Science, Technology, and Society in 17th-Century England* well before that.[1] During my time in graduate school it was one of the very few things touching on history of science that a student of history, particularly English history, could read with any interest, let alone pleasure. The following essay explains how an exchange of correspondence in 1958 led to a continuing reciprocity of interests lasting throughout our careers.

Merton has himself recounted in sociological categories and in detail the most intensive of those episodes.[2] In the process of planning the organization of the *Dictionary of Scientific Biography*,[3] of which I was editor, I remembered that Merton had drawn on the British *Dictionary of National Biography* for biographical information and that lack of uniformity in the data had been a serious impediment. It seemed a possibility that the *DSB* might be assembled in such a way as to afford comparability in the information of sociological significance.

With that in mind, I wrote to Merton on 12 April 1965 saying that the central thrust of articles in the *DSB* was to be scientific, but that at the same time we would like to compile a reference work that would be of maximum use in what the sociology of science would come to be. Accordingly, I asked him to criticize a draft of the letter of instructions we planned to send to contributors with a view to letting us know what kind of biographical information he wished he had found in the *DNB,* and what it might be that future sociologists would seek in our volumes. In my sociological innocence I thought this to be a simple question and expected a straightforward answer.

Nothing of the sort. Merton took the ball and ran with it farther than my colleagues or I had imagined. Computers were in their robust infancy and he saw in the request an unprecedented opportunity to create a computerized data bank for a quantitative historical sociology of science. This was not something to be done lightly or commissioned in a simple letter. It would require a carefully constructed schedule of pertinent biographical information to be supplied, so far as possible, by contributors to the *DSB* even while researching the scientific work of their subjects. To that end, in the year 1965–66 Merton addressed his graduate seminar at Columbia to the task of constructing and testing just such a questionnaire.

A provisional draft was ready in March 1966. With his usual psychological and sociological acumen, Merton had warned that there would be resistance, a possibility that in my enthusiasm I had discounted. So also did Tom Kuhn, who was a member of the editorial board of the *DSB* and whose excitement over the prospect exceeded mine. Others of the Board were less interested, but expressed no opposition. Merton's warning was borne out, however, once we met with the advisory committee charged to oversee preparation of the *DSB* by the American Council of Learned Societies, the sponsor and proprietor of the work. To Kuhn's dismay, and even more so to mine, the committee, composed mostly of senior colleagues, was ambivalent. Reactions were drastically divided between enthusiasm and hostility, with the latter predominating. Several members of the editorial board also now confessed to severe reservations. Apart from unvoiced skepticism about the very discipline of sociology and the possibility of quantitative history, the fear was that being confronted with such a questionnaire would turn off many contributors and that by overreaching we would jeopardize the realization of the *DSB* itself. I dared not press ahead, apprehensive lest similar objections might indeed be encountered in the larger community of prospective contributors.

The product of Merton's seminar was thus stillborn. Such was his generosity that our confidence in each other continued unabated. "We were ahead of our time," was his conclusion about what I still regret as a missed opportunity, specially so now that a very different sociology of science purports to address the content of science itself instead of scientists.

Notes

1. *Osiris Studies in the History and Philosophy of Science,* vol. 4, part 2 (Bruges, 1938).

2. "The Sociology of Science: An Episodic Memoir," in *The Sociology of Science in Europe,* ed. Robert K. Merton and Jerry Gaston (Southern Illinois University Press, 1977), pp.36–47.

3. Sixteen volumes, 1970–1980 (New York: Scribners).

EIGHTEEN

Mertonian Theses*

—⚏—

I t may be appropriate to begin the discussion with an apology for
being the one to have written it, for I am neither a sociologist nor a
scientist but a historian, and must leave evaluation of Merton's work
as sociology to the professional journals. What the editors of *Science*
asked to have set out, however, is the interest it may hold for scientific
readers, and I felt privileged to accept the commission for a reason they
may not have known, which is that historians of science have learned
more about scientists from Merton than from any other sociologist. We
appreciate the extensive and accurately documented use he makes of the
historical literature of science, both primary and secondary, and stand in
awe of his knowledge of our subject.

Having thus begun somewhat personally, perhaps I may be permitted
an anecdote to illustrate the unexpectedness of the kind of thing we have
learned. Some years ago, probably in early 1958, Merton sent me an off-
print of what I have since found to be the most eye-opening single piece
that he has written, his presidential address to the American Sociological
Association on "Priorities in Scientific Discovery." It starts by noting (pp.
286–287) "the great frequency with which the history of science is punc-
tuated by disputes, often by sordid disputes, over priority of discovery." As
I read on, dismay overtook amusement at the parade of eminent scientists
arguing and frequently quarreling with each other, not over what the
truth was, but over who had it first, Newton or Leibniz, Newton or Hooke,
Cavendish or Watt or Lavoisier, Adams or LeVerrier, Jenner or Pearson or

---

* Essay review of Robert K. Merton, *Sociology of Science: Theoretical and Empirical Investigations* (University of Chicago Press, 1973) in *Science,* vol. 184 (10 May 1974), pp. 656–
660. Reprinted in I. Bernard Cohen, ed., *Puritanism and the Rise of Modern Science: The
Merton Thesis* (New Brunswick and London: Rutgers University Press, 1990).

Rabaut, Freud or Janet. Sometimes the great men themselves abstained from contending in the lists of professional recognition for title to their intellectual property only to have their claims championed by disciples or compatriots. All too clearly the particular instances that Merton adduced in a number of variations on the theme of intellectual possessiveness could have been multiplied almost indefinitely.

In a note of acknowledgment to Merton, I wrote that, though it seemed surprising that the phenomenon was so nearly universal an accompaniment to scientific discovery, I did wonder whether the matter wasn't a bit trivial. I don't believe I also said "unworthy" but recollect that such a dark thought was in my mind. Only a few years later, when I began to study and teach materials in the social and institutional as well as the more traditional internal and intellectual history of science, did I come to take the full thrust of what he had in fact said, and said clearly and convincingly. It was that such behavior occurs in service to social norms; that norms arise in the life of real communities governing the conduct of their members; that the phrase "scientific community" is, therefore, no mere manner of speaking about some shared pleasure in the study of nature but refers to an effective social entity; and that, within its membership, which is bounded professionally and not geographically, two main sets of norms constrain behavior and do so in ways that conflict, the one enjoining selflessness in the advancement of knowledge, and the other ambition for professional reputation, which in science accrues from originality in discovery and from that alone. The analysis exhibits the scientific community to be one wherein the dynamics derive from the competition for honor even as the dynamics of the classical economic community do from the competition for profit, and neither of those statements is in any way incompatible with agreeing that the competitors characteristically like their work and choose it for that reason.

Merton replied a little stiffly though politely to my note, and only now on reading other essays do I learn, and suppose I should feel comforted, that my initial resistance was not mere obtuseness, but an instance of a methodological fallacy against which he warns, the supposition that the social importance of a phenomenon is a measure of its sociological significance (pp. 59–60). I find further, and less comfortably, that the instinct to trivialize what is demonstrably significant is a signpost familiar to sociologists, and that it points to distaste for the facts and hence to wishfulness and the substitution of sentiment for analysis (p. 384). The episode (out of modesty I should like to say trivial but, such is Merton's way with a norm, no longer dare do so) brings home to me the distinctive feature of his touch, which is to situate behavior, most often intellectual behavior, in its sociological context, but without thereby robbing it of individuality.

Some ten years later Merton welcomed the appearance of James Watson's *The Double Helix* (1968) for the epitome it gave of the inwardness of a scientific investigation into a strategic problem, and for the confirmation it afforded of the competitiveness among scientists entailed by the premium on being first with a solution. In his appreciation of Watson's candor, Merton dismisses the squeamish reaction that it shows contemporary science to have been corrupted by the scale, pressure, and contagion of a world that is too much with us. In fact it has been ever thus and ever an illusion, fostered by the myth of lonely, leisurely, disinterested contemplation of objects of simple curiosity, that science in olden times—say prior to World War II—was somehow better and purer than life. Indeed, what with the increasing prevalence of research in groups and teams, and perhaps also with the imposition of institutionally induced civilities, the occurrence of multiple discovery, though no less characteristic of science than in the past, has produced relatively fewer priority disputes in recent times than in its heroic ages.

Extending his investigation in the immediate sequel to the "Priorities" paper, Merton advanced the startling hypothesis (p. 356) that "the pattern of independent multiple discoveries in science is in principle the dominant pattern rather than a subsidiary one. It is the singletons—discoveries made only once—that are the residual cases, requiring special explanation." Moreover, the special explanations that he adduces virtually explain away the common notion of the unique and individual discovery in science. In every case of an apparent singleton that he has examined, Merton detects an example of rediscovery of something not fully seized, or else of work unpublished, suspended, or forestalled, and thus potentially when not actually duplicated. Lest the case here seem a little forced, he points out tellingly that scientists habitually live in the fear that such will happen, knowing it in their hearts to be the common fate.

A historian for his part must acknowledge that the argument, even when pushed to the extreme, makes sense of his vaguer feeling that the creations of science pertain to the scientist in a manner different from the relation of the work of art to the artist: that there is a sense in which the problem finds its scientists and that we would thus have had the law of gravity and the laws of motion even without Newton, but would not have had Hamlet without Shakespeare. Consistently enough, Merton supports his hypothesis sociologically by empirical evidence rather than by cognitive or logical considerations. He and his associate, Harriet Zuckerman, have inventoried 264 cases of discovery and found 179 doublets, 51 triplets, 17 quadruplets, six quintuplets, eight sextuplets, one septuplet, and two nonaries.

What, then, of the role of scientific genius? For it is attractive that Merton never denies the reality or importance of greatness in the gifted person or dodges it in the study of faceless aggregates. On the contrary, he dispels the notion that social explanation of a process derogates from the single man's part in it. What he finds, also empirically, is that scientists commonly reckoned to have been great participated in an altogether larger number of multiple discoveries than their lesser colleagues. They did a lot more science and were connected professionally with a much larger number of other scientists. A detailed study of much of Kelvin's published work detects some 32 instances of multiple discovery involving 30 other scientists. A scientific genius is measurably great, then, in the sense (though not necessarily only in this sense) that his contributions are equivalent to those of a considerable number of lesser lights.

That these findings should have been unwelcome to many scientists is not surprising. To have some conflict in inner values exposed to the inspection of others is never exactly pleasant, however salutary, and though Merton did not mean to be insulting, the spectacle that he spreads upon the record flatters only the very few who were self-denying about their claims to priority. No one likes to have his behavior labeled, the less so if it be aptly done, and Merton calls the elation (in part illusion) of the moment of discovery the Eureka syndrome (p. 401). He goes to some lengths to explore the deviant or repressive behavior which the premium on originality produces—plagiarism or falsification in the extreme instances, which are rare in science; cryptomnesia in the common instances of an investigator who forgets, not only that someone else had the fine idea before, but often that he himself did at some earlier stage. To study all this frankly would be healthy, Merton argues in a paper on the advantages to be anticipated from further study of the phenomenon of multiple discovery. More concretely, such investigation could be expected to bear on the problem of creativeness in science, on the role of the scientific milieu, on the comparative methodology of science, on the relation of research establishments to society at large, on the psychology of science, and finally on planning and science policy. On the last score Merton advances the intriguing possibility that multiple research and discovery may not in fact be wasteful and redundant if the problems are important, but may instead be functional, in that repetition and reinforcement will affect the situation more strongly than would single statements.

Among the other papers collected in this volume are a chapter, "The Puritan Spur to Science," from the doctoral dissertation that Merton completed in 1935—*Science, Technology and Society in Seventeenth-Century England* (in *Osiris Studies,* 1938)—and the preface that he composed for the reprinting of that monograph in 1970. Not many a thesis furnishes fuel for a controversy lasting as long as its author's career, much less bidding (as this one is beginning to do) for immortality. It was Merton's contention, advanced at a time when science and religion were supposed to be categorically antithetical each to the other, that on the contrary the Puritan values of diligence, rationality, practicality, asceticism, self-denial, civic spirit, and service to God through the work of this world (for this purpose the study of His works in nature) had the strong tendency of stimulating and validating the scientific enterprise among the generation that founded the Royal Society. In effect the argument utilized the categories of Max Weber's famous *The Protestant Ethic and the Spirit of Capitalism,* where they served the analysis of the religious legitimation of capitalism, and transferred them to an analysis of the religious legitimation of science.

Merton's treatment of the Puritan ethos rests upon the same kind of evidence as did Weber's, an essentially psychological insight into literary sources and biography. It touched us on the quick at first reading in much the same way, at least it did those of us who still recognize our own springs of action in Puritan modes of behavior. It was also open to the same kind of objection on the part of critics of a more literal or Catholic turn of mind than Weber, Merton, or we who remain persuaded that there is something important here that has yet to be fully, made out. For clearly there are difficulties, and grave ones. Economic enterprise for profit has flourished in many a milieu untouched by Puritanism, and a theory of capitalism that requires excluding Florentine bankers has its problems. By a similar token, it is very easy to find notable Roman Catholics among European scientists of the seventeenth century, and the embarrassment may be complemented by citing eminently antiscientific statements on the part of certain Puritan divines.

If it is correct that some sort of intuitive recognition (or resistance) was responsible for the initial and continuing interest in Merton's argument for a stimulating effect of Puritanism on science, then the inability of scholarship either to settle the question or drop it may arise from a discrepancy between that fact and the way in which the case was presented. For Merton made it depend on a statistical analysis of the membership of the Royal Society in its early days. His contention that the persons who predominated scientifically in that body were in significant degree of a Puritan persuasion in religion has not, in my judgment, withstood criticism based on chronology, counting, and biography. During the interval of more than 30 years between the publication and the reissue of his monograph, Merton kept largely silent about the question, preoccupied as he was with the other investigations reported in this book, and even more so with sociology at large. But now that he has answered his critics, the preface (pp. 173–190) in which he does so fails to meet their case. The trouble begins with the fancy that his own book written long before is by somebody else. There are passages in other writings in which Merton likes to make a point by doing sociology right out there in the open before the reader's very eyes, and usually the device is self-deprecating and entertaining in effect, but here it distracts attention from the issue. Neither will it do to refute Lewis Feuer's *The Scientific Intellectuals* (1963) as if that specious book were representative of the skeptics, much less worthy of them. Not much more pertinent is his disclaimer of having argued that Puritanism was an indispensable incentive to scientific work, for I doubt that he has often or seriously been charged with so simplistic a statement. Most generally, however, he rejects the identification of his Ph.D. thesis with the "Merton thesis" on Puritanism and science (though his own emphasis on eponymy in the "Priorities" paper should have convinced him that it is not called that for nothing), and observes a little plaintively that his monograph concerns the whole cultural, social, and economic context of science and technology, and that it adduces much evidence about the importance of mining, navigation, and military needs in the early work of the Royal Society, devoting more pages to these matters than to Puritanism.

That is true, but what Merton does not reckon with—though he recognizes it—is that precisely the question of Puritanism in science served to keep the interest in his monograph alive. In a way, he has himself justified the concentration of his critics on the subject, even if inadvertently. For it is the topic of the most interesting of the few essays on sociology of science that he included in his earlier collection, *Social Theory and Social Structure* (third edition, 1968). Moreover, Storer has chosen the chapter on Puritanism, and not those on economic or military involvement of science, for incorporation in the present work.

Perhaps, therefore, Puritanism in science came home to Merton himself as well as others, and I trust it will not be inconsistent with the profound respect I feel for the whole corpus of his work to say that in treating this question, as perhaps elsewhere on occasion, he sometimes seems to me to misjudge the location of his own greatest strengths. He often emphasizes that sociological research is an empirical undertaking, and though I am not sure he ever says that it is itself science, such is certainly the implication. The reader is often reassured that careful counting has gone on behind the scenes (I have no doubt it has) and that hypotheses are ever being tempered by evidence (I have no doubt they are). It is not there, however, it is rather in his insight into motivation and behavior, individual and collective, that Merton is at his best and deepest. The important quality he brings to his work is psychological acumen, not quantitative rigor, and that makes it a work which is humane, exciting, and inceptive rather than decisive and conclusive. Notice how the latest papers in several lines that he has opened to inquiry end by enumerating the hares he has started rather than the results he has reached. Notice also that the various topics he has pursued in the sociology of science concern the internal functioning of the scientific community rather than its relations with the external structure of society at large, whither an innately statistical and quantitative forte might more naturally carry a sociologist. Perhaps that is why he did not catch (or set) fire over the military and economic associations of science in the seventeenth century even though he did write about them dutifully.

As for the Weber-Merton analysis of Puritanism itself, carrying it further has been frustrated in part by the feebleness of the statistical approach. For one thing, the numbers involved in counting bluenoses on the science side are very small. For another, the form in which the question has been discussed has required that Fellows of the Royal Society be called Puritans or not, and this at a time when Puritanism was already a century old and had permeated (as it still does) the values of people whose religion was pallid or even different. Another sort of analysis is what is needed, one which Merton would be very well qualified to give but has not given. It would be a combined psychology and sociology of the comparative social dynamics of Puritanism, capitalism, and science, and would be independent of whether particular capitalists and scientists were themselves Puritanical in the religious part of their lives, or whether particular Puritans were capitalistic or scientific in the economic or intellectual parts of their lives. Then the intersections and interactions in which Puritanism, capitalism, and science have reinforced each other in their thrust to change their worlds might become more manifest. As a historian, I am convinced that they did. I am convinced that all three pertain to the forward march of history, to the forces that have modernized society, and that all worked corrosively against the complex of Catholic, feudal, and scholastic forces—if that is the appropriate trinity of traditionalism— wedded to preserving the past rather than to making the future.

To study that would require enlarging the boundaries of the problem. At the other end of Merton's sociological range, however, a methodological restriction obtains which also impedes this particular inquiry, though in a different way. In practice, and maybe in principle, his is a sociology of scientists, taking no account of the content of their work as a factor in the social and institutional relations among them. He makes no distinction, to cite the example that is important here, between the mathematical and the experimental, or, better, between what Thomas S. Kuhn has called the classical and the Baconian sciences in the seventeenth century (it is not quite the same distinction, but no matter). In the article on history of science in the new *Encyclopedia of the Social Sciences* (1968), and also in the 1973 Sarton lecture before the AAAS, Kuhn has suggested that if one were to concentrate on the latter sciences, and consider how largely Baconian were the collective interests of the Royal Society, the association of Puritanism with the initial impetus and early activity might be made a good deal stronger. I agree, for the economic and technological topics that Merton discussed and wants noticed would thereby be integrated into the argument, which might then turn on the sociology of the Puritan aftermath rather than the theology of Puritan belief. It would be more convincing for the change in focus.

So much for the parts of Merton's work that a historian is best qualified to discuss. I hope it is not parochial to suggest that scientific readers not already familiar with his writings might wish to begin with those papers and issues, and then go back to the first two sections of the book, where the emphasis is more on sociology itself than on its object in knowledge or science. They will thus have become accustomed to his mode of analysis in relation to their own affairs, and will find it no less illuminating in this more arcane area. One of his recurrent preoccupations has been the reason for so studied a neglect of science on the part of other sociologists. He does not himself claim credit for having finally reversed the situation (though in fact no one has a better right to it). Circumstances have done that, and in a manner that fulfills one of the shrewdest of his predictions, which was that interest in the subject would develop when and only when science itself came to be regarded as a social problem or a source of social problems. For difficulties, strains, dysfunctions, and dangers are what attract the interest of social scientists, not mere importance or success.

A second set of concerns is the degeneration of scientific or cognitive disagreement or conflict into the political and sectarian strife of schools and factions, wherein the question ceases to be "Is that right?" and becomes "Why did he say that?" An important paper, "The Perspectives of Insiders and Outsiders," was largely inspired by the recent proliferation of scholarly and sociological enterprises linked to the aspirations of blacks, women, and other segments of society that feel impeded or abused. Merton discusses the matter from the standpoint of whether it is necessary to belong to such a group in order to have knowledge of it, and concludes with an injunction to openness and tolerance all the more welcome for being sociological.

Grouped in the last section are certain recent investigations that touch scientists in their actual careers more closely than any of the foregoing. Merton there brings under scrutiny the procedures by which recognition is awarded and work evaluated together with the equity of the results, and related to that, considers the effects of age and seniority in the life of the scientific community. It may be that his sense of whimsy is sometimes a little like James Reston's in his occasional Sunday column, a serious man jesting about a recalcitrant subject. Merton calls his paper on the reward system "The Matthew Effect in Science," the allusion being to the statement in the first Gospel, "For unto every one that hath shall be given, and he shall have abundance. . . ." Recognition is the common coin in which scientific rewards are paid, and as in other communities, treasure is unevenly distributed. It is the already prominent scientist whose name becomes associated with projects on which he works with junior colleagues, not the younger people who need the credit. Such collaboration is to their advantage in another way, however, since the notice that the findings attract will also depend on the fame of a name and not just on their importance, though on that too. For beginners the value of working with outstanding people is no mere matter of public relations. The guidance and example, Merton does not hesitate to say the character, of eminent scientists are instrumental in increasing the productivity and effectiveness of those fortunate enough to study with them, particularly in their identification and choice of problems. A significant proportion of Nobel laureates have been trained under older Nobel laureates, whose influence in science becomes then a function of their standing and not mainly of the research that won them the prize.

A more extensive paper, this one written with Harriet Zuckerman, inquires into the refereeing system in scientific publication, an aspect of the institutionalization of science that is coeval with the earliest societies and journals. Comparison with the humanities and social sciences exhibits a very low rate of rejection in science compared with humanistic disciplines; the harder the subject the lower the rate. (That is not his terminology. Always preferring to make a sociological statement, he says of the softer subjects: "This suggests that these fields of learning are not greatly institutionalized in the reasonably precise sense that editors and referees on the one side and would-be contributors on the other almost always share norms of what constitutes adequate scholarship" [p. 472].) It should, at any rate, prove startling to my historical colleagues to learn that leading journals in our field reject 90 percent of the manuscripts submitted to them compared with 24 percent rejected by physics journals. An intensive study of the archives of the *Physical Review* reveals that the judgment of referees is not significantly affected by the age, standing, or institutional affiliation of referees or contributors, and permits the conclusion that physics is well served by the system.

The final paper investigates another factor in science about which much myth and gossip have clustered but little research. Also written with Zuckerman, it is called "Age, Aging, and Age Structure in Science." Among the questions considered is that most famous or infamous one, on which views change as scientists grow older, whether indeed scientific invention is a secretion if not a secret reserved to youth. But I think it will be consistent with the main purpose of this article, which is to draw the attention of scientists to the interest Merton's work holds for them, not to give that answer away, nor even to say what he makes of the question. I shall observe only that this, his most recent paper in the sociology of science, is evidence that one sociologist, in what he will not mind my calling his maturity, need fear no weakening of his powers to see what is deep in things commonly mistaken for obvious, and to make the best and most humane of good sense out of the most unexpected of problems.

—⚬— 19 —⚬—

Remarks on Social Selection as a Factor
in the Progressivism of Science

AS WILL APPEAR, this memoir, unlike most of my published writings, is the fruit of meditation, and to some degree introspection, rather than research. The considerations it advances developed in the course of teaching the history of science and thinking about its relation to general history. In what measure the analysis stands up is for a reader to say. The question of how the elements appear to me now, a full generation since 1968 when I wrote the piece, had better be reserved for a postface.

One thing may be noted at the outset. It will be obvious that Robert Merton exerted a powerful influence on my thinking. Sociology of science has indeed developed into a flourishing subdiscipline since the days when he, as it were, summoned it into existence. That it should be so bears out his prediction that sociology of science would flourish only when and if the role of science in society itself came to be seen as problematic instead of simply beneficial. That is precisely what happened as a carryover from the counterculture of the 1970s. The consequence, however, is not what Merton imagined or, I think, approved. For in large part the sociology addressed to science is constructivist in the sense that it seeks to exhibit the role of social relations in forming the content of science. In his view the function of sociology was to elucidate the dynamics governing the conduct of scientists.

NINETEEN

Remarks on Social Selection as a Factor in the Progressivism of Science[*]

—ɯ—

The purpose of the present essay is to invite consideration of modes in which science may have been functionally related to political and social progress in the course of modern history. The discussion does not touch on the material and technological role of science. I believe that these factors, tremendously important as they undoubtedly are, could be brought within the scope of the argument, but am not presently prepared to do so. What follows is not, therefore, intended to take issue with Marxist or materialist positions in the historiography of science and does not confront them on their own ground. At most, the concluding section advocates an alternative way of looking historically at relations between science and its social environment, one based on considerations concomitant with rather than reducible to technological and economic factors and one that does not presuppose any causal philosophy of history.

I

So far as history has a direction more significant than the mere succession of events in time, that direction is generally taken to be forward toward liberalism, democracy, socialism and an accompanying amelioration of the human condition. Whig history is out of fashion, but there must be very few historians who do not in their own sensibilities evaluate events according to whether they participated in or impeded progress in that sense. Opinions seem to agree, moreover, about the reality of dynamic

[*] Reprinted from *American Scientist* 56, 4 (1968), pp. 439–450. This paper was originally delivered before the XXIIth International Congress of the History of Science held in Paris between Aug. 25 and Sept. 1, 1968.

connections between the growth of science and progressive political and social development.

It will evoke the universality of the assent to this last proposition to call to mind the political and social significance of the signal philosophies of science since the seventeenth century, when modernity in science as in history first became clearly recognizable. One has to go right back to Hobbes to encounter a philosopher who, reasoning positively upon science, drew conclusions that have since been thought politically regressive. Even in the case of Hobbes, however, it would be difficult to call his view of the state and of sovereignty anything but prescient. Thereafter, philosophers have consistently taken science to be a body of knowledge relevant and important ideologically to the progressive side of political and social issues. The matter appears to particularly good advantage in the interplay between philosophy and politics throughout modern British history. At the time of the Whig Revolution John Locke made the case for empiricism in philosophy, for constitutionality in government, and for toleration in society. In the nineteenth century the intellectual content of the reform movement was known as philosophical radicalism, and John-Stuart Mill considered that the case for representative government and for personal liberty depended on first establishing the validity of the inductive method in science. In the twentieth century Bertrand Russell has occupied the positions that derive from British empiricism in relation to the science, mores and politics of our own times.

Nowhere outside of Britain has either politics or philosophy exhibited an equally coherent development, but neither have their relations differed practically in other contexts. Indeed, it is curious that, disagree though philosophers have done in their accounts of knowledge, the three main persuasions in the philosophy of science, namely British empiricism, Kantian idealism, and French *idéologie* merging into positivism, took similar positions with respect to the historic issues of politics. Throughout the nineteenth century, liberal intellectuals in Germany could find in the Kantian tradition prescriptions for peace, freedom, and individuality. In France positivism followed its own injunction and graduated into a program for intellectual emancipation and political reorganization while Saint-Simonianism for its part projected engineering into socialism. In Russia westernizers in literature and philosophy appealed to science, and in the United States the pragmatic variant on empiricism answered to the active and egalitarian character of American democracy. The correlation emerged most clearly in Dewey's philosophy of education.

The reciprocal may be equally impressive, for in no context nor even in any significant instance does the student of political and social philosophy encounter an appeal to science in support of a conservative position. Whether it be Burke, Gentz or Chateaubriand; Maistre, Carlyle or Taine, the argument is from history, law or religion, from language, nationality, or myth, and not from science. Perhaps it is also relevant that the term philosophy itself refers to a different kind of intellectual effort in these instances, one from which epistemology is largely absent.

The evidence of philosophy confirms, then, the impression that, insofar as science relates to the social and political process, it pertains by its nature to progress and the programs of the left. The proposition is inherently persuasive, and there must be something in it. The problem is to know precisely what. For when the historian turns to the kind of reconstitution to which he is accustomed, that is to the narration of the development of scientific knowledge itself (if he is a historian of science), or to an account of political events and social structures (if he is a general historian), the mode in which science has in fact participated in the historical movement toward liberalism and democracy appears very unclear. Abstracting from economic and technological factors (though I recognize how partial the argument thereby becomes), one might specify various ways in which science could be supposed to have figured in political and social development. It might have done so directly, by virtue of its content, or through the political actions of its increasingly influential community, or in both these ways. It might have done so indirectly, through its influence upon the social structure and the dispositions of culture.

Let us consider the several possibilities.

II

As to the first alternative, it might appear obvious *a priori* that the content of the natural sciences can have had no intrinsic relevance to political or social questions. In fact, however, the history of science has been attended by repeated efforts to deduce from scientific constructs and methods rules for political organization, social arrangement, and ethical prescription. Movements of the type of biological romanticism, natural theology, social Darwinism, Spencerian individualism, scientific socialism, and ethical naturalism have been an important feature of intellectual history. What may be doubted is whether science has thereby participated effectively in the history of liberal thought and polity and reciprocally whether they have enhanced the growth of science. My own conclusion is that neither of these effects occurred: that on the contrary purportedly scientific programs of politics, society, and ethics have usually proved authoritarian rather than libertarian in their consequences since what they looked to was precisely the authority of science, and further that they have generally redounded to the disadvantage of science since their proponents have often subjected science and scientists to ethical or political constraints.[1]

Turning to the second partial possibility, i.e., that the scientific community, a growing segment both of society at large and within that of the intelligentsia, has normally been a liberal force in actual politics, here too the conclusion is a negative one. That it should be so may seem somewhat more surprising and require justification. Not that one expects scientists to busy themselves in politics, but insofar as intellectuals generally have been politically active and articulate, one looks to find their weight exerted toward the left. That may still be true of preponderant political opinion among scientists throughout modern history. There is no way to tell, for it has not on the whole been true of their actions, as I still believe it to have been of other groups among intellectuals. This conclusion may not be very welcome, and perhaps it will be well, therefore, to set out explicitly the sectors of evidence from which it mainly derives.

It derives from reasonably extensive studies of the interactions of science and politics in revolutionary and Napoleonic France together with more discursive readings about the political behavior of the scientific community in the nineteenth century and in the contemporary world. In respect of the former, the conclusion is quite definite. The half-century when France held the scientific leadership of the world to a degree unmatched by any nation before or since, that is to say from the Turgot Ministry until the July Revolution, was also the time when science and public affairs first assumed their modern importance for one another. During that period it must be said of the scientific community, and of it alone among definable professional or intellectual communities—i.e., artists, writers, philosophers, doctors, historians, and political economists—that it pressed into the service of each successive regime and received back increasing institutional benefits from each government in turn. The reform ministry of Turgot, the final ministries of the old regime, the constitutional experiment from 1789 to 1792, the dictatorship of the Committee of Public Safety, the reactionary dispensations of Thermidor and the Directory, the military receivership of the Consulate and its transformation into Empire, the restored Bourbon monarchy—against none of them did scientists as a group or scientists as men of intellectual conscience raise their voices in active political dissent. It was not the scientists, it was the social scientists who sustained the legend of Turgot; it was not the scientists, it was the moral philosophers whom Napoleon had to reorganize out of the Institute; it was not the scientists, it was jurists and writers who resisted the perversion of patriotism into terror or later into Bonapartism; it was not the scientists, finally, who remained faithful to the Emperor who had associated them with himself from Egypt onwards, it was a few old Jacobins in the Hundred Days.

Not that scientists were indifferent to their own political advantage: they accepted the suppression of the Academy and went on from participation in war production to impress their necessities so successfully on the state that they emerged a professional rather than a privileged community housed in the type of educational, technical, and academic institutions in which scientific research and careers have developed in Europe ever since that time [3]. But unlike other intellectuals, and whatever their private views, they remained indifferent as a group to those distinctions which the progressive political tradition makes between liberty and tyranny.

There is no convincing evidence that science has in more recent times borne a different relationship to political dissent. Among scientists no general rallying occurred to Oppenheimer or to Vavilov anymore than it did to Lavoisier. Was there any significant emigration from Germany in the 1930's among scientists who were not Jewish? Recollection suggests that there was not, and it is merely a European myth that moral scruples among those who remained retarded or sabotaged development of atomic weapons: the chagrin among German physicists upon the news of Hiroshima was at having been ignorant that it could be done.

In all countries in the post-war years disaffection has been growing between politicians and intellectuals both with regard to the use of force in international affairs and with regard to the internal proceedings of governments. It seems to be artists, writers, and students who take the chances and the lead, however, not the scientists. Yevteshenko says that it is the poets whom the dictators have to fear, and so it seems to be. One of many recent European journalistic surveys of American politics observes that it is surprising to what a slight extent students in the technical schools take part in the agitation against the government's conduct of the Vietnam war. It may be that scientists share emotionally in this alienation of the intelligentsia from the men in government. If one were to judge merely from conversations with many mathematicians or physicists, one would assume they do. Yet it never occurs to them that they are in a peculiarly strategic position among liberals and intellectuals, in that they alone could do something directly to stay the hands of statesmen and to lighten the menace of destruction that weighs from many sides. They could do so in all countries simply by refusing the state their services. They have it in their power to do that, though no sooner is so naive an observation advanced than it meets with the recognition that for obvious reasons nothing of the sort will happen, and it is meant less as a serious suggestion than as a reminder that science can be seen in different guises politically. There is what most scientists themselves see when they look into what they take for the mirror of history: the reflection of peaceful benefactors of humanity bringing about the progress in welfare and enlightenment foretold by Bacon. In the recent past of the historiography of science, this was the vision which George Sarton had in mind when he imagined in science a cosmopolitanism amounting to the new humanism. There is, on the other hand, the cast of characters for Durenmatt's *The Physicists,* and it is not only in the light of nuclear weapons that persons who are not scientists have felt there to be a ruthlessness about research.

One need not go to that emotional extreme, however, in order to conclude that the function of science in relation to the state historically has not been to liberalize it. That function has been rather to enhance the powers of the state and in return to draw advantage from the state for science.

III

Evidently, therefore, it is in contexts other than direct intellectual or political influence that the historian must look for the leavening of the social process by science, and this necessity puts him at a disadvantage, for the areas are those that normally fall professionally to the sociologist. Indeed, of all the desiderata before us in the study of the scientific enterprise, it does seem that the development of an historically directed sociology is the most needed. Perhaps a historian may venture certain remarks, however, about impressions that he would like to have explored, corrected, or possibly dismissed as mere suspicions by systematic study. We have already had important help from sociology, after all, and it would be ungracious not to look for more.

We know something, for example, about the types of social background from which in certain historical instances scientists have been largely recruited. The studies of Robert Merton and others on the relation of the scientific movement to Puritanism in 17th-century England have not produced agreement about the theological issues, but they have certainly brought out and ought perhaps to have been addressed in the first place to a subsequent feature of English history, to wit that science developed almost entirely outside the Anglican Establishment [10, 13, 5]. Until the reforms of the later nineteenth century took effect, Oxford, Cambridge, and the public schools had little part in it, their function having

been the classical education of the governing class. Comparable consequences emerge from a study of the educational background of American
scientists conducted shortly after the last war. The institutions proportionally most productive of scientists were the small formerly denomina-
tional colleges of the Middle-West drawing upon lower middle-class
groups in the heartland of rural self-help. Proportionally the least productive were Roman Catholic institutions, southern institutions, and the
larger private universities of the East Coast, these three groups having in
common that they educated students from milieux oriented toward the
continuation of traditional (though very different) ways of life [8].

So far as I am aware, comparable studies do not yet exist for other
countries. Yet possibly we know enough of the lives and circumstances
of scientists elsewhere to venture a very weak generalization: it is that
scientists have seldom arisen in groups nowadays called Establishments. The number of aristocrats who have contributed anything but
patronage to science must be small indeed. But small also is the number
who derive from the high bourgeoisie in modern France, a class as satisfied with their lives as have been Prussian Junkers, English noblemen, or
southern planters. It appears, on the whole, that science has proved attractive to groups outside such privileged circles as offering careers that
are definite, consequential, and likely to yield results commensurate with
effort and intelligence, in a word careers open to brains and discipline.
Obviously, however, science is a possibility only in communities prosperous enough to afford education and disposed to value it. The number of
scientists who have come from real deprivation must be smaller than that
from the aristocracy.

These specifications are so obvious and broad that they have often
been taken to characterize the provenance of capitalists, entrepreneurs,
and other dynamic elements in modern history. May there not, however,
have been other factors beside or within these that might if explored yield
information on the special quality of scientific careers? The factors I have
in mind are those of personality and individual culture.

Maybe it is irresponsible even to broach the subject, for if the sociology
of science is underdeveloped, its psychology is scarcely born, and what
follows derives from an impression for which the excuse must be that it
is not in my mind alone.[2] That impression is that there exists a widespread
instinct outside science (whether justly or not is beside the point) that scientists on the whole are likely to be of a rebarbative disposition, especially
those of a theoretical and mathematical bent.

Evidence may be adduced from a domain that does not normally attract the interest of sociologists or historians of science. Observers ac-

quainted with American university life will be aware that students on most campuses distribute themselves among voluntary social organizations usually called fraternities. I believe it to be generally true—though I may be mistaken and a survey could easily be made—that those regarded as most desirable on a social scale of values are least likely to number science students among their members, and further that among science students mathematicians are least likely to belong to any such organization and most likely to lead what their contemporaries and the opinion of society at large regard as anti-social lives.[3] If this is true, or nearly true, why is it so? For it is not merely family privilege, social position, or attendance at private school which are reflected by membership in the more fashionable fraternities or clubs, although those factors have an important place. Beyond mere snobbery, however, are the qualities of congeniality, athletic prowess, and personal address, of poise, charm, and easiness which make the company of some people more sought after than that of others, and the absence of which is likely to be felt by young people as a disadvantage in themselves and in one another. Since the Renaissance, indeed, these are the qualities which, whether created by natural good fortune or social advantage, it has ever been the ideal of humanist education to elicit. For if there is any connection between characteristics of personality and educational or professional choices and preferences, it is unlikely to be peculiar to the American environment.

A recent study of the educational background of professional humanists in the United States, undertaken to complement that on scientists, reveals an interesting contrast. In significant degree humanists—historians, philosophers, classicists, students of art and literature—have had their education in precisely those Ivy League universities which, despite their high academic standing and the activity of their faculties in scientific research, have been among the less effective institutions in attracting undergraduates into science [7]. Their main function socially has been the education of young men who have gone on to form a leading element in finance, law, medicine, and the administration of business and government. Perhaps, therefore, it is not unreasonable to say that the liberal arts as subjects of study appeal to those elements in society disposed to the conservation of tradition and the operation of established patterns and institutions. A much more explicit case could be made of the relation between classical studies and the education of the governing class in Britain and France over a period of several centuries. In some ways, the American universities may be more interesting, however (though only for this argument—I mean no chauvinism). While educating the sons of wealthy families, they have also been more open to scholarship students from many levels and sections of the country and have been a means of conducting young men from very different backgrounds into leading positions.

Professor Merton's further work on priorities in scientific discovery bears on motivations in the scientific community, his analysis turning on the immense importance which attaches to innovation and discovery [11, 12]. His findings have two aspects. On the one hand, far from being professionally self-effacing and modest, as the avowed ethic of disinterested knowledge would have it, scientists are individually ambitious and constitute a community motivated by competition for honor and recognition. On the second aspect, Professor Merton insists less strongly, but it may be of even greater significance for the relation of scientific knowledge to the surrounding culture. Until the advent of science in the seventeenth century, organized learning looked to the past for the knowledge that mattered. The humanities still do so. Not so the sciences, wherein the knowledge actively sought is that which, in the searcher's eye, does not yet exist. If it did, his search would be spoiled. Erudition is not held in much honor among scientists. Even a wide and catholic mastery of many branches of an existing science confers little distinction compared to the creation of some novel finding. There is no equivalent to the esteem in which humanists hold scholarship and literacy. Is not this tacit disagreement about the very locus of culture, whether it is there in the past or to be attained in the future, what is responsible for the disjunctions noticed by C. P. Snow in his essay *The Two Cultures and the Scientific Revolution?*—a piece of writing which in my opinion has been treated too cavalierly by both scientific and humanistic opinion [14].

What I wish to suggest, then, is that there may be a functional relationship between the orientation of liberal studies toward the past and their appeal to persons whose disposition is to continue the past into the future, and reciprocally between the orientation of scientific studies toward the future and the instinct of scientists that new knowledge is what counts, thereby to render the future different from the past. The question is one that mingles historical or social with personal past and future. It is reasonable to suppose that persons who are at ease in a culture, whether because of social and economic privilege or because of personal temperament and congeniality or through some combination of social and individual factors, should be drawn to the sympathetic study of the history and elements of the civilization in which they feel at home. It is also reasonable to suppose that persons not so congruently situated in the life around them, whatever may be the blend of personal and social factors constituting their sense of disadvantage, should prefer (intelligence permitting or even compelling) to lose themselves or the world or prove themselves to the world in impersonal problems, should take little comfort in what is known and find more scope in what is not, and should wish, sometimes even aggressively, to contribute to making the future different from the past, their own and that of the world.

(Whether or not this statement is anywhere near the mark, I see no inconsistency between it and the proposition with which scientists usually resist analysis of their motivations, i.e., that they genuinely like their work and enjoy solving scientific problems.)

Clearly these preferences do not operate along a spectrum of political positions from left to right. They are matters of cultural disposition, and appear by taking a different sort of cut into the relations of personality with society and culture. It seems to me, indeed, that factors relevant to determining the composition of the scientific community itself also explain our awareness that, neuter though that community may have been in politics, nevertheless the increasing orientation of all our modern sensibility to progress and the future pertains to science, though in what measure as cause and what as effect I should not like to try to say.

IV

Such an analysis, if it were ever to be established as something more useful than an idea, would need to be capable of bearing on rather small elements and combinations of personal fortune and temperament in each individual and on large numbers of individuals in whole populations of scientists. I put it that way partly in virtue of a remark with which Thomas S. Kuhn concludes his recent *Structure of Scientific Revolutions* [9], to the effect that it would be well to set the evolution of science into the context of a Darwinian historiography and to see its course as a circumstantial development from primitive beginnings rather than, in the fashion of scientific pedagogy, as a linear or exponential progression towards the goal of the science of today.

The notion is capable of development. Stephen Toulmin has advocated it in a recent article [15], and a few years ago at a meeting of the American Historical Association,[4] I ventured to suggest that one might borrow the notion of selection in a qualified way as useful in relating the filiation of scientific concepts in historic succession to the circumstances of the surrounding social environment, that is to say the internal to the external history of science. The idea of selection seems applicable when it is wise to substitute contingency for purpose as the mode in which a highly articulated and internally self-sufficient chain of events, such as the replication of species or the filiation of scientific theories, may have been related to aspects of a context from which it cannot be isolated but wherein the central chain of development bears no direct or visible relevance to events in the world at large except over long periods. One might adapt the term social selection to a Darwinian or circumstantial historiography of science. One might consider the succession of scientific theories, the sequence which is currently written in the internal history of science, to be analogous to heredity, in consequence of which mechanics in the seventeenth and twentieth centuries form visibly the same subject. One might, on the other hand, consider circumstance to be analogous to the environment. Just so, for example, did industrial growth favor the study of machines and social development the professionalization of engineering mechanics. Just so does the operation of power politics in our own day select talent and resources into study of high energy physics at the expense of problems that arise in other sectors.

Returning to the subject of this paper, the relation of science to progress, one might approach it in detail by considering how the intellectual opportunities and internal institutional dynamics of science have operated to select into scientific careers those individuals suited by mind and temperament to flourish there. Approaching the relations of science to the social environment in that fashion might reveal a number of beliefs about science to be as unfounded as the notion that it has been a liberal force politically. Among them is one to which I should like to address a concluding paragraph, for I believe it to be a myth and think the reality more interesting. I mean the ivory tower or garden of Eden vision in which it is supposed that science flourishes most naturally amid peace and quiet and settled times when the scientist, not himself much interested in politics and perhaps not very adept personally, is secure from the distractions of a world in turmoil.

In historical fact science has never been a delicate plant. Compare the relative productivity of England and France in the seventeenth century, the one wracked by civil war and political strife, the other stabilized at the zenith of its monarchy. Even more strikingly compare the vigor of English science in Newton's time to its pallor in the eighteenth century in the safe society of the Whig Supremacy. Consider the relative scientific impoverishment of bourgeois France in the nineteenth century compared to what had gone before and the sparseness of science in an America remote from danger until the twentieth century. The violent events of revolutionary and Napoleonic France, on the other hand, establishing forms in which the very disciplines of professional physics and biology took modern shape, exhibit a positive relation between socio-political stress and scientific vitality. So also does the intermingling of national frustration and scientific eminence in 19th-century Germany. But undoubtedly the most impressive spectacle is that in which we still participate, the geometric rate of scientific growth under the stimulus of the cold war and the space race—in the United States, in the Soviet Union, and now in many countries. There is every reason to think it is the same in China. The evidence, in a word, exhibits a mutually stimulating relation between political and scientific revolution in modern history—new times, new men, new powers. There is no contradiction here with the proposition that it is not the scientific community itself that is politically dynamic, but the culture in which it flourishes, and in order to observe what may be the orientation to tradition of such a culture, we have only after these twenty years of explosive science to look around us at a whole generation in revolt, scornful of tradition as never before in western history.

Notes

1. This conclusion is one to which I have come in the course of other writings. See particularly [1] chapter 8, [2], and [4], pp. 151–201, 342–351.

2. Since preparing this paper for publication, I have had my attention called to Liam Hudson's study [6] of the correlations of personality factors among English schoolboys with their choice of and success in science and liberal arts subjects at the University. This original, suggestive, and very amusing study now seems to me to bring a kind of oblique confirmation to certain of the remarks ventured in the next few paragraphs.

3. There is evidence in Knapp and Goodrich [8] of an inverse correlation between the presence of Greek letter fraternities on a given campus and the productivity of the institution in the undergraduate education of scientists, pp. 37–39.

4. At Washington, D.C., in December 1964.

Bibliography

1. C. C. Gillispie, *Genesis & Geology* (Cambridge, Mass.: Harvard University Press, 1951).

2. ———, "The Encyclopédie and the Jacobin Philosophy of Science," in Marshall Clagett, ed., *Critical Problems in the History of Science* (Madison: University of Wisconsin Press, 1959).

3. ———, "Science and Technology," *The New Cambridge Modern History, 9* (1793–1830), 118–145 (Cambridge: At the University Press, 1965).

4. ———, *The Edge of Objectivity* (3rd Printing, Princeton: Princeton University Press, 1966).

5. A. Rupert Hall, "Merton Revisited," *History of Science, 2* (1963), 1–16.

6. L. Hudson, *Contrary Imaginations* (London: Pelican, 1966).

7. R. H. Knapp, *The Origins of American Humanistic Scholars* (Englewood, N.J.: Prentice-Hall, 1964).

8. R. H. Knapp and H. B. Goodrich, *Origins of American Scientists* (Chicago: University of Chicago Press, 1952).

9. T. S. Kuhn, *The Structure of Scientific Revolutions* (Chicago: University of Chicago Press, 1962).

10. Robert K. Merton, "Science, Technology, and Society in Seventeenth-century England," *Osiris, 4* (1938), 360–632.

11. ———, "Priorities in Scientific Discovery," *American Sociological Review, 22* (1957), 635–659.

12. ———, "Singletons and Multiples in Scientific Discovery," *Proceedings of the American Philosophical Society, 105* (1961), 470–486.

13. T. K. Rabb, "Puritanism and the Rise of Experimental Science," *Cahiers d'histoire mondiale* (1962).

14. C. P. Snow, *The Two Cultures and the Scientific Revolution* (Cambridge: Cambridge University Press, 1959).

15. Stephen Toulmin, "The Evolutionary Development of Natural Science," *American Scientist, 55,* (December 1967), No. 4, 456–471.

—⚋—

Postface

REREADING THE ABOVE in 2005, thirty-seven years after it was written, I have the sense that the first two sections are sound. The discussion of the demographics of American science in Section III, however, was based on the Knapp-Goodrich study of the pre-World War II generation cited in note 6. I am not confident that the findings apply as well, or at all, to the population of American scientists in the succeeding generations. It is true that the contribution of Princeton University and others like it to the education of mathematicians and scientists still occurs primarily and mainly at the graduate level, and that the vast majority of undergraduates major in the humanities and the social sciences. A small but appreciable number are drawn to mathematics and physics, however, and a somewhat larger proportion to molecular biology and genomics.

As for other countries, the generalizations were impressionistic rather than statistically based. I think they were right for Britain and for 18th-century France. They may not be for 19th-century France, however. An authoritative and still to be published study of the public role of science by Robert Fox notes a significant bourgeois element in the population of scientists. I simply do not know enough about Germany or Russia to venture an opinion.

As for the paragraphs on psychological factors, my intuition has not changed. Nor would I modify the concluding paragraphs of Section III and the applicability of the concept of social selection outlined in the conclusion.

Social Science and Probabilistic Analysis in Physics

LIKE "THE ENCYCLOPÉDIE and the Jacobin Philosophy of Science" (No. 6 above), this paper was composed for delivery at a conference, in this case a week-long symposium held at the University of Oxford in July 1961.[1] Alistair Crombie organized the affair under the auspices of the International Union for the History and Philosophy of Science. The proceedings were similar to those at the Madison meeting in 1957. Discussion from the floor followed prepared remarks of one or several critics assigned to comment on each paper.

Excellent people participated. The occasion did not draw them together in the same way, however. It scarcely could. The members of the symposium numbered one hundred and sixty-one. Represented were universities, academies, and museums from most of the countries of Europe. Apart from those who had participated in the Madison meeting, only a handful from the United Kingdom and France aspired to be historians of science. The majority were established scientists, philosophers, historians, museum directors, and officers of institutions of various sorts. Many such attended only one or two sessions. Perhaps too the accommodations in an unheated late-medieval Oxford college during a damp, chilly English July with few bright intervals were less conducive to congeniality than the open, warm, and friendly campus of the University of Wisconsin in gorgeous late-summer weather. Still, the papers for the most part were very interesting and the volume that collects them holds up well.

After completing *The Edge of Objectivity* (1960), I conceived the notion of studying the history of probability, and applied successfully to the National Science Foundation for a grant for that purpose. The paper that follows was the firstfruits. Having become involved in 1963 in organizing and then editing the *Dictionary of Scientific Biography*, I never did complete a book on the subject. Nevertheless, what I learned in that year, in which I audited the late William Feller's undergraduate course on probability, informed much of my later work. I could not otherwise have written *Pierre-Simon Laplace, a Life in Exact Science* (1997) or the parts dealing with Laplace, Condorcet, et al. in the two volumes on *Science and Polity in France.*

My paper was not well received in Oxford. Critics were appreciative enough of the middle passages on Laplace, Poisson, Quetelet, and Cournot. The central point, however, was the conjecture that John Herschel's account in the *Edinburgh Review* for June 1850 of Adolphe Quetelet's *Theory of Probability as Applied to the Moral and Social Sciences* may well have been what suggested to Maxwell the law of distribution of velocities which inaugurated the use of statistical methods in the dynamical theory of gases. My principal critic, Mary Hesse, dismissed this suggestion out of hand. She did graciously acknowledge in a footnote what I had not known. In the interval between the Oxford meeting and publication of its proceedings, it had come to light that Maxwell in a letter to his biographer, Lewis Campbell, expressly writes of having read Herschel's essay.[2] Hesse still considered that to be of no importance, however. It was then thought unlikely, not to say inconceivable, that a technique could be transferred from social to physical science. All the discussantes agreed with her.

I confess it to be a gratification that historians of physics since then are in accord that the Herschel review did contribute, and in an important manner, to Maxwell's development of the kinetic theory of gases.[3]

Notes

1. *Scientific Change: Historical Studies in the Intellectual, Social, and Technical Conditions for Scientific Discovery and Technical Invention from Antiquity to the Present*, A. C. Crombie, ed. (London, 1962: Heinemann), pp. 431–453. In 1962 I delivered this paper with minor modifications in Paris as a lecture at the Palais de la Découverte, where it was printed as a pamphlet with the title *Les fondementes intellectuels de l'introduction des probabilités en physique* (1962).

2. L. Campbell and W. Garnett, *The Life of James Clerk Maxwell* (London, 1882), pp. 142–143.

3. See C. W. F. Everitt, "James Clerk Maxwell," *Dictionary of Scientific Biography* 9 (1974), p. 219, and the citations in his n. 67, p. 228.

TWENTY

Social Science and
Probabilistic Analysis in Physics*

—⚭—

On 21 September 1859, Maxwell appeared before the British Association for the Advancement of Science meeting in Aberdeen and read a paper, "Illustrations of the Dynamical Theory of Gases."[1] He introduced it as an essay of the atomic hypothesis in the special case of the kinetic consideration of gases, the velocity of the particles rising with temperature. The argument demonstrates the laws of motion among an "indefinite number of small, hard, and perfectly elastic spheres acting on one another only during impact,"[2] and proceeds to compare the properties of such a system to the experimental gas laws. The yield is interesting. One set of equations, for example, entails the surprising consequence that friction is independent of density. Perhaps the most encouraging deduction confirmed Avogadro's law based on chemical information that the number of particles is the same in unit volume of different gases. But what physics prizes most highly is the analysis which Maxwell here employed.

In form, the essay consists in a series of propositions analytical of the laws of motion in the hypothetical system. Maxwell established his mode of reasoning in Proposition IV: "To find the average number of particles whose velocities lie between given limits, after a great number of collisions among a great number of equal particles." Maxwell makes it appear

---

* The original title was *Intellectual Factors in the Background of Analysis by Probabilities in Physics*. It appeared under that title in the published proceedings of the conference, *Scientific Change: Historical Studies in the Intellectual, Social, and Technical Conditions for Scientific Discovery and Technical Invention from Antiquity to the Present*, A. C. Crombie, ed., (London, 1962: Heinemann), pp. 431–453. In 1962 I delivered a French translation with minor modifications in Paris as a lecture at the Palais de la Découverte, where it was printed as a pamphlet with the title *Les fondementes intellectuels de l'introduction des probabilités en physique* (1962).

upon analysis "that the velocities are distributed among the particles according to the same law as the errors are distributed among the observations in the theory of the 'method of least squares'."[3]

In the second part of the paper, Maxwell distinguishes between the motion of translation of the system as a whole, and the "motion of agitation" wherein the "collisions are so frequent that the law of distribution of the molecular velocities, if disturbed in any way, will be re-established in an appreciably short time."[4] Thus did Maxwell inaugurate the science of statistical mechanics. The law of errors had itself been introduced as a rule of procedure by Legendre, and clarified geometrically by Gauss before Laplace put it on a rigorous footing in the *Essai analytique des probabilités,* the *Summa* of the early history of probabilities.[5] And the purpose of the present memoir is to survey the intellectual and philosophical elements which preceded and permitted Maxwell's novel application of a probabilistic analysis, not just as theretofore to games or affairs, but to a dynamical problem of matter in motion.

A peculiarity distinguishes the prior intellectual history of probability. The recourse to example was more immediate and frequent than in other modes of mathematical reasoning. Nevertheless, it appears to have been a branch of analysis undernourished by worthy materials. Nor was this for lack of concern in application among the protagonists. Indeed, the school of rational mechanics held high hope of the calculus of probability as an instrument of exact science adaptable (in the complexity if not the contingency of the human condition) to civil and moral matters and to their improvement by administration. Condorcet undertook his probabilistic essay of electoral procedures at Turgot's behest.[6] Laplace from the very beginning of his career divided his main efforts between celestial mechanics and probabilities, the one line concerned with the real world and the other with our procedures for knowing about it. "La courbe," says his famous summary in the *Essai philosophique des probabilités,* "décrite par une simple molécule d'air ou de vapeurs est réglée d'une manière aussi certaine que les orbites planétaires: il n'y a de différence entre elles que celle qu'y met notre ignorance. La probabilité est relative en partie à cette ignorance, en partie à nos connaissances."[7] The dichotomy guided his earliest investigations, and oriented—or perhaps it restricted—his thinking about civil applications toward repairs that might be worked in our ignorance of causes of events.

To "déterminer la probabilité des causes par les événements" was the object of Laplace's earliest memoir of a philosophic character. He describes it as a "matière neuve à bien des égards et qui mérite d'autant plus d'être cultivée que c'est principalement sous ce point de vue que la science des hasards peut être utile dans la vie civile."[8] Uncertainty in knowledge bears either on events or on their causes. If an urn is known to contain a certain number of black and white slips in a given ratio, and the probability is required of drawing a white one, then the cause is known and the event uncertain. But if the ratio is unknown, and after drawing a white slip one is to say the probability that it is as p to q, then the event is known and the cause unknown. All problems of the theory of chance might be assigned to one or other of these alternative classes. The second, that of inverse probabilities, was the more interesting to Laplace, and his memoir treated it by establishing and exemplifying a rule known under the name of Bayes, who had put it forward less cogently in 1763. It will be clearer to quote the terms in which Laplace couched it in the *Essai analytique:*

> Si un événement observé peut résulter de n causes différentes, leurs probabilités sont respectivement comme les probabilités de l'évènement tirées de leur existence; et la probabilité de chacune d'elles est une fraction dont le numérateur est la probabilité de l'évènement dans l'hypothèse de l'existence de la cause, et dont le dénominateur est la somme des probabilités semblables, relatives à toutes les causes.[9]

Other major preoccupations of the *Essai analytique* are apparent in germ in this and in accompanying memoirs—technical problems which led to generating functions; the choice of a correct value among observations (which led to the law of errors); the evaluation of "espérance morale" in games and decisions (out of which economists would later make the principle of marginal utility); the confidence that demographic information might warrant in particular instances. The *Essai* is a summary of all this, notable for the virtuosity of technique and the unity and steadiness of vision in which Laplace held the subject. Nor did he even then move beyond the causal to a statistical treatment of events, and that is the most important limitation to this thinking. His estimates bore upon the chances that we are or are not mistaken about the cause of phenomena and not upon configurations in the data themselves. He studied to know, not the mean about which the barometer varies, but the probability that the fluctuations have a constant cause. And in the case of minor inequalities in planetary motions, the magnitude of such a probability encouraged him to establish the cause.

Nevertheless, it was primarily Bayes's principle of inverse probability that permitted Laplace his hopes for mathematics as an instrument of social and political amelioration,[10] and the sectors of polity in which he thought to help were electoral procedures, decisions of representative bodies, credibility of witnesses, and the reliability of judicial tribunals. In every case the intent was to know causes from events in order to correct false ones. By "false" causes Laplace meant those that produce events unconformable to principles of morality and justice. "*Vérité, justice, humanité*," he told the *École normale* in the lecture on probability which concluded his course in 1795, "voilà les lois éternelles de l'ordre social qui doit reposer uniquement sur les vrais rapports de l'homme avec ses semblables et avec la nature; elles sont aussi nécessaires à son maintien que la gravitation universelle à l'existence de l'ordre physique...."[11] And though Laplace developed it last, the analysis of judicial decisions aroused greater interest than did other problems. Indeed, it became almost a test-case of mathematics applied to morality, and his procedures may be taken as a paradigm of the entire programme.

In the last article of the first edition of the *Essai analytique*, Laplace treated the judgment of a tribunal pronouncing between two contradictory opinions as he had the problem (already analysed) of the testimony of several witnesses about the extraction of one number from an urn containing only two. If p is the probability that each judge finds the truth, then given r judges the probability of a unanimous verdict will be

$$\frac{p^r}{p^r + (1-p)^r}.$$

One may compute p from the ratio of unanimous verdicts i to the total number of cases n. Then from the relation

$$p^r + (1-p)^r = \frac{i}{n}$$

it may be shown that in the case (say) of a panel of three judges,

$$p = \frac{1}{2} \pm \sqrt{\left(\frac{4i-n}{12n}\right)}.$$

Laplace chose the positive root on the ground that it is more natural to assign each judge a greater probability for truth than for error. Let us suppose, then, that we have to do with a court half of whose judgments are unanimous. In that case the probability of the veracity of each judge will be 0.789, and the probability that a verdict sustained on appeal is in fact just will be 0.981 (if the finding is unanimous) and 0.789 (if it is by split vote). "Il y a donc," concludes Laplace, perhaps to exemplify his dictum that probability is only common sense reduced to computation, "un grand avantage à former des tribunaux d'appel, composés d'un grand nombre de juges choisis parmi les personnes les plus éclairées."[12]

An added complication occurred to Laplace for inclusion in a supplement to the *Essai analytique*. Jurors (unlike urn-watchers in this respect) might differ on the fact in perfect good faith. With this consideration in mind, Laplace gave the problem an elaboration which he thereafter addressed to the general public in later editions of the *Essai philosophique*. Taking the condemnation as event, the probability is required that it was caused by the guilt of the accused as opposed to the error of the jurors. As always computation by Bayes's rule presupposed values for the probabilities in play. Laplace assumed that the probability of a truthful juror varies between 0.5 and 1. This may appear a somewhat dubious construction upon his choice of the positive root in the earlier analysis, and Laplace now justified it on the ground that in an ordered society one cannot well suppose jurors more prone to error than to truth. Even so his computations alarm. In a panel of eight members of whom five suffice for conviction, the probability of error came to $^{65}/_{256}$ or more than $^1/_4$. On the other hand, Laplace felt that English criminal procedure weighted the odds too heavily against the security of society. It was his considered opinion in the *Essai philosophique* that a majority of 9 out of 12 produced the nicest equilibrium between the protection of society on the one hand and of innocence on the other.

To Poisson as to his master, judicial probabilities appealed as a critical test of mathematics applied to the moral order. He feared lest it fail, however, through insecurity in the *a priori* values which Laplace had assigned to the chance of guilt in the accused and error in the juror. To repair this deficiency Poisson put a novel approach in hand.[13] In 1825 the French government had begun publishing annual *comptes généraux* of judicial proceedings. Thereby Poisson could look to experience for the numbers. He assigned a probability of guilt from the ratio of convictions to cases over the years 1825–33 inclusive, and computed the probability of a juror or judge finding in error by analysing the voting records of panel and of bench. Poisson took confidence from what he enounced (none too clearly) as the law of large numbers. In practice, this would appear to have been a reference of Bernoulli's theorems to the phenomenon of statistical regularities.[14] Taken over the whole of France, the rate of conviction was notably uniform: 0.61 for the first six years, varying only to 0.62 in 1826 and 0.60 in 1830. During those years a majority of seven to five sufficed to convict. The rule was changed to eight to four in 1831, and convictions declined to 0.54. In 1832 and 1833 instructions went out to judges to consider extenuating circumstances in fixing sentence, and jurors responded by increasing their severity to a rate of 0.59 in 1832 and staying steady there in 1833.

Statistical treatment, then, was the distinctive departure in Poisson's analysis:

> L'objet précis de la théorie est de calculer, pour des jurés composés d'un nombre déterminé de personnes, jugeant à une majorité aussi déterminée, et pour un très grand nombre d'affaires, la proportion des acquittements et des condamnations qui aura lieu très proba-blement, et la chance d'erreur d'un jugement pris au hasard parmi ceux qui ont été ou qui seront rendus par ces jurés. Déterminer la chance d'erreur ou d'acquittement prononcé dans un procès connu et isolé serait impossible selon moi, à moins de fonder le calcul sur des suppositions tout à fait précaires, qui conduiraient à des résul-tats très différents, et, à peu près, à ceux qui l'on voudrait, suivant ces hypothèses que l'on aurait adoptées.[15]

Poisson, indeed, was careful to distinguish between the mathematical and the moral sense of "guilty." It would be more exact, he observed, to substitute "condemnable" for "coupable," since the numbers in play are of those actually condemned, whether or not justly:

> Ainsi, lorsque nous trouverons que sur un très grand nombre de jugements, il y a une certaine proportion de condamnations erro-nées, il ne faudra pas entendre que cette proportion soit celle des condamnés innocents: ce sera la proportion des condamnés qui l'ont été à une trop faible probabilité, non pas pour établir qu'ils sont plutôt coupables qu'innocents, mais que leur condamnation fût nécessaire à la sûreté publique. Déterminer parmi ces condam-nés, le nombre de ceux qui réellement n'étaient pas coupables, ce n'est pas l'objet de nos calculs. . . . [16]

Fortunately, there was reason (Poisson felt) to believe the number small, because of the rarity of pardons and of verdicts which offended public opinion. Even in the mathematical sense, moreover, his statistical proba-bilities appeared far more auspicious for the prospects of justice in France than did Laplace's estimate of the chances in particular cases. Before 1831, for example, the probability that a juror would not be mistaken in his vote was a little better than $^2/_3$ in the case of crimes against persons and ap-proximately $^{13}/_{17}$ in crimes against property. Taking account of appeals, the probability of guilt was 0.98 in convictions sustained of crimes against persons and 0.998 against property when the court of first instance had voted eight to four. Conversely, the probabilities of innocence were 0.72 and 0.82 respectively when the first findings were reversed. Taking into consideration the initial probability of not being condemned, the probabilities that a guilty person would be acquitted were 0.18 and 0.07. This meant that there were about forty innocent parties among 8,000 found guilty of personal crimes at eight to four, and about eighty-eight out of 22,000 similarly condemned for violations of property. In civil cases, to cite one other result (for the years 1831–3), the probability of a correct judgment at first instance was 0.76, that of correct confirmation on appeal 0.948, that of a correct reversal on appeal 0.64, that of a confirmation of a first appeal by a second 0.75.

Nor did Poisson dodge the objection that the proceedings of revolutionary tribunals invalidated his calculations. There were always two roots to the basic equation, and Poisson followed Laplace in taking the positive root as the solution for the sane society in which judges have a greater tendency to justice than to error: "Mais il n'en est plus de même quand les jugements sont prononcés sous l'influence des passions; ce n'est plus la racine raisonnable des équations, c'est l'autre solution qu'il faut employer, et qui donne aux condamnations une si grande probabilité d'injustice."[17]

Poisson's "law of large numbers" sometimes seems to reach out towards a statistical mechanics, but he was not the man to find the way.[18] One paragraph, for example, does suggest that "La constitution des corps formés de molécules disjointes que séparent des espaces vides de matière pondérable, offre aussi une application, d'une nature particulière, de la loi des grands nombres." From any point in the interior of a body, one might draw lines measuring the distance to the nearest molecule. Though small in all directions, that distance might be ten, twenty, or even a hundred times greater in one direction than in another. The distribution of molecules would be very irregular around any point and constantly changing with their vibrations. Dividing an element of volume by the number of molecules therein and extracting the cube root of the quotient, one would have a "mean interval" of the molecules, independent of the irregularity of distribution, and abstracting from any compression produced by weight; this mean interval would obtain throughout a body at constant temperature. But Poisson gave no rule for the distribution of molecules around his mean, and he was in fact thinking not of kinetics but of the solid state, not of a way to compute the variations from a mean, but of his law of large numbers as the key:

De ces exemples de toutes natures, il résulte que la loi universelle des grands nombres est déjà pour nous un fait général et incontestable, résultant d'expériences qui ne se démentent jamais. Cette loi étant d'ailleurs la base de toutes les applications du calcul des probabilités, on conçoit maintenant leur indépendance de la nature des questions, et leur parfaite similitude, soit qu'il s'agisse de choses physiques ou de choses morales, pourvu que les données spéciales que le calcul exige, dans chaque problème, nous soient fournies par l'observation.[19]

Poisson presented the regulative remarks and the numerical results of his investigation of judicial processes in preliminary reports to the Academy of Sciences in 1835 and 1836,[20] and the interest they provoked initiated a philosophical discussion of the prospect for probabilities.[18] To review the main positions will suggest, perhaps, that formation of a statistical mechanics required operations of technique on data in a relation to nature that was by no means obvious. Poisson in his statistical analysis did transcend the Laplacean dichotomy according to which mechanics describes nature causally and probability repairs our ignorance of causes. Nevertheless, Poisson intended no retreat from the strict determinism of nature itself or from causality as the grail of science. And it is ironical that Comte, whose philosophy seems most in keeping with the expanding fortunes of the probabilistic analysis, should have rejected the technique while heaping scorn and contumely upon its advocates; whereas Cournot, who alone advanced the idea of an order of chance *per se* and adopted probabilism as an advance upon universal determinism, never laid hold on the statistical assemblage of information or its rationalization about a mean, the which techniques might have bodied his probabilistic philosophy of science into science itself.

From our perspective it appears as if Comte and his disciples might properly have seized on probability and carried it over from the account it gave of error to an account of fact. In their philosophy science knows only for prediction, not by the light of reality, and predicts only for control and not to say truth. And Comte in his phenomenalism and relativism might well have built whatever there was of uncertainty in the observations right into science itself. Instead, he repudiated probability. He wrote, indeed, with violence. He stigmatized it as the illusion "propre aux géomètres" who should seek to render social science positive by "une subordination chimérique à l'illusoire théorie mathématique des chances." The geometric aberration was even more vicious than that of biologists who would create a sociology as appendage to biology. Both would dispense with the indispensable in foregoing historical analysis. That fallacy had been excusable enough in the time of Jacques Bernoulli, less so in that of Condorcet, and impossible to forgive in Laplace's later years, "when the general state of human reason already began to permit a glimpse of the truly fundamental spirit of a healthy political philosophy." Poisson fared even worse, among "imitateurs subalternes" who are exploiting the just prestige of mathematics in a sort of "manie algébrique, maintenant trop familière au vulgaire des géomètres." The very notion of evaluating probability was radically irrational, except in games of chance where it started and still belongs. It would lead us to reject as unlikely events which, however, are sometimes going to occur. It would lead us to "donner notre propre ignorance réelle pour la mesure naturelle du degré de vraisemblance de nos diverses opinions." And no doubt Comte was himself too close to Laplace to see probabilities in a context other than that of Laplace's philosophy of nature, causal and deterministic.[21]

The temper of Cournot's philosophy is appealing by comparison, and many of the positions which he took have since been firmly occupied.[22] Certainly he intended to lay down a foundation for a probabilistic science, and he identified and clarified elements that later went into the statistical view of phenomena. But it does not appear that his thoughts were heeded or his writings influential. The abortive quality of his career makes a curious study, therefore, and the neglect that was accorded him a sad one. His view of the relations of science and metaphysics was as astringent as Comte's, and far more modest about the function of philosophy. Criticism and analysis of the ideas and procedures of science are the main business of philosophy, which indeed depends on science for materials on which to work. But science does not graduate into philosophy and become historically subservient to a positive sociology and ultimately even obsolete. Science and philosophy retain their identities in a dialogue of matter and form. Cournot appears, moreover, to have been the closest of the French rational school to the empiricism of his British contemporaries, both in knowledge and in spirit. He was the translator of Herschel's *Treatise on Astronomy*, and the inherent compatibility of his interests with Herschel's will appear as significant in relation to his probabilism.

Cournot formed his views from mathematical experience. The problems that attracted him had a common characteristic—they occurred in sectors which invited statistical or probabilistic analysis or both in the combination which would break down in mathematical practice the barrier of principle between physical and civil phenomena. Two of his earliest articles consider the probability that the sequence and inclination of the orbits of comets are determined causally. He treated judicial statistics from a point of view similar to Poisson's, drawing on fuller records in the years after 1835. His first general treatise, *Recherches sur les principes mathématiques de la théorie des richesses*, was a true trailblazer. It is not too much to describe it as the first full treatise of mathematical economics and the link between the definitions of moral expectation in the theory of probabilities and the mathematization of economic science at the hands of Jevons and Marshall. Here again Cournot figured as an intermediary between French rationalism and British empiricism. His moderation emerges to characteristic advantage from his remarks on economics. Most economists, he observed, set their faces against mathematical expression because they took an overly simple view of its value. They expected numerical results, and if they could not have them, regarded formalization as pedantic. But the object of analysis is not mere computation. It is to find relations between magnitudes which are not numerable and between functions that go beyond algebraic expression. Thus, the theory of probabilities will yield important propositions, even though for lack of information it often happens that numbers can be assigned to contingent events only in cases that are merely curious, and notably in that of games. Rational mechanics is no different in principle. It demonstrates theorems. Nor does their interest depend entirely on the numbers, which have usually to come from experiment. So with economics:

> I am far from having thought of writing in support of any system, and from joining the banners of any party; I believe that there is an immense step in passing from theory to governmental applications; I believe that theory loses none of its value in thus remaining preserved from contact with impassioned polemics; and I believe, if this essay is of any practical value, it will be chiefly in making clear how far we are from being able to solve, with full knowledge of the case, a multitude of questions which are boldly decided every day.[23]

In the expansion of his interests from the mathematical, Cournot moved from comets and judicial statistics through quantification in political economy to the composition of *Exposition de la théorie des chances et des probabilités*. That work advances the ideas which Cournot elaborated into a full philosophy. Its propositions are bound to arrest the attention of anyone interested in the intellectual history of probability. Probabilism was to Cournot what positivism was to Comte, the keystone of a philosophy, notable if not notorious. He, too, took issue with Laplace, not, however, by dismissing probability with dogmatic petulance, but by promoting it and assigning chance an objective standing in the world of phenomena. For probability is no mere "calculus of illusions":

> Pendant longtemps on n'a guère appliqué le calcul des chances qu'à des problèmes sur les jeux, problèmes purement spéculatifs ou d'un futile intérêt pratique, et à des faits de statistique sociale dont les causes se dérobent par leur complication à toutes investigations mathématiques, et pour lesquels nous n'avons d'autres données que celles de l'expérience. On s'est peu occupé de l'adapter à des questions de philosophie naturelle, questions pour ainsi dire de nature mixte, où l'on aurait pu espérer de confronter les données de l'observation avec des relations fournies par la théorie.[24]

That chance and order both subsist in objective reality was Cournot's point of departure. Nor did this proposition imply any retreat from causality, but rather a quite novel appreciation of its plurality. Every effect has a cause and may in its turn become the cause of a subsequent event. Such a chain of successive causes and effects forms a linear series of events. There may be any number, an infinite number, of such chains of events running through the world quite independently one of another. One causal chain leads the tile to fall from the roof. Another leads a man to walk by at just that instant. And the chains cross in this, a fortuitous but not an uncaused blow on the scalp. "Events brought about by the combination or conjunction of other events which belong to independent series are called *fortuitous* events, or the results of *chance*."[25] And these intersections are what open the domain of physical phenomena to the calculus of probability.

Mathematical probability is connected to phenomena that happen thus by chance through the notion of physical impossibility. It is physically impossible, for example, that a cone should stand stably on its apex, that a balance should be perfectly precise, or that a sphere should be struck so shrewdly on a line through its centre that no rotation would result. The chances against such an event are logically identical with the improbability of a man in blindfold drawing the one white ball from an urn containing besides an infinity of black ones.

> Consequently it may be said more briefly, in the accepted language of mathematicians, that a physically impossible event is one whose mathematical probability is infinitely small. . . . Thus mathematical

probability becomes the limit of physical possibility, and the two ex-pressions may be used interchangeably.[26]

Cournot developed the comparison into a distinction between what he called the subjective and objective senses of probability. Laplace had confined probability to the former sense, and thereby told—or fore-told—only half the story. Given the hypothetical being invoked by La-place, possessing a sensorium and a mind capable of reporting and know-ing all the causes and effects—given those capacities, mathematical probability would disappear for lack of an object. This would not follow from Cournot's conception of nature, wherein even a perfect intelligence would still require its services in computing intersections of indepen-dent causal chains. For such a being mathematical probability would fill the office of experiment among ourselves. Far from being a calculus of il-lusions, therefore, mathematical probability seemed to Cournot precisely that application to phenomena which most widely justified the ancient saying, *Mundum regunt numeri.* And it is not the least interesting aspect of his philosophy that he should have associated it with the limitations of mechanics:

> We have no basis for believing that we can give an account of all phenomena simply by means of the ideas of extension, time, and motion, that is, in a word, simply by the continuous magnitudes on which the measurements and calculations of mathematics rest. The acts of living, intelligent, and moral beings are by no means ex-plained, in the present state of our knowledge, and there is good rea-son to believe that they never will be explained by mechanics and mathematics. Consequently, these acts do not take their place in the field of numbers on the same basis as geometry and mechanics do. Yet they find their place in this field in so far as the notions of combination and of chance, of cause and of fortune, are superior to geometry and mechanics in the order of abstractions, and apply to phenomena in the domain of living things as to those that pro-duce the forces which activate inorganic matter; to the reflective acts of free beings as to the inescapable determinations of appetite and of instinct.[27]

Is it only in retrospect that this is suggestive? Only in the light of sub-sequent statistical methods which in fact owe little to Cournot? It may be so. No filiation can be established, at any rate, between the first practice of statistical mechanics, and this, the philosophy which systematically explored the role of chance in natural phenomena. In his classification of the sciences, Cournot himself—his views are nothing if not unexpected—regarded political economy as quantifiable and therefore ranking after physics and before biology. Biology would ever remain inaccessible to mathematics, divided in principle from physics and politics and all exact science. And though it is surprising for a mathematician thus to have begun his philosophy in political economy and ended it in vitalism, it seems clear that its inconclusiveness turned on a simple failure. He never hit upon the analytical devices and numerical regularities which might have converted his hypothetical play of chance into a computable order of chance. He never reasoned on the mean, or exploited the law of error, and only occasionally referred in passing to the work of Quetelet, who did just that, and who pieced together the techniques.

It conditioned the pre-history of probability in physics that its protagonists should have had their footing in astronomy and their eye upon society. Founder of the Observatory in Brussels, Quetelet went to Paris in 1823 to study under Laplace, Fourier, and Poisson, became more interested in Laplace's course on probabilities than in those on astronomy, and returned to make his observatory a centre rather of statistical than celestial observation. His career might be taken as a medium for studying the migration of techniques. Popularization fell among his responsibilities, and in 1828 he published *Instructions sur le calcul des probabilités*. The chief interest consists in two chapters explaining Fourier's method of approximating a mean in statistical information by the rule of least squares.[28] He had already in his earliest mathematical paper printed a geometrical study of what he did not then recognize as the binomial curve. From the outset, therefore, he disposed of the two essential elements of which the combination would yield his analytical technique. But he did not as yet see their application to mobilizing the statistical information which all the while he made it his vocation to assemble.

Quetelet's summary of the special statistical studies of those earlier years remains the most famous of his writings—*Sur l'homme et le développement de ses facultés*.[29] The sub-title is *Essai de physique sociale*, and the work wears two aspects, a tabular and a programmatic. Book I contains vital statistics: birth rates, taken with reference to population and to marriages; variations thereof with climate, class, economic condition, and habitat; ratios of male to female in town and country, by legitimate and illegitimate status, and according to the absolute and relative ages of parents; death rates by climate and locale, by profession, by age-group, and by sex; scattered information on the incidence of still-birth and certain diseases; estimates of densities of population and rates of growth. Quetelet took his figures where he could find them. He looked to Malthus, Sadler, and Porter, to the publications of scientific societies, to a British peerage, to the French judicial statistics on which Poisson had drawn, to parish registers and urban bureaus, to his own Belgian census returns and to those of Prussia. In Book II Quetelet turned to anthropometry. He gave figures on height and weight and their proportions according to age, sex, and circumstance, and introduced manpower measurements of the force exercised by arm and thigh. Book III deals with the effects of the moral and intellectual faculties. Quetelet chose the drama as a cultural phenomenon to be handled statistically, and made estimates of the comparative productivity of the French theatre. It figures among the findings that tragedy is a secretion of youthful dramatists and comedy of aged ones. What most struck the moralists of the age were the figures bespeaking the regularity of incidence of crimes in any modes. In this connection Quetelet observed that the difference between speaking of a propensity to courage and a propensity to theft is that the law defines and the state identifies instances of the latter. We have the figures.

All this constituted the prologue to a social physics, his hopes for which Quetelet placed in the identification and specification of man thus measured, man the mean. The flyleaf to Volume II bears an epigraph from Laplace on probabilities: "Appliquons aux sciences politiques et morales la méthode fondée sur l'observation et sur le calcul, méthode qui nous a bien servi dans les sciences naturelles." Statistics contain the evidence for regularities that exhibit the service to law of moral phenomena. There are two conditions for a social science, two conditions and an instrument that brings it within reach of quantification. The first condition is that we should study "les qualités de l'homme qui ne sont appréciables que par leurs effets,"[30] and the second that we should abandon as inaccessible to science consideration of man as an individual and take him in the mass. And what gives authority for this procedure is precisely the instrument by which we reduce the laws to numbers. "Le calcul des probabilités montre que, toutes choses égales, on se rapproche d'autant plus de la vérité ou des lois que l'on veut saisir, que les observations embrassent un plus grand nombre d'individus." But the inaccessibility of the individual derogates not a whit from the rigour of the laws. Physics offers the analogy of statical analysis, which employs the centre of gravity rather than considering all the points distributed through the figure of a body. In just such wise will the *homme moyen* figure in the laws of social physics:

> L'homme que je considère ici est, dans la société, l'analogue du centre de gravité dans les corps; il est la moyenne autour de la-

quelle oscillent les éléments sociaux; ce sera, si l'on veut, un être fictif pour qui toutes les choses se passeront conformément aux résultats moyens obtenus pour la société. Si l'on cherche à établir, en quelque sorte, les bases d'une *physique sociale,* c'est lui qu'on doit considérer.[31]

Nevertheless, social physics was at best a qualitative science in *L'homme et ses facultés,* for Quetelet had devised no useful way of assigning numerical values to the faculties or dimensions of his human mean, who as yet was nothing more determinate than the average man. For example, in his discussion of the weight of a population of 10,000 souls (or bodies), Quetelet multiplied the average weight at certain ages by the proportion of the population within those limits, and computed the mass of men and women between the years of thirty and forty, forty and fifty, etc. The population of Brussels he estimated at 4,572,810 kilograms, which was to say four and a half times the weight of a cube of water ten metres square. Indeed, the entire human race would not balance thirty-three cubes of water 100 metres on a side.

So much for *Sur l'homme et ses facultés.* The phrase "social physics" was little of a novelty, and the vein is humanitarian and uncritical, not to say naïve. Ten years later Quetelet published a further general work, *Lettres à S.A.R. le Duc régnant de Sax-Cobourg-et-Gotha sur la théorie des probabilités appliquée aux sciences morales et politiques.*[32] The title seems to promise even less than *Sur l'homme.* At this date one does not expect to find an important treatise masquerading under the conceit of letters to an enlightened princeling. The book contains neither new compilations nor mathematical novelties, and it must be admitted that not all the examples satisfy. Nevertheless, this volume is no mere exhortation. Rather, it is a synthesis, a real synthesis in which probabilities and statistics met in that relationship of language and subject which opened the way to a statistical account of nature. It seems to me, indeed, that the difference between this social physics and Maxwell's statistical mechanics is in the objects rather than the methods. The latter distributes velocities instead of deaths or births about a mean, but the distribution is similar and the mean is what counts. Consider Quetelet himself on his conception of statistics as a science:

Collecting statistics is generally very well understood. But it is not so with the definitions of this science: there is almost always a tendency greatly to confine its domain. I think that the definition which I propose, and which moreover varies little from that given by many modern scientific men, sufficiently circumscribes the attributes of statistics, for it not to be confounded with the historical sciences or the other political and moral sciences which is the nearest approach to it. *It only considers a state during a determined period: it only collects the elements connected with the life of this state, applies itself to make them comparable, and compares them in the manner which is most advantageous with a view of showing all the facts they can reveal.*[33]

The argument turns upon the distinction which Quetelet now introduces between an average and a mean, or as he says an arithmetical mean (that having been his criterion of *l'homme moyen*) and a true mean. Since writing *Sur l'homme* he has found out how to represent the distribution of deviations from the "true"—i.e. most probable—value. The technique resolved the situation under study into two sorts of occurrence, complementary and equal in probability, which might come about in groups distinguished by the combinations—for example, the drawing of black and white balls two at a time, or five at a time, or 999 at a time. Quetelet chose the deaths of men and women in a city where equal numbers die, and showed that the combinations of male and female deaths taken in groups of one, two, three, and up to thirteen may be represented by Pascal's triangle, which was to say by the binomial expansion.

This was rather a case of recognition than of discovery. A memoir of 1844 contains Quetelet's first construction of this "échelle de possibilité."[34] Like the least square rule, which he had expounded in 1828, a "courbe de possibilité" figures in Quetelet's earliest work, without his having then imagined its identity with the "binomial law" by which he would situate his mean amid the data.[35] *Sur l'homme* itself contains tables of meteorological observations and of human stature which might have been graphed to represent the distribution of positive and negative deviations from the mean. But Quetelet did not then see the opportunity: "Je ne déterminai d'abord pas la véritable nature de la courbe qui se rapportait à ces lois (relatives aux facultés de l'homme); mais je reconnus plus tard que c'était la fameuse formule de binome de Newton."[36]

What was novel, therefore, was not the mathematics, but the analytical idea that exploited it and permitted the combination of a distribution according to the binomial expansion with the least square rule for prediction of probable error. The latter is the basis of Quetelet's "modulus of precision," a measure of the contraction of the curve of possibility towards the mean. That diminution varies as the square root of the number of observations (or possibilities). And now perhaps we may identify the elements of Quetelet's synthesis. Hitherto (unless I am wrong in this) Pascal's triangle or the binomial expansion had governed combinations of events equal in chance and opposite in nature, whereas the least square rule had governed the distribution of mistakes and inaccuracies. The very phrase "law of errors" restricted application to departures from the truth or target. Thus, the crucial idea that permitted Quetelet his synthesis was that measurements themselves may be taken as events subject to equal chances of error in excess or in defect. The difference in sign disappears when the magnitude is squared to compute the probability. One must cite his favourite illustration. He took it from the thirteenth volume of the *Edinburgh Medical Journal*, where there appeared a compilation of the chest measurements of 5,738 Scottish soldiers of various regiments. The figure for the puniest was 33 and for the huskiest 48; and

> The mean of all these measurements gives a little more than 40 inches as the average circumference of the chest of a Scotch soldier: this is also the number which corresponds to the largest group of measurements; and, as theory points out, the other groups diminish in proportion as they recede from it. The probable variation is 1.312—a value of which we should not lose sight.
>
> I now ask if it would be exaggerating, to make an even wager that a person little practised in measuring the human body would make a mistake of an inch in measuring a chest of more than 40 inches in circumference? Well, admitting this probable error, 5,738 measurements made on one individual would certainly not group themselves with more regularity, as to the order of magnitude, than the 5,738 measurements made on the Scotch soldiers; and if the two series were given to us without their being particularly designated, we should be much embarrassed to state which series was taken from 5,738 different soldiers, and which was obtained from one individual with less skill and ruder means of measurement.[37]

In uniting the distribution of binomial combinations with the law of errors, Quetelet may all unwittingly have done more for positivism than its founders, for this step abolished in practice the distinction between phenomenon and error, between nature and science, or (in Cournot's terms) between subjective and objective probability. No difference of treatment remains between a magnitude and a measurement, between an error of nature and an error of science. Thus, he liberated the analysis from its relativity to Laplace's ignorance of causes, and in effect opened the way to a probabilistic description of nature itself. "A chemist," he writes,

> would have the greatest chance of arriving at results in conformity with truth, were he perfectly sure of his analyses. In drawing a glass of water from a pure spring, the atoms of oxygen and hydrogen would be found in an infinite number, and consequently in the ratio fixed by the Creator in the composition of water. This case corresponds with the extraction of an infinite number of balls from the urn.[38]

And in another passage, "The urn, then, which we interrogate is Nature."[39]

How far did Quetelet see down this, the vista he helped to open? Not far, perhaps—his gaze was fixed on social physics, not on physics, nor did he effect the transfer to mechanics of the bell-shaped curve. One who did see further was Herschel, who appreciated the prospect from the standpoint of British empiricism. Quetelet's *Letters on Probabilities* commanded little attention until Herschel took the English translation as occasion for a splendid essay in exposition of the entire subject of probabilistic analysis, an article which the *Edinburgh Review* published at the head of the journal in July, 1850.[40] I do not know whether Maxwell read and was instructed by this fine piece. He may well have been. Herschel republished it in a collection of his occasional essays on philosophy of nature in 1857, two years before Maxwell introduced statistical mechanics to the British Association. But I have nothing to go on beyond this *post hoc* relationship and Maxwell's well known familiarity with Herschel's work and admiration for his standing. And even if the connection were established, the transmission of ideas in 19th-century science occurred along a front of communications broader than person-to-person filiation of influences.

In any case, Herschel's essay brought before the public explicitly that appreciation of the role of probabilities on which statistical mechanics did in fact depend. His own background of interests and values suited him to play the mediator—his studies of the distribution of stars in space, his support for liberal rationalization in political and economic affairs, his natural theology which recognized the contingency of the physical creation. In this last he was at one with Cournot, who had translated his *Astronomy* and who shared his astronomical and economic interests. Nor did Herschel's interest in Quetelet begin with the *Letters on Probability*. They had corresponded for some years, and Herschel actually prompted

Quetelet to define the law of distribution from which he constructed his first curves. But with all this, what fundamentally informed Herschel's essay was his agreement with the philosophy of science of John Stuart Mill.

"Experience," he begins, "has been declared, with equal truth and poetry, to adopt occasionally the tone, and attain to something like the certainty of Prophecy."[41] Out of this confidence in the uniformity of past and future, Mill has constructed a philosophy of logic, a philosophy which amounts almost to a "discovery" in showing that all reasoning is from particulars to particulars, and that the business of inductive science is to determine what are the really relevant circumstances, what the uniformities which it then calls laws of nature. But what security have we for the truth of anything asserted about a fact that we have not observed? What measure is there of such security as may be found? For inductive philosophy contents itself "with *practical,* as distinct from *mathematical,* certainty in all physical inquiry, and in all the transactions of life."[42] Probability is only the expression of that security and that certainty. Its theory gives the measure of it. It is, therefore, "as a practical auxiliary of the inductive philosophy that we have chiefly to contemplate this theory."[43]

Into its procedures, "its delicate and refined system of mathematical reasoning," no metaphysics intrudes and no preoccupation with causation.[44] Cause means only the occasion for occurrence of a result with greater or lesser frequency. It may just as well consist in taking away some obstacle as in an efficient agent. Nor is the result variable in intensity or extent proportionally to the degree of the cause. A result is only an event that does or does not happen, of which the theory studies the frequency, whatever may be the physical or moral agencies at work within the system. Herschel knew the strain on ordinary habits of thinking about the civil and physical world imposed by so radical a reversion to common sense. To mitigate it was the purpose of his essay:

> There still remains behind, however, this inquiry—which we have
> known to occur as a difficulty to intellects of the first order—*Why*
> do events, on the long run, conform to the laws of probability?
> What is the *cause* of this phenomenon as a matter of fact? We reply
> (and the reply is no mere verbal subtlety), that events do not so
> conform themselves—the fact to the imagination—the real to the
> ideal—but that the laws of probability, as acknowledged by us, are
> framed in hypothetical accordance with events. To take the simplest
> case, that of a single contingency—the drawing of one of two balls,
> a black and a white. We suppose the chances equal, in theory; but,
> in practice, what is to assure us that they are so? The perfect simi-

larity of the balls? But they need not be similar in any one quality but such as may influence their coming to hand. And, on the other side, the most perfect similarity in all visible, tangible, or other physical qualities cognisable to our tests is not such a similarity as we contemplate in theory, if there remain inherent in them, but undiscernible by us, any such difference as shall tend to bring one more readily to hand than the other. The ultimate test, then, of their similarity in that sense is not their general resemblance, but their verification of the rule of coming equally often to hand in an immense number of trials: and the observed fact, that events *do* happen according to their calculated chances, only shows that *apparent* similarities are very often *real* ones.[45]

One would not, indeed, wish to argue that Maxwell *had* to read this essay. It is more impressive, perhaps, to take it as an epitome of the adaptability of probability in the medium of British liberalism and empiricism to important preoccupations of 19th-century science. Let me simply, by way of conclusion to what is meant most tentatively as an introduction, allude to two of the many further notes which Herschel strikes in his exposition, and which echo in familiar fashion.

The first may suggest the reflection that neither did Darwin really need to read Malthus. Herschel moves from the passage just quoted on the relation of probability to the world to a consideration of what we mean by tendencies observed in experience. Were we to ask why the strong win out and the weak get nothing, we would reply that this does not happen in every case, and that although we cannot go into the dynamics of every competition, nevertheless we see what happens in a number of instances and observe a "visible enough *tendency* to the defeat of the weaker party." Contrariwise, when we say that success is proof of ability, what we mean by ability is simply some undefined collection of qualities which has a tendency to success. We may not be able to identify all the qualities, or in very confused or obscure circumstances any of them, and have to fall back on our experience of the tendency itself:

And it may further happen that this tendency, which we are driven to substitute in our language for its efficient cause, may be so feeble—whether owing to the feebleness of the unknown cause, its counteraction by others, or the few and disadvantageous opportunities afforded for its efficacious action (general words, framed to convey the indistinctness of our view of the matter)—as not to become known to us by long and careful observation, and by noting a preponderance of results in one direction rather than another.

And thus we are led to perceive the true, and we may add, the only office of this theory in the research of causes. Properly speaking, it discloses, not causes, but tendencies, working through opportunities—which it is the business of an ulterior philosophy to connect with efficient or formal causes; and having disclosed them, it enables us to pronounce with decision, on the evidence of the numbers adduced, respecting the reliance to be placed on such indications.... [46]

And the second passage to be quoted evokes the approximation of social to physical science—or vice versa in the case of statistical mechanics—in a common method:

Whether statistics be an art or a science ... or a scientific art, we concern ourselves little. Define it as we may, it is the basis of social and political dynamics, and affords the only secure ground on which the truth or falsehood of the theories and hypothesis of that complicated science can be brought to the test. It is not unadvisedly that we use the term Dynamics as applied to the mechanism and movements of the social body; nor is it by any loose metaphor or strained analogy that much of the language of mechanical philosophy finds a parallel meaning in the discussion of such subjects. Both involve the consideration of momentary changes proportional to acting powers—of corresponding momentary displacements of the incidence of power—of impulse given and propagated onward—of resistance overcome—and of mutual reaction. Both involve the consideration of time as an essential element or independent variable; not simply delaying the final attainment of a state of equilibrium and repose—the final adjustment of interest and relations—but, in effect, rendering any such final state unattainable. ...

Number, weight, and measure are the foundations of all exact science; neither can any branch of human knowledge be held advanced beyond its infancy which does not, in some way or other, frame its theories or correct its practice by reference to these elements. What astronomical records or meteorological registers are to a rational explanation of the movements of the planets or of the atmosphere, statistical returns are to social and political philosophy. They assign, at determinate intervals, the numerical values of the variables which form the subject matter of its reasonings, or at least of such 'functions' of them as are accessible to direct observation; which it is the business of sound theory so to analyse or to combine as to educe from them those deeper-seated elements which enter into the expression of general laws. We are far enough at present from the actual attainment of such knowledge, but there are several encouraging circumstances which forbid us to despair of attaining it.[47]

Notes and References

1. James Clerk Maxwell, *Scientific Papers*, ed. W. D. Niven (2 vols. Cambridge, 1890) I, 377–409.

2. Ibid. I, 377.

3. Ibid. I, 380-2.

4. Ibid. I, 392.

5. P. S. Laplace, *Œuvres* (14 vols., Paris, 1878–1912) VII, 353.

6. S. D. Poisson, *Recherches sur la probabilité des jugements* (Paris, 1837) 3.

7. Laplace, *Œuvres*, VII, p. viii.

8. "Mémoire sur la probabilité des causes par les événements," *Mémoires de mathématique et de physique, presentés . . . par divers savans*, VI (1774) 612–56, p. 622.

9. Laplace, *Œuvres*, VII, 183.

10. *Mémoires de mathématique et de physique*, VI, 652-3.

11. Laplace, *Œuvres*, XIV, 173.

12. Ibid. VII, 469–70.

13. Poisson, *Recherches sur la probabilité des jugements* (Paris, 1837).

14. Ibid. p. 7

15. Ibid. pp. 17–18.

16. Ibid. p. 6.

17. Ibid. p. 26.

18. For example, Ibid. pp. 7–8.

19. Ibid, p. 12.

20. *Comptes-rendus . . . de l'Académie des Sciences*, I (1835) 473–94; II (1836) 377–80, 395–400.

21. Auguste Comte, *Cours de philosophie positive*, ed. E. Littré (6 vols., Paris 1869) II, 255 n. (27ᵉ léçon); IV, 366–368 (49ᵉ léçon).

22. F. Mentré, *Cournot et la renaissance du positivisme* (Paris, 1908); Jean de la Harpe, *De l'ordre et du hasard* (Neuchâtel, 1936), M. H. Moore, ed. and trans., *An Essay on the Foundations of our Knowledge by Cournot* (New York, 1956); E. Callot, *La philosophie biologique de Cournot* (Paris, 1960).

23. A. Cournot, *Researches into the Mathematical Principles of the Theory of Wealth* (tr. Nathaniel Bacon, New York, 1927) 5.

24. Cournot, *Exposition de la théorie des chances et des probabilités* (Paris, 1843) 261.

25. Cournot, *Essay on the Foundations of Knowledge* (Moore ed.) 41.

26. Ibid. p. 47.

27. Ibid. p. 50.

28. Joseph Lottin, *Quetelet, statisticien et sociologue* (Louvain, 1912) 118 citing Fourier, *Recherches statistiques sur la ville de Paris*, III (1823) pp. ix–xxxi.

29. 2 vols., 1836.

30. Ibid. II, 114.

31. Ibid. I, 21–2.

32. Brussels, 1846; citations are to the English translation (London, 1846).

33. Ibid. p. 182.

34. "Sur l'appréciation des documents statistiques," *Bulletin de la commission centrale de statistique*, II (1845) 205–86.

35. "Sur une nouvelle théorie des sections coniques considerées dans le solide" (14 October 1820), cited in Lottin, p. 14.

36. Quoted in Lottin, p. 155.

37. Quetelet, *Letters on Probability*, 92–3.

38. Ibid. p. 75.

39. Ibid. p. 20.

40. *Edinburgh Review*, XCII (1850) 1–57.

41. Ibid. p. 1.

42. Ibid. p. 3.

43. Ibid. p. 29.

44. Ibid. p. 2.

45. Ibid. pp. 30–1.

46. Ibid. pp. 31–2.

47. Ibid. pp. 40–1.

CONCLUSION

l'Envoi

—⁂—

B y way of conclusion, and not simply to this collection of essays, it may be appropriate to set forth what the relations of science to other aspects of culture seem to me to have been across the sweep of modern history. The paragraphs that follow are not meant to be an argument. The purpose is merely to pull together and express my own thoughts.

My treatment of science and French polity has been criticized on the grounds that it does not allow for the involvement of science in the political process.[1] It has never been my practice to take issue with critics, nor shall I do so now. Let me simply identify the differences along these lines between Keith Baker, the most eminent among those critics, and myself. Professor Baker has devoted an entire essay in his collection, *Inventing the French Revolution,* to taking me to task for writing as if science and politics had nothing to do with each other.[2] Endnotes cite page numbers in *Science and Polity,* but readers who fail to turn to the back of the book will not know that I have treated at considerable length the ways Professor Baker mentions in which scientists were deeply involved in providing technical advice to government on such matters as rationalization of commerce and development of industry, public works, improvement of agriculture, sanitation and public health, standardization of weights and measure, development and production of munitions of war, instituting an educational system, institutionalizing science itself, and many comparable activities. Little if any of all that had to do with basic science or current theory. In the "Natural History of Industry" (above II, 5) I refer to such engagements as the application to industrial or military problems, not of science, but of scientists.

Professor Baker also kindly arranged a panel discussion of my books during a meeting of the French Historical Studies Society held at Stanford in March 2005. I hope it may there have become evident that disagreements between us may reduce to matters of terminology rather than fact. In writing of science per se, I have in mind the developing body of knowledge concerning the structure, forces, and condition of nature. In writing of politics, I have in mind the process of gaining and exercising the power of government. In my view the activities mentioned in the preceding paragraph and others like them pertain, not so much to politics, as to polity or the public weal in general. Many, though not all, scientists do engage in such technical activity. Their distinctive mission is not that, however. It is searching for knowledge of nature and imparting such knowledge to colleagues, to students, and so far as possible to the public.

In the case of Kepler, for example, his finding that planets revolve about the sun in elliptical orbits had no effect on the governance of the Holy Roman Empire. Nor were his astronomical investigations impeded by application of his talents to astrological prediction and casting horoscopes for Rudolf II, a technical duty. In the case of Newton, proof of the inverse square law of gravitational attraction neither speeded nor delayed the deposition of James II. Nor did it affect his administration of the Mint, which required only rationality and attention to detail. In the case of Lavoisier, the oxygen theory of combustion had nothing to do with igniting the French Revolution. Nor did it figure either in his administration of the Gunpowder Commission or his execution.

Professor Baker's most explicit criticism, if I understand him, is not along such lines, which are those of science proper. It is rather that I fail to allow for the alleged role of science as an authoritative model of rationality serving to legitimate modernization of government. The cardinal examples he cites are the design of the metric system (which, to be sure, I treat at length only in the second of my two volumes), the reform ministry of Turgot, and Condorcet's discipleship in attempting to found social science on the model of natural science. Here I have to say that to what degree rationality is historically indebted to science or science to rationality is unclear to me. I am inclined to think the relationship one of reciprocity rather than derivation. Even if the derivation of political reform from science were to be allowed, however, and one may entertain the hypothesis, what was involved was scientific method, not to say scientistic ideology. Neither one involves actual scientific knowledge any more, and indeed rather less, than does technical activity.

It may be thought that I am taking too narrow a view of science in restricting the term to knowledge of nature and excluding mere technical activity and ideology. Perhaps so. That restriction is what I have in mind, however, in asserting that science and politics are distinct in their inwardness. This is not to say that there are no relations between them, only that the relations are indirect rather than intrinsic. Nor are relations between different undertakings in society and different aspects of existence in the world the less important for being indirect. The example set to reformers by scientific method and the exercise of technical expertise may thus be considered an indirect relation between science and public affairs. Enlarging the scope to embrace science in general, one thinks of Darwinian evolution in consequence of natural selection. Biological heredity and the physical environment have nothing intrinsically to do with each other. It is nevertheless environmental circumstance that selects accidental variations in heredity into the process of the evolution of organic species, including homo sapiens.

Science too evolves, not merely in satisfying scientific curiosity, though that in considerable part, but also in response to needs and opportunities arising in the societal environment. Such response does not form the content of scientific knowledge, but it may and often does determine whether and with what vigor a particular line of scientific investigation is prosecuted. Thus the fissionability of Uranium-235 had nothing to do either with the personalities of Niels Bohr and John Wheeler, whose calculations revealed it, or with the threat that Nazi Germany posed to the world. Absent the latter, however, it is unlikely that any government would have incurred the expense of exploiting the prospect either for energy or for weaponry and also unlikely that high energy and particle physics would have attained the dimensions they did in post-war years. More recently, molecular biology and genomics are fascinating in themselves. But the resources lavished on development of those specialties would scarcely have been forthcoming in so large a measure except for the hope of therapeutic applications and the expectation of genetic engineering. Those are examples of indirect connections between public affairs and the development, as distinct from the content, of scientific knowledge.

An infinity of others could be adduced. Beyond that, the common locution "scientific discovery" is misleading. Science like other aspects of culture is created, not discovered as if it lay there concealed under layers of ignorance. But with respect to the relation of the creation to its creator there is a difference between science and other aspects of culture. We say "creative writing" and "creative art." We do not say "creative science" in any symmetrical mode. The difference in usage bespeaks our sense that the work of art pertains to the writer or artist in a way that the work of science does not pertain to the scientist. Hamlet, the Mona Lisa, and the Mass in B-Minor would not exist if Shakespeare, Leonardo da Vinci, and Bach had never lived. It is different with science, even the greatest science. If Newton had fulfilled predictions at his premature birth and died in infancy, the planets would still move subject to the inverse square law of gravity. Although no one else would have written the *Principia,* it could be argued convincingly that others would then or soon have written down everything in it that really mattered to later physics. Much the same is true of nearly all the great contributions to modern science. For scientists find their problems but also problems find their scientists, and usually not only one of them. Evidence is the frequent occurrence of simultaneous so-called discovery of many an important phenomenon, often attended by disputes over who had it first. It is all very well to say that convergent social forces are the explanation of simultaneous discovery, but pace social constructionists, they do have to converge on something out there in nature.

What, then, pertains to the scientist and not largely to the state of knowledge? Style, no doubt, and also motivation. Style may be the man in science too, but it is far more difficult to know the man of science that way. We can recognize Lavoisier's clarity of mind in his prose and in his reasoning. We can if we know both chemistry and French. We can discern the play of some quizzical turn of mind in Maxwell's physical models, but unless we understand electro-magnetism, we have to illustrate it from his verse. We are ready to be purified by the austerity of Einstein's taste in theory, but unless we know quite a lot of physics, we have to experience it vicariously on the basis of a few utterances much repeated together with what other physicists can tell us.

No one would wish to deny the personal element that goes into the making of scientific knowledge. Michael Polanyi enlarged upon it in eloquent manner.[3] The problem for the humanist is to know what difference the personal element in the creation of a piece of science makes to the body of knowledge once it has left the hand of its creator, which is immediately. No trace remains in astronomy of Kepler's determination to fit the planetary orbits into a cosmic nest composed of the five regular polyhedrons. The planetary laws are independent of Kepler's geometric passion, and no one since has followed the route by which he eventually learned them. Nor have organic chemists needed to repeat Kekulé's vision of a snake biting its own tail in order to model the structure of the benzene ring. In short, the personalities and motivations of scientists—Kepler's and Kedulé's inspirations are extreme examples, obviously—are full of interest. It is a biographical and historical interest, however, and none the less important for that, but not a scientific interest.[4]

National as well as personal styles also obtain in the doing of science.

French science has characteristically tended to be Cartesian and mathematical, British science Baconian and experimental, German science philosophical and theoretical, and American science technological and industrial in scale. But those differences in taste and practice have had no effect on the long run acceptability of the findings. The role of cultural change more largely has certainly affected the outlook of scientists as of everyone. The reciprocity of rationalism and science in the Enlightenment has already been mentioned. The shift from an encyclopedic to a positivist mode in French science in the revolutionary and Napoleonic period was paralleled in the conduct of public life. The early twentieth century saw the breaking of classical molds in art, literature, and physics alike. It is, however, difficult to think that artists, writers, and physicists read or look at one another's work in deciding to set off on a new tack when the climate of opinion changes. Something deeper must be at work. To the best of my knowledge, which may be insufficient, the occurrence of such tectonic shifts in sensibility has been recognized by scholarship, but their dynamics has not been elucidated in convincing fashion.

Finally, as to the historical process in general, I agree with two professors almost a century apart in the University of Cambridge, one Scottish, the other English. In his inaugural lecture as Professor of Experimental Physics in 1871, James Clerk Maxwell asked rhetorically "Is the student of history and of man to omit from his consideration the history of those ideas which have produced so great a difference between one age and another?" In the Introduction to his *Origins of Modern Science* (1950) Herbert Butterfield asserted that the Scientific Revolution of the sixteenth and seventeenth centuries "outshines everything since the rise of Christianity and reduces the Renaissance and the Reformation to the rank of mere episodes, mere internal displacements, within the system of medieval Christendom."

Published early in my career, *The Edge of Objectivity* (1960), a history of scientific ideas, begins its account of the Scientific Revolution with the instance of Galileo and the Law of Falling Bodies before reaching back to Copernicus, Kepler, and Vesalius. Recent scholarship has traced roots of modern scientific practice to earlier times: of experimental method and inductive reasoning to Robert Grosseteste in the thirteenth century, of statics to rediscovery of Archimedes, of kinematics to 14th-century scholastics, of optics to the Arabs, of chemical procedures to alchemy, of technology to innovations of craftsmen and builders in the High Middle Ages, of the quest for power over nature to magic and hermeticism. As background in the Renaissance one may also cite the behavior patterns of a Brunelleschi, a Leonardo, a Michelangelo, a Christopher Columbus, a Vasco da

Gama, a Prince Henry of Portugal. Theirs were doings animated by the for-
ward-looking instinct that later formed a Galileo, namely that knowledge
finds its purpose in action and action its reason in knowledge, that if a
problem can be solved it should be solved, that if a constructive thing can
be done it should be done.

Among the persons just named are two voyagers and an impresario.
For clearly history of science should encompass the voyages of discovery,
which initiated a process of geopolitical dynamism that eventually acquired the *vis viva* to carry spacecraft to the moon. What relates the early
voyages and their sequels to science is that they were bearers of the civilization that created modern science and that came to dominate the world
until yesterday, for good and for ill. Further, what differentiates them
from the random travels of antiquity and of other civilizations, which
entailed only adventure, legend, or commerce, is that they were always involved with, though perhaps not motivated by, the problem of how the
world is made, in a word with knowledge. Still, only in the Scientific
Revolution did all the above activities and attitudes, schematized in complementary fashion by Bacon and Descartes, come together into a set of
investigations that formed the modern picture of the world we live in,
wherein we are neither the center nor (though this was for long implicit)
the raison d'être. Traditional minds draw small comfort from Pascal's reflection that, though man be but a reed, the weakest thing in nature, he is
a thinking reed.

My book incurred some criticism at the time of publication for the lack
of a clear definition of objectivity. Since then it has met with occasional
disfavor on the (not altogether consistent) grounds that, whatever objectivity means, no such attitude is attainable in this world of culturally conditioned existence and irreconcilable ideologies. There is some justice to
the former charge. If I were re-writing the book today, I would not insist
so patently on objectivity as the common feature of the successive theories that carried the day. I should be more inclined to use terms such as
externalization of nature to describe the central cultural tendency of
science, and alienation to evoke its consequence in sensibility. But though
I might be able to compose a subtler book, it would not be a different one
thematically. It still seems to me that science has exacted a price, and the
price is that anthropomorphic considerations of goal, purpose, suitability,
and wish be eliminated from its formulations and statements. I still think
that broadly speaking the intellectual history of science finds its sequence
or direction in that sometimes painful process, which (witness intelligent
design) may never be completed. For my taste, however, the price is worth
paying. A scientific world view has no place for an overweening, all pos-

sessive Dr. Faustus. The Enlightenment has fallen into disrepute in post-modernist writings, but I am at one with 18th-century writers and philosophers in holding that science has been and is a potent, if not om-nipotent, weapon in the never-ending battle against ignorance, superstition, dogma, and material deprivation.

Other civilizations, notably those of China, India, and Mezo-America, developed indigenous and interesting systems for dealing with natural phenomena. Nevertheless, beginning in the Scientific Revolution and un-til very recently, modern science, rooted as it was in antiquity, has been a creation of European civilization and of it alone among all those that have seen the world. Together with the technology that attends it, science is, moreover, the only aspect of western civilization that the others have fully assimilated. Western religions and philosophies, western arts and letters, have met with mixed receptions at best in other parts of the world. To the extent that western political and social systems have displaced traditional forms, it was in the wake of imperialism or military conquest. Beginning with the Japanese, however, Asian polities have deliberately acquired science, if only in self-defense at the outset. They operate it just as well as do Americans and Europeans and no differently. Thus, though created out of personality and in culture, science is not then bound by personality or culture. It is impersonal and general. I can think of nothing else of which that may be said.

Perhaps it is a reflection that justifies describing science as objective, a body of knowledge made by persons, but made about the world and not about themselves.

Notes and References

1. *Science and Polity in France at the End of the Old Regime* (Princeton, 1980), fol-lowed by *Science and Polity: the Revolutionary and Napoleonic Years* (Princeton, 2004).

2. *Op. cit.* 1 (Cambridge: Cambridge University Press, 1990), pp. 153–167.

3. *Personal Knowledge* (Chicago: University of Chicago Press, 1958).

4. I should like here to record my entire agreement with Steven Weinberg in a fine article, "Physics and History," *Daedalus* (Fall, 2005), pp. 31–39.

APPENDIX

Also by the Author

—ɯɯ—

Genesis and Geology: A Study in the Relations of Scientific Thought, Natural Theology, and Social Opinion in Great Britain, 1790–1850. Harvard Historical Studies, Vol. LVIII (Cambridge, Mass.: Harvard University Press, 1951) xv + 315 pp., 2nd printing, 1969; Harper Torch Book Edition, 1959. New edition, with foreword by Nicolaas Rupke and a new preface, Harvard University Press, 1996. Chinese translation, 1999.

Edited, *A Diderot Pictorial Encyclopedia of Trades and Industry: Manufacturing and the Technical Arts in Plates Selected from the Encyclopédie . . . of Denis Diderot* (New York, 1959: Dover Publications). 2 volumes; 485 plates; xxx + 920 pp.

The Edge of Objectivity: An Essay in the History of Scientific Ideas (Princeton: Princeton University Press, 1960), 562 pp. Oxford University Press edition, 1960. Translations: Japanese, 1966; Greek, 1975; Korean, 1981; Italian, 1981; Polish, 1991; Romanian, 2000. Reissued with a new preface, Princeton University Press, 1990.

Les Fondements intellectuels de l'introduction des probabilités en physique. (Paris: Palais de la découverte, 1963). 27 pp.

Editor, *Dictionary of Scientific Biography* (New York: Scribners, 1970–80). 16 volumes.

Lazare Carnot, Savant. With an Essay by A. P. Youschkevitch. (Princeton: Princeton University Press, 1971) xiii + 359 pp. French translation (Paris: Vrin, 1976).

Science and Polity in France at the End of the Old Regime (Princeton: Princeton University Press, 1980) xii + 602 pp. Italian translation, Il Mulino, Bologna, 1983. 2nd paperback printing as *Science and Polity in France, the End of the Old Regime* (2004).

The Montgolfier Brothers and the Invention of Aviation, 1783–84, with a Word on the Importance of Ballooning for the Science of Heat and the Art of Building Railroads (Princeton: Princeton University Press, 1983) xiv + 212 pp., 11 plates and 70 illustrations. French translation, Actes Sud, 1989.

The Professionalization of Science: France (1770–1830) compared to the United States (1910–1970) (Kyoto: Doshisha University Press, 1983). 40 pp.

Administrator, *The Princeton Mathematics Community in the 1930s.* An Oral History Project. Interviews with Albert W. Tucker, et al.; Interviewers, A. W. Tucker, William Aspray. Ed. Frederic Nebeker. (Princeton: Trustees of Princeton University, 1985).

Monuments of Egypt, the Napoleonic Edition: The Complete Archaeological Plates from "La Description de l'Égypte." Edited with Introduction and Notes by Charles Coulston Gillispie and Michel Dewachter (New York: Princeton Architectural Press, 1987) xxx + 47 pp.; 426 plates; Map. 2nd ed., 2 vols. boxed, 1988). French translation, *Monuments d'Égypte*, 2 vols. boxed (Paris: Éditions Hazan: 1988). Italian translation, 1990. 3rd Printing, 1991; 4th Printing, 1994.

Pierre-Simon Laplace, 1749–1827, a Life in Exact Science (Princeton: Princeton University Press, 1997) xii + 323 pp. With contributions by Ivor Grattan-Guinness and Robert Fox. Paperback ed., 2000.

Science and Polity in France, the Revolutionary and Napoleonic Years (Princeton: Princeton University Press, 2004) ix + 751 pp. Illustrated.

Index

www.ingramcontent.com/pod-product-compliance
Lightning Source LLC
Chambersburg PA
CBHW081339190326
41458CB00018B/6051